CONCEPTS OF SPACE IN GREEK THOUGHT

PHILOSOPHIA ANTIQUA

A SERIES OF STUDIES
ON ANCIENT PHILOSOPHY

FOUNDED BY J. H. WASZINK AND W. J. VERDENIUS

EDITED BY

J. MANSFELD, D.T. RUNIA
J.C.M. VAN WINDEN

VOLUME LXV

KEIMPE ALGRA

CONCEPTS OF SPACE
IN GREEK THOUGHT

TUTA SUB AEGIDE PALLAS · 1683 ·

CONCEPTS OF SPACE IN GREEK THOUGHT

BY

KEIMPE ALGRA

E.J. BRILL
LEIDEN · NEW YORK · KÖLN
1995

The paper in this book meets the guidelines for permanence and durability of the Committee on Production Guidelines for Book Longevity of the Council on Library Resources.

Library of Congress Cataloging-in-Publication Data

Algra, Keimpe.
Concepts of space in Greek thought / Keimpe Algra.
p. cm. — (Philosophia antiqua, ISSN 0079-1687 ; v. 65)
Includes bibliographical references and index.
ISBN 9004101721 (hard)
1. Philosophy, Ancient. 2. Space and time. I. Title.
II. Series.
B187.S7A54 1994
114'.0938—dc20 94-33784
CIP

Die Deutsche Bibliothek - CIP-Einheitsaufnahme

Algra, Keimpe:
Concepts of space in Greek thought / by Keimpe Algra. -
Leiden ; New York ; Köln : Brill, 1994
(Philosophia antiqua ; Vol. 65)
ISBN 90-04-10172-1
NE: GT

ISSN 0079-1687
ISBN 90 04 10172 1

PRINTED IN THE NETHERLANDS

in memoriam matris

TABLE OF CONTENTS

ABBREVIATIONS

CAR	E. Maass, *Commentariorum in Aratum Reliquiae*, Berlin 1958 (= repr. 1898)
DG	H. Diels, *Doxographi Graeci*, Berlin 1958 (= repr. 1879)
DK	H. Diels / W. Kranz, *Fragmente der Vorsokratiker*, 3 vols., Berlin 1960[10]
FDS	K. Hülser, *Die Fragmente zur Dialektik der Stoiker*, Stuttgart 1987-1988
FHSG	W. W. Fortenbaugh/ P. M. Huby/ R. W Sharples/ D. Gutas, *Theophrastus of Eresus: Sources for his Life, Writings and Influence*, 2 vols., Leiden etc. 1992
LS	A. A. Long / D. N. Sedley, *The Hellenistic Philosophers*, 2 vols., Cambridge 1987
LSJ	H. G. Liddell / R. Scott / H. Stuart Jones, *A Greek-English Lexicon*, Oxford 1953 (= 1940[9])
RE	*Paulys Real-Encyclopädie der klassischen Altertumswissenschaft*, herausg. von G. Wissowa, Stuttgart 1894 etc.
SVF	J. von Arnim, *Stoicorum Veterum Fragmenta*, 3 vols., Leipzig 1903-1905; vol. 4, indexes by M. Adler, Leipzig 1924

INTRODUCTION

This book is a (considerably) revised and enlarged version of my 1988 doctoral dissertation which was entitled *Concepts of Space in Classical and Hellenistic Greek Philosophy*. It attempts to study the various ways in which Greek philosophers, mainly in the classical era (the fifth and fourth century BCE) and in the Hellenistic period (especially in the third century BCE), treated the problems connected with physical space. This period is of immense importance to the history of western philosophy, because it witnessed the development of the great philosophical and physical systems that were to dominate the intellectual discussions for many centuries to come: Platonism, Aristotelianism, Stoicism and the atomist systems of the Abderites and Epicurus. Also for the student of the history of 'the concept of space' this period has great significance. The theory of the receptacle in Plato's *Timaeus*, Aristotle's theory of place in *Phys.* Δ, but also the early Stoic and Epicurean concepts of space and void played an important role in physical discussions in later antiquity and for many centuries afterwards. To the extent that such later developments may be considered the *Rezeptionsgeschichte* of these early concepts of space, it will be clear that detailed studies of these concepts and of the specific problems they involved may be of more than merely intrinsic interest. This is in no way to belittle the importance of later developments in Greek thought. Indeed, recent scholarship (in particular the studies of Samuel Sambursky and Richard Sorabji) has shown that the contributions of several Neoplatonist thinkers have been both interesting and influential. Although occasionally such later developments will be briefly discussed—as in chapter 5, which deals with various ancient interpretations and elaborations of Aristotle's theory—the present book will focus primarily on the physical theories of the fourth and third centuries BCE.

The six chapters of this book are to be regarded as six more or less separate studies. As a whole, the book does not pretend to offer a complete 'history of the concepts of space and place' for the period with which we are dealing. The decision to select three particular physical systems (viz. those of Plato, Aristotle and the early Peripatetics, and the Stoics) for closer study is motivated because the way in which these philosophies deal with the problems of space and place involves a number of problematic features which are still in need of further clarification. In addition, the theories of some other philosophers (Melissus, Democritus, Strato, Epicurus) are given a more concise treatment (in chapter 2) either because they are less problematical or because they have already been dealt with adequately in the scholarly literature. Accordingly, our studies should be regarded as 'contributions to the history of a problem', not as an exhaustive survey.

In spite of the more or less self-contained character of the individual chapters, they are linked by a common methodological approach. This may call for some further comment. Unlike other problems of early and classical Greek natural philosophy—like the question of the shape of the earth, or the question whether or not physical reality is compounded of four elements—the problem of space has not as yet stopped worrying philosophers. The fact that space is on the one hand an ineliminable part of the furniture of the physical world—or at least an ineliminable aspect of the way in which we experience that world—whereas on the other hand it proves extremely difficult to reach a consensus about what it actually is, has fascinated many thinkers from the times of Parmenides to the present day. Even in our century—in which in other respects the separation between physics and philosophy seems to be almost complete—the revolutionary developments in theoretical physics, such as the introduction of the General Theory of Relativity, have given rise to a fresh philosophical interest in problems concerning space and time. This raises the question whether it is legitimate—and whether it would be fruitful—to interpret the ancient materials in the light of modern developments. In fact some important historical studies in the field (Jammer, Sambursky) have to a greater or lesser extent been undertaken from such a 'modern' perspective.

Rewarding though such an approach may be in some cases, it is not what I have opted for primarily in the present book. Instead I have tried to further our understanding of a number of ancient theories of space by studying them as much as possible in their historical context. The word 'context' should here be taken in a very broad sense. On the one hand it refers to the more general philosophical or methodological framework within which ancient theories of space took shape. Viewing the ancient material in its context in this particular sense involves trying to determine the peculiarities of the way in which ancient thinkers approached 'the problem of space' rather than looking primarily for common ground between their theories and modern counterparts. What, in other words, were the questions *they* asked, which factors determined the nature of the specific problems *they* had to deal with, and in what way were their theories embedded in a specific philosophical and historical context? Thus I hope to be able to show that the problems Plato experienced in specifying the way in which the sensible world participates in the intelligible world rubbed off to a large extent onto his 'theory' of space as propounded in the *Timaeus*. Similarly, a number of problematic features of Aristotle's theory of place (such as his views on natural place, discussed in the first section of chapter 5, or the question of how his 'classical' conception of place relates to others which are also to be found in his works) can be better understood when we take account of the nature and consequences of Aristotle's dialectical method as reflected in the text of the *Physics*. As a result of this 'contextualist' approach, the reader will sometimes find himself faced with what may at first blush appear to be mere excursions into areas that are fairly remote from the main subject (Eleatic ontology, Platonic metaphysics, Aristotelian methodology, Stoic modal logic). Yet I hope to be able to show that in the end these sections do contribute to our understanding of the various theories of space and place.

The word 'context' may also be taken to refer to the way the ancient tenets were transmitted. Also in this sense of the word I have tried to study the ancient material 'in its context'. After all, we often have only second-hand knowledge of ancient theories of space, i.e. we have to rely on ancient testimonies rather than on *verbatim* fragments. This makes it all the more important to find out how and why our ancient secondary sources include such testimonies, and to try to establish what this may tell us about the reliability of the information they provide. In cases where our dependence on such secondary sources is notoriously strong (early Peripatetics, early Stoics, Posidonius), I have accordingly devoted quite some space to an analysis of the nature and status of the sources.

If our 'contextualist' approach may lead us to stress what was peculiar to individual theories and concepts of space, rather than to underline the common ground between them, it might in return be asked in what sense the different philosophers who have been discussing the problem of space throughout history were all still talking about the same thing. This problem will be dealt with, and the approach taken will be defended, in the first chapter ('Writing the History of Concepts of Space'). There I shall provide an outline of how, to my mind, the history of 'the concept' of space should be envisaged. Moreover, it will be shown that this view has important consequences for the way in which the historian should apply his descriptive vocabulary. The second chapter ('*Topos, Chôra, Kenon*: Some Case Studies') will show how terminological problems—especially the absence of a one-word equivalent of 'space', and the different possible applications of the term *to kenon*—to a large extent determined the outlines of the philosophical debate about space in the period we are dealing with. As such its observations are in some ways preliminary to the discussions which are to occupy the subsequent chapters. As a *parergon*, our inquiry will allow us to add some qualifying remarks concerning the correctness of the many 'automatic' translations which are to be found in the scholarly literature. The main part of this book, the chapters 3–6, deals with three important stages in the history of Greek philosophies of space, which need no further introduction here viz. the theory of the receptacle of Plato's *Timaeus* (chapter 3), Aristotle's theory of place and early Peripatetic reactions to that theory (chapters 4 and 5), and the Stoic theory of space (chapter 6).

As I indicated, this book may be regarded as a revised and enlarged version of my 1988 dissertation. I should therefore specify here to what extent the contents of the two books overlap and to what extent the present book presents information that is new. The present chapter 1 is a partly rewritten and considerably condensed version of the first chapter of the dissertation. The original chapters 2 and 3 recur in this book in an only slightly altered form: I have made a number of small corrections and I have tried to update the footnotes by adding references to recent scholarly literature. The fourth chapter of the present book is based on the fourth chapter of the dissertation, but parts of it, in particular the section on Aristotle's dialectical method (section 4.4), have been rewritten to a large extent. The whole fifth chapter ('Problems in Aristotle's Theory of Place and Early Paripatetic

Reactions') is new. Yet I should acknowledge that I was fortunate to be able to lecture on themes taken from its first section (Aristotle's conception of natural place and natural motion) to academic audiences at the universities of Cambridge and Groningen. In addition, section 3 (on Theophrastus) covers a subject which I discussed earlier in Algra (1992). Chapter 6 is a considerably rewritten version of chapter 5 of the 1988 dissertation. In addition, it now contains a section on Posidonius, which partly covers the same ground as Algra (1993).

In order to make this book accessible also to Greekless readers, I have restricted the use of untransliterated Greek as much as possible to the footnotes. But sometimes, especially in chapter 6, the philological component of my research was so predominant or so important that Greek phrases had to be incorporated into the main text (although in those cases I have usually added translations). Nevertheless, key terms (like *topos, chôra, kenon*) have been transliterated throughout the text.

Quotations from ancient sources within the main text have of course been translated into English. In those cases where I thought the original Greek text was likely to be relatively inaccessible, or where I thought the reader might find it useful to be able to compare the wording of the Greek text, I have added the latter in a footnote. Except when indicated otherwise, Greek texts are given as printed in the modern standard editions. These are, to limit myself to the most important authors, the editions of Oxford Classical Texts (OCT) for Aristotle, Plato, and Diogenes Laertius, the Teubner editions for Sextus Empiricus and Plutarch (with the exception of *SR* for which I have used Cherniss' text in the Loeb edition), and the Berlin edition of the *Commentaria in Aristotelem Graeca* for the Greek commentators on Aristotle. For citations from the Presocratics I have, as a rule, used the text as printed by Diels/ Kranz, *Fragmente der Vorsokratiker*, Berlin 1960[10] (abbreviated as DK). Unless otherwise indicated, translations are my own, although I should of course acknowledge my debt to existing English translations, especially to the revised Oxford Translation of the works of Aristotle.

Over the years I have benefited from the comments and suggestions, made in various forms and on various occasions, of many scholars and friends. In particular I would like to thank Jaap Mansfeld, David Runia, Richard Sorabji, David Furley,

David Sedley, Thomas Kuhn, Bertram van Winden, Paul Mercken, Malcolm Schofield, and Han Baltussen. At various stages of the project Mark Howard and Patty van Hees tried their best to correct my English. It goes without saying that they are not responsible for the Batavisms they did *not* catch. Gonni Runia did an excellent job producing the final desk-top version. But most of all I would like to thank Henri van de Laar for his invaluable help in proof-reading the typescript, checking references, compiling the *Bibliography*, and preparing the *Indices.*

CHAPTER ONE

WRITING THE HISTORY OF CONCEPTS OF SPACE

Introduction

A recurring feature in many 'histories of philosophy' is the tendency to 'upgrade' the ideas of thinkers of the past by linking them up with modern notions. Readers of handbooks in the history of philosophy are well acquainted with such diachronic 'pairings' and the concomitant talk about 'precursors'. Democritus and Bohr, Parmenides' spherical Being and recent notions of spherical space: they are not uncommonly encountered in happy juxtaposition.

It is of course commonplace to remark that such comparisons, though they may in themselves be stimulating, are also quite often fairly misleading. And indeed fellow-historians are usually quick to point this out. The problem seems to be that people who make comparisons of this kind unduly act *as if* different, and historically widely separated, concepts of an atom or of matter share a common essence in virtue of which the proponents of such different concepts are in an important way still 'talking about the same thing'. With the *proviso* that historians acting like this need not be consciously committed to any well-defined ontological or epistemological version of essentialism, we may in these cases well speak of unwarranted 'essentialist' presuppositions. Since, however, such 'anachronistic comparisons' almost by their own nature force us to check their accuracy, their flaws are usually easily recognized and for that reason they turn out to be rather harmless.

Another guise in which essentialist presuppositions in the sense just described may crop up in the historiography of philosophy—and, of course, in historiography in general—is far less easily detectable. They may affect the very basis of historiographical accounts, the concepts with which the historian describes his subject. In studying e.g. historical concepts of space or matter many scholars start out with some fairly specific idea at the back of their heads about what space or matter is and what it is not.

Thus when they ascribe spatial concepts to ancient philosophers they run the risk of investing those ancient concepts with incorrect or too specific connotations.

Now in those cases where all this is done in an explicit manner—as e.g. when Crombie expresses doubts about whether Plato's receptacle 'is space', because, as he believes, 'space is simply a system of relations whereas Timaeus' 'third force' seems more like a thing'[1]—we are at least in a position to assess the 'argument' and, if necessary, to prove it wrong. The problem is, however, that more than once underlying essentialist presuppositions figure in a less overt way.

As an example we may again adduce the problem concerning the interpretation of the receptacle passages in Plato's *Timaeus*. Among the most prominent exegetical controversies concerning Plato's receptacle is the question whether it should be described as space or as matter. As we shall see in chapter 3, one of the reasons why interpreters have opted for the *exclusive* identification of the receptacle with matter, or the *exclusive* identification with space, is the assumption that these two descriptive labels are actually incompatible. This alleged incompatibility is in its turn based on some, on closer scrutiny unwarranted, presuppositions concerning the 'real nature' of matter and space in general, viz. that matter is corporeal and that space is an independent (absolute) three-dimensional extension. In this case a very specific meaning is ascribed to a descriptive term, not on grounds of an inquiry into the way in which the corresponding concept actually figures in its historical setting, but on grounds of an implicit idea on the part of the historian about what the labelling term *must* mean. In other words, the label slides from a descriptive into an explanatory role. The fact that, contrary to what we saw in the case of 'anachronistic pairings', this is done implicitly—in most cases even unconsciously—makes the procedure only the more dangerous.

It will be clear that what I have labelled 'essentialist' presuppositions, appearing either in the guise of anachronistic pairings or in the guise of too specific translations or descriptions, may lead to bad historiography. Hence, before we go on with the proper business of our historical investigations, it seems worthwhile to try to answer the two following questions:

---

[1] Crombie (1963) 218.

(1) In which sense were the philosophers who have been theorizing about space (or place) throughout history all talking about the same thing? What, in other words, constitutes the common ground between various divergent conceptions of space or place?

(2) To what extent should the answer to question (1) affect the way in which the historian uses the descriptive labels 'place' and 'space'?

These questions will accordingly constitute the subject of this first chapter. Two objections might at first sight be raised against prefacing our historical studies by such a rather theoretical account. In the first place, it might be argued that the difference between the 'descriptive' and the 'explanatory' use of labels is in fact only gradual. Historical descriptions, it might be argued, are always to some extent explanations and it is illusory to think that such a thing as an 'objective description' exists. In many ways the historian's own thoughts, prejudices, conceptual apparatus etc. will rub off onto his historical descriptions. This trivial truth, however, does not relieve us of the burden of being as conscious as possible of such processes, and of trying to frame some methodological rules of thumb which may safeguard us from at least the most obvious historiographical blunders. Secondly, it might be objected that the historian should not waste his time over such methodological or philosophical issues and that he had better get on with his proper business, the investigation of 'concrete' historical problems. However, as I have already indicated and as I shall try to show in the further course of this chapter and this book, a number of failures, misunderstandings and obscurities to be encountered in the scholarly literature on our subject are arguably due to the fact that an awareness of the problem just referred to has been disappointingly rare among scholars writing about 'the concept of space' in Greek philosophy. At any rate the number of loose, inadequate or erroneous descriptions of ancient spatial concepts is large enough to warrant our bringing up the issue here.

This first chapter will be structured as follows. In section 1.1 I shall try to answer question (1) by showing how the history of 'the' concept of space or 'the' concept of place is to be envisaged, and in what sense philosophers who were committed to different concepts of space or place were all still talking about the same thing. Next, section 1.2 will try to answer our question (2) by showing how the results of the investigations of 1.1 may get a methodological relevance. The way in which the historian should use his descriptions and translations will be sketched and as negative examples a number of cases will be adduced where what I called essentialist presuppositions have led to confused or incorrect historiography.

1.1 *Concepts of space and place within different theories*

1.1.1 It will be convenient to try to answer the question whether different concepts of place share any common ground by applying the notion of 'meaning'. After all, what we intend to do is to use the results of the present section in order to determine what historians, using 'place' or 'space' as descriptive labels *mean* or *should mean.* At the same time such an approach will enable us to adduce some useful results of the contemporary debate about theory change and meaning variance. It should be made clear from the outset, however, that I certainly do not want to suggest that inquiries in the history of ideas should be based on previously and independently established philosophical theories of meaning. On the contrary, I would rather maintain that for the historian such theories serve at most as hypotheses, to be confirmed or refuted by his case studies. Also in the present study *philosophy* plays this modest role of *ancilla historiae* and the treatment of theoretical or philosophical issues will accordingly be limited to that which serves the interests of our inquiry.

If we then reformulate question (1) of the previous section and ask whether and to what extent the meanings of terms like 'place' and 'space' in different theories are connected, roughly three possible answers present themselves. The meanings are (a) radically disconnected, (b) fully identical, or (c) in some ways connected, in others not. Possibility (a) would of course amount to the radically relativist position concerning meaning variance and theory change which is now familiar from philosophers like Thomas Kuhn and Paul Feyerabend.[2] If the history of concepts of space were actually such as possibility (a) presupposes, it would in

---

[2] Cf. Kuhn (1962) and Feyerabend (1962); see also P. Feyerabend, 'On the meaning of scientific terms', *JPh* 62 (1965) 266–274, and *idem*, 'Problems of empiricism', in: R. Colodny (ed.), *Beyond the Edge of Certainty* (Englewood Cliffs, NJ 1965) 145–261.

fact consist of a series of different and unrelated concepts figuring within different and incompatible theoretical frameworks. For the historian this would of course involve a fundamental problem of translation and description: his descriptive tools would turn out to be misleading on any account. For whether he takes his descriptive labels from contemporary scientific vocabulary, or from common parlance—which, as is well known, Feyerabend considered to be a 'wrong theory'—in both cases he would turn out to be applying a basically incommensurable vocabulary.

Apart from possible *philosophical* objections against this position,[3] there is the general *historical* objection that in practice the notion of radical meaning variance—either between different individual theories or between 'conceptual schemes'—turns out to be so clearly at odds with common intuitions and with scientific and historical practice that it may count as a *reductio ad absurdum* of the strong relativist position. In this respect the history of concepts of space constitutes no exception. Defendants of different concepts of space have usually shown themselves quite able to take the concepts of their opponents for what they really were and though e.g. such a famous clash of competing views on space as takes place in the Leibniz-Clarke correspondence may at times have much of a *dialogue de sourds*—because both parties vigorously stick to their positions—the combatants nevertheless try their best to find their ammunition not only in their own system and concepts, but also in that which they criticize. They are, therefore, not fully and utterly at cross-purposes.[4] In other words, the strong relativist answer to the first question posed in our introduction,

---

[3] Thus, e.g. it has been put forward that the main idea on which Kuhn's views on the history of science—which are very similar to Feyerabend's—rest, the idea of a conceptual scheme, is vague or incoherent, on which see D. Davidson, 'On the very idea of a conceptual scheme', *Proceedings and Addresses of the American Philosophical Association* 57 (1974) 5–20; repr. in: D. Davidson, *Inquiries into Truth and Interpretation* (Oxford 1984) 183–197. It has moreover been argued that the very notion of incommensurability is incoherent as well, in that the very act of establishing that two theories are incommensurable presupposes that they can at least be superficially compared, so that they turn out to be not that incommensurable after all.

[4] Cf. Koyré (1957) 246: 'to a large extent, they simply repeat, or elaborate the same arguments—philosophers in general ... seldom, if ever, convince each other, and a discussion between two philosophers resembles as often as not a *dialogue de sourds*—and yet they come more and more into the open, and the fundamental issues come more and more to the foreground'.

amounting to the thesis that philosophers who framed different theories about space were *in no relevant way* talking about the same thing, has to be rejected.

Possibility (b), on the other hand, would amount to a radical version of what I called 'essentialism'. Maybe this radical variety of essentialism does work in the case of *some* theoretical concepts.[5] However, when we are dealing with concepts of *space*, it is equally hard to square with the historical evidence as its opposite, the radically relativist position. For it is too rigid to account for the various ways in which the terms 'space' and 'place' have *de facto* been related throughout history. Moreover, the differences between the various concepts at issue are simply too great to allow them to be subsumed under one common 'essence'.

We seem to be in need, therefore, of a 'theory of meaning' which is in accordance with actual historical evidence in evading the vicious consequences of complete meaning relativism and in being, at the same time, 'vague' enough to account for a variety of ways in which the same terms occurring in different theories may, in spite of their different meanings, still be related. The outlines of such an alternative theory of meaning have been drawn by Dudley Shapere, who, in several philosophical studies and from a number of different perspectives,[6] has thrown light on the relationship between the meaning of scientific terms and theory change.[7]

---

[5] One of the more radical versions, Putnam's so-called 'causal theory of reference' (according to which the fundamental identity of terms throughout theory change is secured by their having the same *reference*) may be applicable to those cases where, within one and the same theoretical framework, e.g. new particles are discovered and 'dubbed'. Cf. H. Putnam, 'The Meaning of 'Meaning', in: H. Putnam, *Philosophical Papers*, 2, Mind Language and Reality (Cambridge 1979) 215–272, and other essays in the same volume. For a most helpful survey of the issue and for further references see Hacking (1983) chapters 5 ('Incommensurability') and 6 ('Reference').

[6] Especially his 'The Structure of Scientific Revolutions' (review of Kuhn's study), *PhR* 69 (1960) 376–385; 'The Paradigm Concept', *Science,* vol. 172 (1971) 706–709; Shapere (1966); 'Space, Time, and Language—An Examination of Some Problems and Methods of the Philosophy of Science', in: B. Baumrin (ed.), *Philosophy of Science: The Delaware Seminar*, vol. 2 (New York, etc. 1963) 139–170 and 'Reason, Reference, and the Quest for Knowledge', *Philosophy of Science* 49 (1982) 1–23. All these studies are now conveniently printed together in Shapere (Dordrecht 1984).

[7] Note, by the way, that because of all the essentialist connotations of the term 'meaning' Shapere would rather favour the abolition of talk about meaning in the context of the history and philosophy of science altogether.

Shapere takes the path of assuming that meanings can be similar, or comparable, in some respects, while also being different in other respects. The similarity exhibited by different concepts of 'electron', e.g., are to be accounted for by the Wittgensteinian notion of *family resemblance.* The main elements of the position defended by Shapere—as well as his reply to the possible objection whether adducing so vague a notion as family resemblance does not amount to just *labelling* the obscurity rather then removing it—may become clear from the following quotation. Speaking about 'the concept' of 'electron' as used, within different theoretical contexts, by different physicists (Stoney, Feynman), Shapere notes:

> What counts as an electron is determined by the best beliefs available at a given time; rather than there having (necessarily) been a constant 'meaning' throughout the tradition from Stoney to Feynman, it is more accurate to say that the 'concept' of 'electron' is just the family of uses of that term which are related in an ancestry-and descent (or cousinhood) relation by a chain-of-reasoning connection. I take this to be an application to scientific change of Wittgenstein's notion of 'family resemblance', except that here the basis of that 'resemblance' is not left as a vague metaphor: the relationship is governed by reasons, and is clear to the extent that it is clear, at a given stage of scientific development, what is to count as a reason in some specific case. Similarly what we are *talking about* (referring to) throughout a 'tradition' in science need not be one specific something; what we are talking about at any given stage is determined by the best information on the subject (e.g., of electrons) at that stage, and is related, if it is at all, to what people previously talked about in talking about that subject by a chain-of-reasoning connection. .... Nothing else is available to us to determine what we 'mean' or are 'talking about', and nothing else is needed to explain our activities; in particular we need not suppose a something-we-know-not-what (an 'essence') whose nature we seek.[8]

There are a few crucial items in this quotation which I would like to select for further comment. In the first place the distinction made between 'what we are talking about throughout a tradition',

---

Cf. Shapere (1966) 85. I shall here nevertheless stick to the (conventional) use of the word 'meaning', not because I believe that this notion is indispensable for the purpose of seeking out central features of scientific theories (like Shapere, I do not), but because my ultimate concern is with the meanings of descriptive terms in historical studies.

[8] Shapere (1984) xxxiv.

i.e. '*the*' concept of electron, which need not be a specific something, and 'what we are talking about at any given stage', i.e. the *specific* concept of electron, or rather: *a* concept of electron, figuring in one specific physical theory. This distinction allows us to maintain that, in a way, theories about the electron are talking about the same thing and, at the same time, that in saying this we are not implying that the meaning of the term 'electron' within two such different theories must be identical.

Secondly, Shapere here seems to envisage only diachronic differences between theories. However, the application of his theory is by no means necessarily restricted to these cases. Also the connection between different uses of the same term in competing, and hence synchronic, philosophical systems in antiquity may be accounted for in terms of a family resemblance of the kind sketched by Shapere. As I shall try to show, the different concepts of space endorsed by different Greek philosophers are a case in point.

Thirdly, Shapere is here suggesting that at each successive stage there is a consensus: people are agreed about what are 'the best beliefs available at any given time' and they will readily change these beliefs as soon as others which are arguably better have become available. But this element does not appear to be essential to Shapere's theory either. In the exemplary case adduced in the text quoted (electrons) it may well have been true that people were agreed at each different stage about what were the 'best beliefs available'. Still, this does not seem to be how science in general works. Often, of course, competing theories or competing systems coexist for some time. This is notoriously true for ancient times when physics in general was highly speculative, and it is *a fortiori* true for ancient discussions concerning such highly theoretical concepts as space and time.

Finally, it should be noted that it is the idea of chain-of-reasoning-connections which disposes of the problem of incommensurability. For each 'new' use of a term may be plausibly supposed to be a reasoned departure from at least part of the contemporary or previous uses of that particular term.[9] Thus, in the course of time, the meaning of the term 'electron' may have gradually shifted to

---

[9] The chain of reasoning need, of course, not be linear: it may show ramifications or a network pattern.

such an extent as to have, in the end, hardly any more features in common with the meaning of the term at its first use (if such a first use can be traced down anyway).[10] By implication, theories which are chronologically widely apart may use one and the same term in ways which are so different that we can hardly—if at all—point to a *specific* common denominator. Still, the connection of family resemblance, as secured by the chain of reasoning, may be intact so that we may still speak of, e.g., two 'concepts of electron'. In a number of cases the chain-of-reasoning connection leading to ever different concepts may in principle extend indefinitely. However, as we shall see, in the case of the history of concepts of space things are different. Before this subject came under the sway of the developments in mathematics in the 19th and early 20th century, but most certainly in the case of the history of ancient concepts of space, the chain of reasoning was conspicuously limited by the confinements of ordinary experience and common language. This is what I shall try to show in the next subsection.

1.1.2 I shall concentrate here on concepts of space that have been defended throughout the history of western thought before the beginning of the twentieth century, i.e. in what we might call the pre-Einsteinian era. If we disregard the difference between subjective-idealist (or relativist) theories of space on the one hand and realist theories of space on the other, and if we just concentrate on the different *functions* a concept of space may fulfil within a physical theory, I think we are justified in recognizing roughly three main kinds of spatial concepts. In other words, different physical or metaphysical systems from antiquity onwards have been investing physical space (and, as we shall see, *mutatis mutandis* also place) with one or more of the following three different functions. Space may figure as

(a) a kind of prime stuff or 'reservoir of physical possibilities',[11] or

(b) a framework of (relative) locations, or

---

[10] Compare in this respect the way in which the mathematization of space in the 19th and 20th century gave rise to theories of space on a non-Euclidean basis, a development which constituted an important departure from common experience (on which see below, section 1.2.2).

[11] I borrowed this term from Hussey (1983) xxxi.

(c) a container, the 'fixed stage where things play out their comedy',[12] a space *in* which things are and *through* which they can move, to paraphrase Epicurus.[13]

The (a) kind of space is present in theories which tend to focus on the extension of individual physical things, without regarding this extension as separate or separable. We may add, however, that this conception of space arguably has its roots in ordinary thinking and speaking. Thus, when we say that a particular thing 'occupies so and so much room' we may mean nothing more than that it has this particular size and, by using an expression like 'to occupy', we need not necessarily imply that this extension has its own independent existence. Arguably, Descartes, who thought extension to be the ultimate defining characteristic of substance (the *res extensa*), and who for that reason identified extension, matter and space, was thus committed to an (a) kind of space. Also the early Newton seems to have been committed to such a concept of space.[14]

Those who defend a concept of space of the (a) kind are likely to run into a number of difficulties when it comes to explaining locomotion. For since space of the (a) kind is not separate—rather, it is a constitutive element of (and so closely tied to) the particular body at issue—it will be unable to play any role of importance in an account of the location or the motion of individual physical objects. It was for this reason that Descartes was forced to maintain, apart from, so to speak, his metaphysical concept of *space* and the connected concept of what he called *internal* place, a concept of

---

[12] The quotation is from Bunge (1978) 134.

[13] Cf. *Ep. Hdt.* 40: εἰ ⟨δὲ⟩ μὴ ἦν ὃ κενὸν καὶ χώραν καὶ ἀναφῆ φύσιν ὀνομάζομεν, οὐκ ἂν εἶχε τὰ σώματα ὅπου ἦν οὐδὲ δι' οὗ ἐκινεῖτο, καθάπερ φαίνεται κινούμενα·

[14] Such a more or less Cartesian—or perhaps Platonic—concept of space is apparent in his (early) *De Gravitatione et Aequipondio Fluidorum*, in: A. Rupert Hall & Marie Boas Hall (edd.), *Unpublished Scientific Papers of Isaac Newton* (Cambridge 1962) 106–107: ' ... non opus est ut effingamus aliquam substantiam non intelligibilem dari cui tanquam subiecto forma substantialis inhaereat: sufficiunt extensio et actus divinae voluntatis. Extensio vicem substantialis subiecti gerit in qua forma corporis per divinam voluntatem conservatur. ... Inter extensionem et ei inditam formam talis fere est Analogia qualem Aristotelici inter materiam primam et formas substantiales ponunt'. Cf. McGuire (1982) esp. 177, where concerning the role of space in this early treatise McGuire remarks that 'So conceived, uncreated and extended space can be viewed as a 'receptacle' within which God himself imposes the formal nature of material things'.

external place which had a surprisingly Aristotelian ring, in that it defined the place or location of an individual thing in terms of (the surface of) its immediate surroundings.[15]

In general, in so far as space of the (a) kind can be that *in which* something else is present, it is not a that-in-which in the strictly local sense, but rather in the sense in which—to use Aristotelian vocabulary—a quality may be said to be present, or to inhere, *in* a particular substance.[16] In chapter 3 we shall see that this was probably the primary sense in which the receptacle of *Timaeus* 48e ff. was called by Plato the 'that-in-which' (*to en hôi*).

Perhaps the most renowned among the concepts of space of the (b) kind is that which was defended by Leibniz. On this view, space should not be accorded any separate reality. Rather, it should be regarded as the sum total of the ordering relations that hold between physical entities. In everyday speech this concept of space (or rather the concomitant notion of place) has its counterpart in expressions conveying the location of particular things in terms of their surroundings, as in 'a fish is swimming in the water' or 'Sokrates is staying in Athens'. In antiquity we seem to encounter a rather similar conception of space in Theophrastus' critical elaboration of Aristotle's theory of place.[17]

Newton's concept of absolute space is no doubt the most famous representative of the (c) kind of space, his concomitant notion of *relative* space rather belonging to the (b) kind. Also this concept of absolute space might be regarded as an extension of one of the senses in which the word space is used in common parlance. We use 'space' in this sense e.g. when we say that a body moves 'through' space, and the concomitant notion of absolute place

---

[15] R. Descartes, *Principia Philosophiae* II, 14, ed. Adam/ Tannery, vol. 8, 48: 'Locum autem aliquando consideramus ut rei, quae in loco est, internum, & aliquando ut ipsi externum. Et quidem internus idem plane est quod spatium; externus autem sumi potest pro superficie quae proxime ambit locatum'. Arguably, he was forced at the same time to use the concept of motion in various senses since the, so to speak, *metaphysical* concept of motion (the motion by means of which God individuates physical bodies) is by no means identical to the concept of motion used when Descartes is attributing motion to physical bodies once individuated. See on these and connected problems Anderson (1976).

[16] Compare, for example, the quotation from Newton's early treatise *De Gravitatione* written out above, n. 14, which has it that 'extensio vicem substantialis subiecti gerit *in qua* forma corporis per divinam voluntatem conservatur'.

[17] On Theophrastus' views on place see below, ch. 5, section 5.3.

when we say that body (A) 'occupies' the place formerly held by body (B). Though perhaps not as fully developed and thought over as the Newtonian concept of space, also the void-space of the ancient atomists, at any rate the Epicurean version, belongs here, as is apparent from the quotation from Epicurus cited above (note 13).

Two further observations on this classification of different types of concepts of space are in order. First, in the unreflected practice of everyday speech the (a) and the (c) kind may well be used interchangeably as they both identify space with extension. There are situations in which it is of no great importance whether we think of that extension as being inseparable from a particular body, or as having an independent existence. Therefore, an alternative way of classifying the kinds of concepts of space might be to distinguish between those conceptions which take space as a certain volume or an extension (= our (a) and (c) concepts) and those which take space as a system of relations (our (b) kind).[18] This will do as long as it is realized that the space-as-extension view may appear in different and even incompatible guises. It is perhaps owing to the fact that our (a) and (c) concepts of space are (and were) not always clearly distinguished in common parlance that, as I shall show in chapter 3, in the *Timaeus* Plato could use his *chôra* in both roles.

Secondly, in so far as the three kinds here discussed are all ultimately rooted in common experience and—if only for that reason—seem to presuppose a Euclidean geometrical framework, they do not embrace modern concepts of space, i.e. concepts of space developed, broadly speaking, from the second half of the 19th century onward. The space of the General Theory of Relativity, for example, might at first blush seem to belong primarily to the (a) kind. For also according to the principles of the General Theory of Relativity space is not something which exists independently of that which fills it and for that reason we might say that it is in a rudimentary sense comparable to e.g. the Cartesian concept of space.

---

[18] This distinction was e.g. advocated by Taylor (1928) 668: 'space may be envisaged primarily as a network of relations .... Or you may think primarily of space as metrical, as that *of* which a given body occupies just so much'.

This, however, should not blind us to the fact that the Einsteinian space-time differs in one fundamental respect from all 'classical' concepts of space (whether of the (a), the (b) or the (c) kind): it has hardly any connection with everyday experience or with common parlance.[19] In fact it could only get its outlines after the revolutionary developments in nineteenth century mathematics had established the equivalence, at least from a conceptual and logical point of view, of Euclidean and non-Euclidean geometries, thereby depriving Euclidean geometry—which is still usually regarded as the geometry of ordinary thought—of its privileged position. Moreover, in addition to its non-Euclidean character, also the close interrelation it assumes exists between space and time,[20] and its presupposing a physics in which the notion of a field plays a basic role, are features which place this modern concept of space miles apart from its 'classical' namesakes. One should be careful always to keep this in mind when drawing comparisons between certain ancient or at any rate 'classical' concepts of space and the space-time of the General Theory of Relativity.[21]

---

[19] Cf. Einstein's famous address to Newton in his 'Autobiographisches' (1949) 30–31: 'Newton verzeih' mir; du fandst den einzigen Weg der zu deiner Zeit für einen Menschen von höchster Denk- und Gestaltungskraft eben noch möglich war. Die Begriffe die du schufst, sind auch jetzt noch führend in unserem physikalischen Denken, obwohl wir nun wissen, dass sie durch andere, *der Sphäre der unmittelbaren Erfahrung ferner stehende* ersetzt werden müssen, wenn wir ein tieferes Begreifen der Zusammenhänge anstreben' [my italics].

[20] On the difference between the four-dimensionality of Einsteinian spacetime and the perspectives of classical physics cf. Einstein (1949) 56: 'Auch der klassischen Mechanik liegt das vierdimensionale Kontinuum von Raum und Zeit zugrunde. Nur haben im vierdimensionalen Kontinuum der klassischen Physik die 'Schnitte' konstanten Zeitwertes eine absolute, d.h. von der Wahl des Bezugsystems unabhängige, Realität. Das vierdimensionale Kontinuum zerfällt dadurch natürlich in ein dreidimensionales und ein eindimensionales, sodass die vierdimensionale Betrachtungsweise sich nicht als *notwendig* aufdrängt'.

[21] In arguing that the 'space' of pre-atomist Greek thinkers like Parmenides was spherical and (therefore?) non-Euclidean (Cornford speaks of pre-Euclidean), Cornford (1936) fails to distinguish properly between a spherical space (i.e. the space occupied by a finite spherical cosmos) with a Euclidean geometry—which, as I believe, was the common view of many ancient Greek philosophers—and the modern conception of a finite space with a non-Euclidean geometry (apart from this, it may of course well be doubted whether there is such a thing as a Parmenidean conception of space in the first place, but that is not what interests us here).

1.1.3 Before we go on to survey some main kinds of concepts of place that have been defended throughout history, I shall first say something about the criteria that usually govern the choice between 'space' and 'place' as potential descriptions of ancient concepts.

In general, there seems to exist a kind of inarticulate consensus according to which 'space' is a more general term than 'place'. We may observe that 'place' is rather used in a relational setting (place being, either potentially or actually, the place *of something*) whereas 'space' rather refers to an underlying frame of reference or to the sum total of all places,[22] though, as a glance at the relevant items in a dictionary will establish, the rules are not without exceptions. For the sake of convenience, however, I would propose to adjust our descriptions to the most common philosophical and everyday usage and to use 'place' exclusively in a relational setting (place as the place *of something*) and use 'space' exclusively as referring to a larger underlying extension, or frame of reference, or receptacle, which constitutes the sum total of all places. We may here note in advance—this subject will be further discussed in chapter 2—that, if the labels 'space' and 'place' are thus defined, there does not exist, contrary to what is often supposed in the scholarly literature, a one-to-one correspondence between either of these descriptive labels and the Greek terms *topos* and *chôra*. In accordance with the classification of concepts of space provided in the previous section, we may now draw up the following list of basic kinds of concepts of place. Place may figure as:

(a) the (material) extension of a physical body
(b) the relative location of a physical body
(c) the extension (in the sense of: part of an absolute extension) occupied by a physical body.

An example of place of the (a) kind is, of course, Descartes' *locus internus*.[23] As I already remarked in connection with Descartes' concept of space, such a concept of place cannot serve to localize a particular body, nor can it play any role in the account of locomotion. The 'classical' Aristotelian concept of place, the one which issues from the dialectical discussion of *topos* in *Phys.* Δ, was a

---

[22] Cf. e.g. Leibniz in his 5th letter to Clarke (in: Robinet (1957) 144): 'espace est ce qui resulte des places prises ensemble'.
[23] On which see above, n. 15.

concept of the (b) kind. Also Descartes' *locus externus*[24] as well as Leibniz' concept of place[25] belong here. A concept of absolute place of the (c) kind is, as might be expected, to be found in Newton's famous *scholium* on space and time in his *Principia*: he there defines place as a part of space, sharing in the characteristics of space so that corresponding to his absolute space there is an absolute place (and corresponding to his relative space a relative place): 'Locus est pars spatii quem corpus occupat, estque pro ratione spatii vel absolutus vel relativus'.[26]

Just as we saw that the threefold classification of space concepts could be reduced to two main kinds: space as extension or volume on the one hand and space as a system of relations on the other, so too the concepts of place may be reduced to two kinds: place as an extension (whether separate or not) or place as a relative location.

At this point I should add a final observation. I have here to some extent simplified matters by representing the three kinds of concepts of space in their most straightforward form, viz. as concepts of *physical* space. It is in this form that they still clearly show their rooting in common experience. Yet it should be kept in mind that in later antiquity we witness a number of what might be called *metaphysical* applications of these various conceptions of space—applications that are in fact much further removed from everyday usage. The neoplatonist systems of e.g. Syrianus and Damascius are interesting cases in point.[27] Here space and place are accorded a role in the process of formative emanation: space is now no longer an inert physical grid or framework by which the order and position of physical bodies can simply be *measured*, but it is conceived as a dynamic formative entity which itself *creates* this order and position. Nevertheless also these novel dynamic

---

[24] On which see above, n. 15.

[25] Cf. Leibniz' fifth letter to Clarke, in: Robinet (1957) 142: '[les hommes] ... considerent que plusieurs choses existent à la fois, et ils y trouvent un certain ordre de coëxistence .... Lorsqu' il arrive qu' un de ces coëxistents change de ce rapport à une multitude d' autres, sans qu' ils en changent entre eux, et qu' un nouveau venu acquiere le rapport tel que le premier avoit eu à d'autres, on dit qu' il est venu à sa place'.

[26] I. Newton, *Philosophiae Naturalis Principia Mathematica*, ed. Koyré/ Cohen/ Whit-man (1972) vol. I, 47.

[27] On these neoplatonic 'dynamic' conceptions of place and space in general (including those of Iamblichus, Syrianus, Proclus, Damascius and Simplicius) see Sambursky (1982) 15–231; and Sorabji (1988a) 204–211.

conceptions of formative space or place can be subsumed under the three kinds we have here distinguished. For they may be of what we called the (a) or (c) kind (in the case of Syrianus who views space as a three-dimensional extension allotting natural or proper places to various kinds of matter), or of the (b) kind (in the case of Damascius, whose conception of 'external' place may be described as the formative order and position of substances relative to others, whereas his 'internal' place is the formative order and position of the parts of a particular substance).[28]

1.2 *Initial and final descriptions*

1.2.1 The position taken in the previous two sections may be summarized in the following thesis. The different concepts of place and space that have been defended throughout history should be regarded as a class of which the members are connected by a kind of family resemblance. As to the concepts of space and place of ancient philosophy, and probably also as to those of later philosophy up to the 19th century, the nature of this family resemblance can be further specified by pointing out that the different concepts all have their roots in ordinary experience and common parlance. Thus an answer is provided to the first question raised in the introduction of this chapter, viz. in which way philosophers who defended different concepts of space were still talking about the same thing.

We are now in a position to answer the second question posed in the introduction, viz. how, accordingly, the historian should use his descriptive labels. That the answer to the first question does not *ipso facto* answer the second can be easily shown. For even if the radical version of meaning relativism—with its concomitant *fundamental* problem of translation—be rejected, we are still confronted with a *practical* problem of translation. Even if, for example, we recognize that the different concepts of space as they have been defended throughout history do not constitute a series of fully disconnected individual concepts, we still have to face the question what we are actually doing, or rather: what we *should*

[28] On Syrianus' theory of space see Sambursky (1982) 17–18 and 53–61 (where the relevant parts of the account of Simplicius are printed and translated). On Damascius' theory see Sambursky (1982) 21–22; Sorabji (1988a) 207–209; Algra (1992) 159–162.

actually be doing, when we translate e.g. Plato's or Aristotle's *topos* by 'space'.

Though an awareness of the problems connected with this question is sometimes encountered among 'ordinary' historians,[29] historians of philosophy in general seem to have been much more alert to the possible ways in which e.g. Aristotle may have given a distorted picture of the history of the Presocratic systems, by using his own terminology to describe them than to the possible ways in which we ourselves run the risk of providing misleading accounts by investing our descriptive terms and translations with too specific or just incorrect connotations.

For our purposes the ideas about meaning variance outlined in section 1.2 may get a methodological relevance by the application of the distinction between *initial* and *final* translations or descriptions.[30] Final translations or descriptions result from a careful study of the context in which the concepts to be translated or described are used (e.g. when, after studying Aristotle's discussion of *topos* in *Phys.* Δ, this concept of *topos* is described as a concept of place of the (b) kind). Initial translations or descriptions are applied more or less automatically because the historian thinks he already knows what the concept at stake amounts to.

We are now in a position to lay down the methodological principle that, as long as we choose an initial translation for reasons of *prima facie* family resemblance between the ancient concept and any modern counterpart, and as long as we keep

[29] Cf. S. Lauffer, 'Die Sklaverei in der griechisch-römischen Welt', *Rapports of the 11th Intl. Congress of Historical Science*, vol 2 (Uppsala 1960) 81 (quoted in Finley (1975) 63): 'The word *Sklave, esclave, schiavo,* which stems from the Middle Ages and originally marked Slavic war captives from eastern Europe can be transferred to antiquity only in an anachronistic way, and that means with misunderstanding. Furthermore, the word brings to mind the Negro slavery of North America and the colonial areas in the most recent centuries, which makes its transfer to the relationships of antiquity even more difficult. The ancient slave is an entirely different social type'.

[30] I am here more or less 'recycling' a distinction made by Moses Finley (1985) 61–62 (on which see the previous note) between *initial* generalizations on the one hand and *final* generalizations on the other. Final generalizations, according to Finley, are the generalizations with which a historian concludes his studies. They are explicit and liable to criticism. They are in fact vital to the continuation of the historical practice in that they are likely to call forward either assent or repudiation. Initial generalizations, on the other hand, are usually inexpressed presuppositions with which the historian starts out.

realizing that this initial translation is *only* an initial translation, nothing is wrong yet. In other words, by translating e.g. an ancient spatial term as 'space' or 'place' we just label it as expressing a concept which is an element of a class, comprising a variety of members which are connected by relations of family resemblance. Subsequently, by a closer investigation of the (theoretical) context in which the ancient term is seen to figure, we may eventually succeed in adequately qualifying this initial translation (by adding specific features and thus turning it into a final translation) or replacing it by a more suitable one.[31] This methodological principle, in combination with the ideas on meaning variance outlined above, allows us to formulate the following three 'rules of thumb':

(1) The historian speaking about e.g. 'the history of *the* concept of space' should be careful not to use 'space' in any specific sense, let alone in any of its modern technical senses, but rather as a kind of common denominator of a class the members of which—though connected by relations of family resemblance—may turn out to be rather dissimilar.

(2) When initially (i.e. without detailed knowledge of the context) translating an ancient term by 'space', the historian is merely labelling the corresponding ancient concept as a member of a rather heterogeneous class. He should be aware, therefore, of the unspecific character of this translation.

---

[31] Of course these considerations do not only apply to the problem of translating scientific or philosophical terms. Also objections which the historian Lauffer put forward in the text quoted above against translating the greek term *doulos* by 'slave' (on which see above n. 26) can be met. We may in general use 'slave' as an initial translation of the word *doulos,* because of some *prima facie* resemblances between the *doulos* described in a particular text and 'our' slave. But in doing so there are two things we should bear in mind. First, the fact that we do translate the ancient term does not imply that we have already decided to invest this ancient *doulos* with all (or even with all essential) features of our 'slave'. Second, in the course of our interpretation it may turn out that the features shared by the ancient and the modern concept are so few or so irrelevant that another translation is to be preferred. This, of course, does not preclude the possibility of the popular stereotype of what it is to be an ancient *doulos,* being, for lack of awareness of possible intercultural differences, in the end largely determined by a number of anachronistic connotations carried by this first translation. However, the professional historian must be keen always to apply his initial translations with a certain *proviso.* Eventually he may, by a closer investigation of the contexts in which the ancient terms were used, succeed in adequately refining, qualifying, or even refuting the first conjectural translation.

(3) As the specific technical load of a concept will in each historical case depend on the role it is made to play by an individual philosopher or physicist, an attempt to describe and translate (in the sense of: giving a final translation) any specific ancient spatial concept should in the end be based on a careful contextualist approach. The fact that we initially describe and translate the concept at issue as 'space' should only be taken as a first orientation and not as an attempt to invest the ancient concept with all connotations of any specific concept of space.[32]

1.2.2 As was indicated earlier, one of the main reasons to pay so much attention to what we are really doing, when we describe and translate ancient concepts, and to take so much trouble to provide some methodological rules (which of course are hardly anything more than a somewhat sophisticated version of the trivial advice 'be careful'), is the fact that implicit 'essentialist' views as to what it really is to be a concept of space are not uncommonly found in the secondary literature. They basically appear in two different guises: as a tendency to provide too specific initial translations or descriptions, or as a tendency to come up with anachronistic 'pairings' of the kind described in the introduction of this chapter. In what follows I shall give some examples of both phenomena. I have tried to select them from typical and responsible studies by well-known historians.

First an example of a rather misleading anachronistic pairing. Consider the following programmatic statement which is to be found in Max Jammer's famous *Concepts of Space* :

> It is the purpose of this monograph to show the development of *the* [my italics] concept of space in the light of the history of physics.[33]

If—in line with the conclusions drawn above—we take the expression '*the* concept' to be elliptical for something like 'a family of concepts of space which play different roles but which still show a certain resemblance' nothing is wrong. However, Jammer's

[32] Note that, from the point of view of historical method, it is of no importance whether that specific concept of space has sophisticated and clear-cut outlines, or whether, in using it, we merely express ourselves in conformity with an unarticulated common sense view.

[33] Jammer (1954) 2.

position—though unfortunately on this point he is nowhere really explicit—seems to be different. As the following example shows he seems to think that if person A and person B each had a concept of space, and if in both cases that was a concept of a *finite* space, the two concepts of space cannot fail to be fairly similar. This allows him to make a comparison between chronologically and theoretically very divergent 'concepts of space':

> [...] space, according to Aristotle is finite, matter being itself finite [...] The idea of a finite physical space, thus understood, is not as absurd today as it must have appeared fifty years ago [...] It is perhaps not wholly unjustified to suggest a comparison between the notion of 'physical space' in Aristotle's cosmology and the notion of Einstein's 'spherical space' as expounded in early relativistic cosmology.[34]

Concerning this quotation the following observations can be made. First, in *Phys.* Δ—which, as the context shows, is what Jammer is here talking about—Aristotle is not providing a theory of *space*, but a theory about what is going on when we localize individual physical bodies, in short, a theory of *place*. So even on this elementary level the comparison turns out to be misleading. This, however is not my main objection. For it might be argued that such an Aristotelian theory of place could easily be transformed into a more general theory of space.[35] What really makes the comparison unacceptable is the fact that even if we were granted to maintain that this Aristotelian space—being in fact nothing but a framework of locations of individual substances—is finite because the Aristotelian cosmos is finite, it is still misleading to represent this finiteness as a common ground between Aristotelian and Einsteinian space. For the Aristotelian and the Einsteinian conceptions of finiteness are of a significantly different kind.

In the case of the General Theory of Relativity the supposed

---

[34] Jammer (1954) 20; the additional point of comparison adduced by Jammer on the same page, viz. 'the idea of 'geodesic lines' determining the geometry of space' suggesting 'a certain analogy to the notion of natural places and the paths leading to them', is too vague to be taken seriously.

[35] In fact, it may be argued that Aristotle's pupil Theophrastus made an attempt to accomplish such a transformation, on which see below chapter 5, section 5.3. For a recent attempt (!) to provide 'An Aristotelian Theory of Space' which is to be regarded as an 'outline of the necessary development of the theory' [scil.: of place] see King (1950), particularly the second part (91 ff.).

finiteness of space-time is supported by mathematical considerations perfectly alien to Aristotle's thought. More precisely the curious Einsteinian conception of a finite universe is based on the distinction made in non-Euclidean Riemannian geometry between the unlimited and the infinite, the 'curved space' adopted by the General Theory of Relativity being both finite and unlimited. Aristotle's finitism, on the other hand, was called forward by metaphysical considerations, in particular by the preeminent status of his notions of 'form' and 'limit'. These very considerations inhibit us from calling his universe 'unlimited'.

So, unless Jammer's remarks are meant to be extremely trivial —merely pointing out that these kinds of space are both, though for different reasons, *finite*,—they are bound to be positively misleading. In general, attempts to invest ancient thinkers with 'interesting modern theories' tend in the end to prove nothing more than that the person at stake held 'the right view, though for the wrong reasons'. In this case, however, owing to his different conception of finiteness, Aristotle cannot even be argued to have subscribed to 'the right' (i.e. presumably: Einstein's) view.[36]

Next, some examples of how essentialist presuppositions may infect historical descriptions or translations. The most common essentialist presupposition is of course the assumption that there are a number of privileged features which are shared by several concepts of space, and which together constitute what we might call the *ideal type* of space.

Thus it has been argued by Henry Mendell that in *Phys.* Δ Aristotle did not provide a theory of *space*, which is in itself correct. One of the arguments adduced, however, is that 'a classical (!) theory of space must at least have two components: space is a continuous extension and it is independent of any substances which occupy it'.[37] What is in fact only one of the *possible*

---

[36] It should come as no surprise that Sambursky's *The Physical World of the Greeks*, a work which is on the whole seriously infected with precursoritis expresses a view quite similar to that held by Jammer: 'Aristotle's combination of geometry and matter to form his concept of place is not unlike the conception of space in the General Theory of Relativity' (page 96). Similar remarks in Düring (1966) 319. Guthrie (1967–81) II, 46 makes an equally unfortunate comparison between Parmenides and Einstein: 'The All of Parmenides is more like the curved and finite Einsteinian space than the Euclidean conception which still dominates our ordinary thinking'.

[37] Mendell (1987) 224.

concepts of space (and what we encountered as only one among the kinds of concepts of space actually held by philosophers) is here in fact transformed into the *prototype* of all *real* concepts of space. Consequently it is implied that what I just labelled space of the (b) kind (e.g. Leibniz' space or Theophrastus' elaboration of Aristotle's theory of place) does not belong to the class of concepts of space, at least not to the class of concepts of *classical* space, whatever that may be.[38]

Similar objections apply to the way in which E. Hussey, in his recent commentary on the third and fourth books of Aristotle's *Physics*,[39] explains the absence of any positive elaboration of a concept of *space* (as distinct from *place*) in Aristotle's *Phys.* Δ. According to Hussey, two factors are primarily responsible for this absence, viz.

(1) the fact that theories of a three-dimensional self-subsistent space are rejected by Aristotle, and

(2) the fact that the other main role to be played by a concept of space, viz. the role of a receptacle, a reservoir of physical possibilities, is in Aristotle's system covered by *matter*.[40]

What is implied by these lines is that what would in fact be the only possible theory of space on Aristotelian lines (viz. an elaboration of his theory of place into a theory of space of the (b) kind outlined above) is said to be not *really* a theory of space, since it obviously does not fit Hussey's ideal type of space. On the other hand, Aristotle's concept of *matter*, which Aristotle himself took pains radically to distinguish from a concept of *topos*, is here said to be *in reality* a concept of *space*, a fact of which Aristotle unfortunately was unconscious. Thus a phenomenon in Aristotle's physics is explained on the basis of presuppositions which are, for

38 It might be that in referring to the 'classical' concept of space Mendell had the Newtonian concept in mind, which of course is often called that way. However, it was only to become *classical* in classical (post-Newtonian) physics; so it is not clear at all why we should expect any ancient theory of space to be classical in this sense.

39 Hussey (1983) xxviii–xxxii.

40 Hussey (1983) xxxi; cf. also page xxix, where, however, Hussey is obviously claiming too many metaphysical applications for Aristotelian matter: 'Space might still be needed as a metaphysical guarantee that spatially existing bodies, locomotion, and spatial relationships will continue to be possible—but, in so far as it is needed in this (sic!) role, Aristotle does have a 'space'[viz. matter, KA]'.

one thing, simply wrongheaded, and which, for another, are quite alien to Aristotle's own thought.

More or less the opposite presupposition concerning the 'real nature' of space is found in the course of Crombie's exposition of the contents of Plato's *Timaeus*. It has already been briefly referred to in the introduction of this chapter. Concerning the receptacle Crombie remarks that 'we begin to wonder whether the word 'space' is not metaphorical. For space *is simply a system of relationships*, whereas Timaeus' 'third force' seems more like a thing' [my italics].[41] Now, although there can be little doubt that Plato's exposition of the receptacle presents huge problems, I am afraid the problem here detected by Crombie is not among them. Space does not have to be of the (b) kind and in those cases where it is not—as in Plato's *Timaeus*—a thing-like presentation is less odd than is here suggested.

As a final example I would like to point to Cornford's argument for preferring 'place' to 'space', as a translation of the term *chôra* in Plato's *Timaeus* : ''Place' would indeed be a less misleading translation than 'space', because 'place' does not suggest an infinite extent of vacancy lying beyond the finite sphere of the universe'.[42] Cornford's fear of a misleading translation in reality betrays a misleading presupposition concerning the real nature of space on his part. Plato was committed to a concept of finite space and so were most if not all Aristotelians and Platonists in antiquity. Although the notion of a finite space may seem rather counterintuitive, because it is difficult to imagine that such a space is not bounded by anything external, it is not intrinsically absurd or self-contradictory, and there is no reason to suppose that the notion of space in itself involves the notion of infinite extension.

Conclusions

The present, preliminary, chapter tried to answer two questions: (1) in which sense were the different philosophers who have been theorizing about space all talking about the same thing, and (2) what are the consequences of the answer to question (1) for the way the historian uses (or should use) his descriptive labels?

---

[41] Crombie (1963) 218.
[42] Cornford (1937) 200, n. 2.

As to question (1), an outline was provided of how the history of concepts of space should be envisaged. It was suggested that the different concepts of space endorsed throughout history are best regarded as a class of which the members are connected by a family resemblance. This family resemblance could be further specified by pointing out that in antiquity (and for a considerable time afterwards) the different philosophical concepts of space were in the end all rooted in the conventions of ordinary thinking and common parlance.

Once this was established, question (2) could be answered by providing some simple methodological rules which help to avoid at least some gross historical distortions. How indeed unwarranted essentialist presuppositions may lead to such historical distortions was shown by a number of examples taken from some well-known historical studies.

As yet we have concentrated on 'our side of the matter', i.e. on what *we* mean or should mean when we call an ancient concept e.g. a concept of 'space'. We did not yet seriously study how the Greeks themselves used their own spatial terms. This will be done in the next chapter.

CHAPTER TWO

TOPOS, CHÔRA, KENON: SOME CASE STUDIES

Introduction

The present chapter will investigate the ways in which the Greeks, both in common parlance and in philosophical language, used their main spatial terms, *chôra*, *topos*, and *kenon*. Since in antiquity neither empirical experiments nor mathematics played any role of importance in philosophies of space, the ways in which spatial terms were commonly applied constituted, so to speak, the raw material of the philosophical debate. Sometimes these spatial terms were used to express clear-cut concepts in a fairly unambiguous manner, as e.g. in Aristotle's dialectical presentation of the four possible candidates (form, matter, separate extension, limits of the surrounding body) for the identification of *topos* in *Phys* Δ. In these cases the philosophical debate might be described as a rather straightforward clash between different concepts of space. More often, however, it was precisely the vagueness of the concepts or the ambiguity of the terms which played a leading role in the debate.[1] This was particularly the case in the period before the rise of the great Hellenistic schools, when philosophers had not yet created a more or less technical vocabulary. The present chapter will provide some outlines of the ways in which terminological problems dominated Greek philosophies of space to a considerable extent, during the period with which we are dealing in this book. As such it will provide a useful basis for the discussions in chapters 3–5.

---

[1] Of course the distinction between *conceptual* and *terminological* unclarity is in practice merely a matter of perspective. Whether or not we will in each case infer conceptual confusion on a philosopher's part from an apparent terminological confusion will be a matter of charity. Thus, as will become clear in the present chapter, *we* may be willing to credit the early atomists with a consistent theory of space—in spite of the fact that it was probably not coherently formulated—whereas Aristotle was obviously unwilling to apply such a charitable interpretation. In the present chapter I shall in general speak as if philosophers who used one and the same term in various manners were *ipso facto* (even if unconsciously) committed to different concepts.

The first section (2.1) will be concerned with the terms *chôra* and *topos*. I shall try to show how the philosophical use of these terms can be connected with their everyday use. A closer look at the rules which apparently governed the choice between the two terms will allow us to criticize the 'automatic' translations often found in the scholarly literature, which assume a one-to-one correspondence between *topos* and 'place' and *chôra* and 'space'. The Greek language did not have a terminological distinction matching the conceptual distinction between place and space.

This is not to say that the philosophers did not differentiate between place and space. At least Plato and Aristotle may be charitably credited with such a conceptual distinction, and we may translate both *topos* and *chôra* as 'space' or 'place' according to the context. Nevertheless the lack of a univocal term exclusively denoting space *qua talis* had two important corollaries. First, it left room for the use of the term *kenon* in this connection (i.e. as an equivalent of 'space'), a phenomenon which created problems of its own. Second it furthered the attempts made by Epicurus and the early Stoics to 'isolate' the notion of space, by inventing new 'technical expressions' for it.

Next, in 2.2, we shall focus on the different applications of the term *kenon*. It will be argued that the typical substantivated use of the Greek term allowed of different applications (without this implying a difference from the purely lexicographical point of view).[2] Moreover, it will be argued that the first philosophically interesting uses of the term *kenon* were determined by the Eleatic notion of not-being.[3] Thus the different applications of the term *kenon* by the early atomists may all be regarded as called forward by different aspects of what may be called 'the Eleatic challenge'. The *prima facie* incompatibility of these different applications induced Aristotle's criticisms and led to attempts, attested for Epicurus and Strato, to refine the rules of application of the spatial terms in general, and of the term *kenon* in particular.

Nevertheless we shall see that also philosophers who were professedly committed to a well-defined concept of *kenon* are still sometimes seen to use different applications of the term *promiscue*.

---

[2] This should be borne in mind when in the present chapter I speak of different 'concepts' or 'conceptions' of the void.

[3] I have left out a discussion of the odd, more or less mythical, concept of κενόν that was endorsed by the early Pythagoreans.

2.1 *Chôra* and *topos*

2.1.1 Of the terms *chôra* and *topos* the former is the one which appears earliest in the written sources. Its basic meaning seems to be 'land/region/ground'. Thus, e.g., in the 8th book of the Odyssey Alcinous can ask Odysseus ὅππῃ ἀπεπλάγχθης τε καὶ ἅς τινας ἵκεο χώρας / ἀνθρώπων ('where did your wanderings lead you and which countries of men did you visit?').[4] In subsequent literature this meaning seems to be common.[5] When applied to a smaller piece of ground *chôra* gets the meaning 'stretch/field/ground/place'. Homer's swineherd puts Odysseus' bow down 'right there on the ground' (θῆκε φέρων αὐτῇ ἐνὶ χωρῇ, i.e. on the ground before his feet),[6] and, when during a chariot race Menelaus is said to be very close to Antilochus, the distance between them is compared to the distance between a horse and its chariot: ὁ δὲ ἄγχι μάλα τρέχει οὐδέ τι πολλὴ / χώρα μεσσηγύς, ('and the horse is running directly in front of it and there is but a small stretch between them').[7]

In those cases where *chôra* should be translated 'place/space' the idea is always that of an extension, whether two- or three-dimensional, which is occupied or which can be occupied. I have not been able to find instances where *chôra* is used in order to refer to a relative location or position. The expression which comes closest to such a use of *chôra* is Xenophon *Cyr.* 4, 5, 37: ἕως ἂν χώραν λαβῇ [τὰ πράγματα] which LSJ translate 'till they are brought into *position*, into *order*'. Though this translation is not incorrect, the use of the verb λαβῇ indicates that *chôra* still has much of its original sense: it is an extension which can be occupied ('taken').

In written sources the term *topos* is not encountered before Aeschylus. To judge from the ways it was generally used, we may infer that it was largely synonymous to *chôra*. The Eumenids in

---

4 *Od.* θ 573.

5 Cf. e.g. Aeschylus *Eu.* 287, where Orestes beseeches χώρας ἄνασσαν τῆσδ' Ἀθηναίαν ἐμοὶ μολεῖν ἀρωγόν ('[I call for] Athena, the goddess of this city/land, to come to my help'; a few lines later, *Eu.* 292, we encounter the word in combination with τόπος in the phrase χώρας ἐν τόποις Λιβυστικῆς ('in the places/regions of the Libyan land'). In Homer χώρα may also mean 'land/shore' as opposed to 'sea': cf. *Il.* Π 68, *Od.* π 352 (where it is said that a ship is στρεφθεὶς ἐκ χώρας).

6 *Od.* φ 366.

7 *Il.* Ψ 521.

Aeschylus *Eu.* 249 tell us that 'every *region* of the earth has been grazed off' by them (χθονὸς γὰρ πᾶς πεποίμανται *τόπος*). In the *Persae* Darius speaks of his army being 'in the *regions* of Greece' (ἐν Ἑλλάδος *τόποις*).[8] In those cases, however, where *chôra* and *topos* are used together, *topos* apparently may denote a *part* of the χώρα, as in Plato *Lg.* 760 c: οἱ τῆς χώρας *τόποι* ('the *places* of the country').[9] On the other hand *topos* may—in contradistinction, as it seems, to *chôra*—also be used to denote relative location or position. This is notably apparent in those contexts where also the term *chôra* is present. Thus in Demosthenes 4, 31 we find ὁ *τόπος* τῆς χώρας which must mean 'the (geographical) *position* of the region/country'.[10]

We may tentatively conclude, then, that in common parlance *topos* and *chôra* were used more or less *promiscue*, the only traceable difference being that, whereas *chôra* appears to have always denoted a certain *extension*, *topos* could also be used just to denote location in relation to the surroundings. Moreover, in contexts where the terms are used together and where they *both* denote extension *topos* seems to have been thought of as smaller than, and part of, *chôra*.

2.1.2 As will be clear from all this, whether *topos* and *chôra* should be translated 'space' or 'place' depends on the context in which they appear rather than on anything like their 'intrinsic meaning'. This is not a trivial thing to notice, for the scholarly literature quite often assumes that there is a one-to-one correspondence between *topos* and 'place' on the one hand and *chôra* and 'space' on the other.[11]

Such a correspondence seems, for instance, to be what is implied by Jammer's statement that 'In the *Phys.* Aristotle uses exclusively the term 'place' (*topos*)'.[12] If we look at the context in which this remark is to be found, it appears that it should probably be understood as qualifying the previous contention that 'Aristotle's

---

[8] Aeschylus *Pers.* 769.

[9] See also the phrase from Aeschylus *Eu.* quoted above, n. 5.

[10] See also Pl. *Lg.* 705 c.

[11] Some examples in addition to those here adduced in the text: Claghorn (1954) 13: 'it was to place (τόπος) [...] rather than to space (χώρα) that Aristotle devoted most of his attention'; Keyt (1961) 291: 'Aristotle finds that Plato identified [...] space (χώρα), place (τόπος), and void (κενόν)'.

[12] Jammer (1954) 15.

theory of space is expounded chiefly in his *Categories* and [...] in his *Physics*'.[13] On the basis of these statements the reader might be tempted to believe that, whereas in the *Phys.* we are dealing with *topos* (= place), the *Categories* use another spatial term, not specified by Jammer, which is equivalent to our 'space'. However, one glance at the relevant passage of the *Cat.* is sufficient to show that there also Aristotle uses the term *topos*.[14]

Similarly, in the chapter of *Aristotle's Physical World* which deals with 'matter and place', Solmsen contrasts Plato and Aristotle with the following words: 'For the *Timaeus* it is, on the whole, space (*chôra*) rather than place (*topos*) which helps to set individual physical things [...] apart from the eternal Forms'[15]. Regardless the overall truth or falsity of this sentence, it is certainly not the case that in Plato the words *chôra* and *topos* have the clearly separated meanings which Solmsen apparently ascribes to them. It may be sufficient in this connection to quote *Tim.* 79d4 where *chôra* should no doubt be translated 'place' : τὸ δὲ θερμὸν [...] εἰς τὴν ἑαυτοῦ χώραν [...] ἰέναι. Since, of course, the existence of a one-to-one correspondence between *chôra* and 'space' and *topos* and 'place' is disproved if we find only one counterexample, this example, stemming from a philosophical context, should suffice to show the potential inadequacy of this kind of 'automatic' translations, at least in as far as we are concerned with the word *chôra*. Moreover, a quick glance at a dictionary will be sufficient to show that also in many non-philosophical contexts the word *chôra* might well be translated 'place' (e.g. in the expression—already referred to above—ἕως ἂν χώραν λαβῇ τὰ πράγματα).[16]

In the case of *topos* the situation is somewhat more complex. As the term is usually applied in a relational setting,[17] it may accord-

---

13 Note that, strictly speaking, it contradicts rather than qualifies that previous contention. This is exemplary of Jammer's slack argumentation in the sections dealing with Aristotle.

14 Amusingly, Patricios (1971) 17–36, who in general paraphrases Jammer, at last (page 26) seizes the opportunity to improve upon his master and to supply the Greek word Jammer has left out. For apparently without consulting Aristotle's Greek text he argues that 'Aristotle's theory of 'space' (*χώρα*) is explained briefly in his *Categories* while his *Physics* is concerned exclusively with the theory of place (τόπος)'.

15 Solmsen (1960) 131.

16 Xenophon *Cyr.* 4, 5, 37; see LSJ, *s. v.*, 3, for more examples.

17 Cf. Sextus' explicit remark, *P.* 3, 134, that ὁ τόπος πρὸς τῷ σώματι

ing to the rules outlined in the previous chapter, it is true, in most cases conveniently be translated by 'place', with the additional qualification that, as in the case of our 'place' (on which see ch. 1) the idea of three-dimensional extension is not necessarily involved. Nevertheless, there are instances when *topos* is used to denote the underlying extension, not of an individual thing, but rather of the whole universe, or of all things.[18] In these cases, of course the translation 'space' is also possible, and often even to be preferred.

We may conclude for the moment that even the limited survey of cases here provided is sufficient to show that, though *topos* may in most cases conveniently be translated by 'place' and *chôra* at least in a number of instances by 'space', our opting for either of these translations will in the end have to be determined by the context in which the Greek terms occur, and cannot take place automatically.

2.1.3 The earliest philosophical uses of *topos* and *chôra* are still closely connected to these common language uses, though of course physics usually endows *topos* and *chôra* with three instead of two dimensions. Plato seems to put the terms more or less on a par, though, as was already pointed out[19] at times *chôra* may denote a larger extension than *topos*. As a third equivalent Plato uses the term *hedra*, a term which in everyday (and poetic) speech could indeed be used as a synonym for *topos* or *chôra*,[20] but which was by and large dropped in later philosophical language,[21] possibly because of its bearing too concrete associations ('bench, seat').

It is interesting to notice that in those passages in *Phys.* Δ where he is giving his own positive views concerning place—i.e. where the localization of a body is basically a localization relative to the

---

νοεῖται οὗ ἔστι τόπος.

[18] Cf., e.g., the dialectical passage in Arist. *Phys.* Δ 208 b 34 ff.: εἰ δ' ἐστὶ τοιοῦτο (scil.: preexisting like Hesiod's χάος) θαυμαστή ἂν εἴη ἡ τοῦ τόπου δύναμις [...] οὐ γὰρ ἀπόλλυται ὁ τόπος τῶν ἐν αὐτῷ φθειρομένων (this passage is substantially echoed by S. E. *M.* 10, 12).

[19] See above, 34; cf. Brisson (1974) 213.

[20] Cf. Aeschylus *Eu.* 11: ἐς τήνδε γαῖαν ἦλθε Παρνησοῦ θ' ἕδρας; Hdt. VII, 37: ὁ ἥλιος ἐκλιπὼν τὴν ἕδραν.

[21] Some exceptions: Arist. *Meteor.* 356 a 4 and Plu. *Quaest. Conv.* 653 ff. (= Epicur. fr. 61 Usener).

immediate surroundings (the outer limit of the containing body) —Aristotle exclusively uses the term *topos*. On the other hand he uses *topos* and *chôra* interchangeably in passages which are clearly of a dialectical nature (i.e. which picture the views of his opponents who thought of place as an *extension*).[22] Thus he seems to have complied with the conventions of common parlance, according to which, as we noticed, *chôra* denoted basically an extension, whereas a relative localization could only (or at least: more naturally) be expressed by the term *topos*. In those cases elsewhere where Aristotle does use the term *chôra*, he is arguably loosely applying a common sense concept of space (space as a three-dimensional extension) which is not his own.[23] Where he refers to his own positive views, he sticks to the use of *topos*, also outside the confinements of the *topos* discussion in *Phys.* Δ.[24]

In the case of the great Hellenistic schools we witness a slight, but not unimportant, change in the use of spatial terms. The Stoics at any rate, but in his own way also Epicurus[25], turned the terms *chôra* and *topos* into more or less technical terms. That these technical uses were, not surprisingly, still rooted in the conventions of common parlance, appears from the fact that both Epicurus and the Stoics used *topos* in a *relational* setting, i.e. as the *topos of something*, so that in both cases the term may well be translated, in accordance with the rules laid down in the previous chapter, by our 'place'. Moreover, they also complied with the convention that when *topos* and *chôra* are juxtaposed *chôra* refers to a larger extension than *topos*. What was new, however, was that both Epicurus and the early Stoics isolated a notion of 'space as such' and that they defined their technical *topos* and *chôra* in relation to that notion of 'space as such'. Interestingly, neither of these schools used *topos* and *chôra* themselves as one-word equivalents of what we would call 'space'—yet another argument against automatically translating e.g. *chôra* by 'space'.

The Stoics referred to space as such as 'that which can be

---

22 Cf. *Phys.* Δ 208 b 7, 209 b 8, and 209 b 15.

23 Examples of such 'loose uses' on Aristotle's part will be extensively discussed in ch. 4.

24 So the statement that 'for both Plato and Aristotle the two terms are nigh synonyms' (Mendell (1987) 214) needs qualification. For a more extensive discussion of Aristotle's use of spatial concepts see below ch. 4.

25 On which see below, 53–55 and ch. 6, 263–281.

occupied by a body' (τὸ οἷόν τε κατέχεσθαι ὑπὸ σώματος). The meanings with which they invested the words *chôra* and *topos* will be discussed at large in ch. 6. At the moment it may suffice to note that, though their *topos* may well be rendered as 'place', the term *chôra* seems to have been given different technical and non-technical senses. At any rate, it cannot simply and automatically be translated as 'space'.

Epicurus used the name *anaphês phusis* for space in its broadest sense, while turning *chôra* and *topos* into technical terms which refer to space *in different contexts*, *chôra* being space as 'room' when bodies are moving through it, whereas *topos* is space when it is occupied by body (i.e. place).[26] Though, as we shall see below (in 2.2.4), Epicurus himself did not always stick to his own conceptual distinctions, they at least in principle allowed his system to stay free of some of the obscurities inherent in the theory of space of the early atomists.

2.1.4 By way of a summary, we may draw the following conclusions:
(1) The terms *chôra* and *topos* could in a number of contexts be used interchangeably, both in ordinary Greek and in their first philosophical applications, though in other contexts the one or the other might be preferred (*topos* in order to denote relative location, *chôra* in order to denote a larger extension than *topos*).
(2) The rules governing the choice between *chôra* and *topos* did—contrary to what is usually suggested in the scholarly literature—not always run parallel to the rules governing our choice between 'space' and 'place' so that 'automatical' translations have to be rejected.
(3) Until the period of the Hellenistic schools neither Greek common parlance, nor the early philosophical Greek had a term which *exclusively* denoted space. In this respect Epicurus and the early Stoics were important innovators.

2.2 *To kenon*

2.2.1 The next spatial term up for discussion, *to kenon*, differs in an important respect from both *topos* and *chôra*. The latter two are

[26] See below, 53–54. For a convenient *exposé* of these matters see now Long/ Sedley (1987) I, 30.

nouns, whereas *to kenon* is a substantivation of the adjective *kenos* ('empty'). In common Greek usage the word *kenos* only occurs as an adjective to form, in combination with a noun, expressions like κεινῇσι χερσί ('with empty hands', Herodotus I, 73) or κενὰς παλαίστρας ('empty wrestling-places', Aristophanes *Nu.* 1054). So it seems that the substantivation of the adjective in philosophical usage constitutes a technical development of ordinary language.

It has been argued more than once that the use of the definite article to in Greek allowing this kind of substantivation contributed much to the development of abstract thought in Greek philosophy.[27] The other side of the medal is—as may be observed e.g. in the writings of Plato, who made extensive use of such substantivated adjectives—that expressions of the form 'to (x)', where (x) is an adjective, may become ambiguous. They may mean either 'the concept of (x)' (or: '(x)-ness') or 'a thing which is (x)'. In the latter case, the adjective is still used qua adjective, while the particular noun which is being qualified is left out. In the former case, the expression bears an 'abstract' sense. Thus an expression like *to dikaion* may mean 'justice' but it may also mean 'the just thing/action'.[28]

I would suggest that a similar phenomenon is responsible for the different uses attested for the term *to kenon.* The occurrences of the term in Greek philosophical texts give us reasons to suppose that *to kenon* could be used to denote

(1) emptiness as such; when used in an extensional (as opposed to intensional) way, i.e. not as a mere *concept, to kenon* might thus refer to mere extension or space;

(2) a specific empty (x); this could be :

  (1) an empty space or place;

  (2) an empty 'thing', e.g. an empty vessel.

This explanation would at any rate account for the fact that in Greek philosophical discussions of the void we encounter basically three different conceptions of *to kenon,* viz.:

(a) *kenon* = space (example: Aristotle's characterization of the Platonic receptacle as *kenon* in *Phys.* Δ 214 a 12 ff.);[29]

---

[27] Cf. B. Snell, *Die naturwissenschaftliche Begriffsbildung im Griechischen,* in: *Die Entdeckung des Geistes* (Göttingen 1975[4]) 205–218.

[28] For a fairly complete survey of the 'meanings' of such common names as *to hosion* in Plato's work see De Rijk (1986) 41–44.

[29] On which see below, chapter 3, 112–113.

(b) *kenon* = empty space or empty place (example: Epicurus' primary use of kenon as the empty space separating the atoms);[30]

(c) *kenon* = an empty thing or an empty part of a thing (e.g. an empty vessel, an empty pocket; example: Strato's theory of empty pores);[31]

At first sight, the separate mention here of the (c) conception may seem rather odd. Indeed, the difference between the (b) and (c) kind tends to be a matter of perspective. In other words, the (b) and the (c) conceptions of void do not necessarily correspond with two different conceptions of space. Nevertheless, the gist of this difference in perspective may be adequately illustrated by the following example provided by David Sedley :

> There should be nothing intuitively abhorrent about the idea of something with a purely negative characterization occupying places and moving—a gap in the traffic, for example. When you carry your thermos flask to work, you would do well to think of the vacuum in it as moving from place to place with it. If you insist instead that the vacuum in it is empty *space* and therefore incapable of moving *in* space, you may have to conclude that throughout your journey the vacuum in your flask is being replaced in a constant stream.[32]

In the present section I want to show how a number of philosophers used two (or even all three) of these kinds of *kenon promiscue.* At the same time it can be shown that this sometimes led to obscurities which were an easy target for critics like Aristotle. Attempts were made, e.g. by Epicurus, Strato, and the early Stoics to forestall these criticisms by making the necessary conceptual distinctions. However, the tendency to slide from one conception of *kenon* into another was apparently so strong that even some of these philosophers may be seen not to have always been faithful to their own conceptual distinctions. I shall first discuss the way the term *kenon* is used by Melissus (2.2.2), and then the usage of early atomists (2.2.3), Epicurus (2.2.4) and Strato (2.2.5). Some views of Aristotle and the Stoics will be referred to in passing.

---

[30] On which see below, 53–56.

[31] On which see below, 58–69.

[32] Sedley (1982) 175; see also Sorabji (1988a) 18 and esp. 78, where it is suggested that a more appropriate name for the underlying immobile vacuum would be 'space', whereas an appropriate name for the mobile vacuum (what I have called the (c) kind) would be 'volume' .

2.2.2 If we discount the fumbling attempts of the early Pythagoreans to come to terms with void,[33] which from a physical perspective are hardly of any importance, the first philosophical problematization of the concept of *kenon* of which we know, appears in the work of the fifth century Eleatic Melissus. In a very interesting fragment (fr. B 7 DK) preserved by Simplicius (*In Ph.* 111, 18 ff.) Melissus offers a number of deductions which purport to show—very much in the fashion of the list of 'signs (σήματα) of Being' offered by Parmenides fr. 28 B 8 DK—that his one Being is in every possible respect immutable. It is eternal, infinite and homogeneous, it does not perish or grow (οὔτ' ἂν ἀπόλοιτο οὔτε μεῖζον γίνοιτο), it does not change its internal order (μετακοσμέοιτο), or suffer (ἀλγεῖ), or move (κινεῖται), and, finally, it is not dense and rare (πυκνὸν καὶ ἀραιόν). I shall here only deal with the latter two characteristics.

The last part of the fragment (§§ 7–10 in the numbering of DK) runs:

> (§ 7) Furthermore, there is absolutely no void. For void is not-being and the nothing could not exist. And it does not move. For it cannot move in any direction. But it is full. For if there were void, it would move into that void, but since there is no void it has nothing to move into.
> (§ 8) And it could not be rare and dense. For it is not possible that the rare is as full as the dense, but the rare is more empty (κενεώτερον) than the dense.
> (§ 9) And this is the criterion for distinguishing between the full and the not-full: if something provides room or receives, it is not full. If, however, it does not provide room or receive, it is full.
> (§ 10) So it must be full, if there is no void. And if it is full, it does not move. [34]

The first argument (7) obviously links up with a number of features of Parmenides' description of Being. Thus Parmenides' Being is 'immobile' (ἀτρεμὲς, fr. 28 B 8, 3 DK) and 'immovable in

---

[33] Cf. Sedley (1982) 176.

[34] (§ 7) Οὐδὲ κενεόν ἐστι οὐδέν· τ ὸ γὰρ κενεὸν οὐδέν ἐστιν· οὐκ ἂν οὖν εἴη τό γε μηδέν. οὐδὲ κινεῖται· ὑποχωρῆσαι γὰρ οὐκ ἔχει οὐδαμῇ, ἀλλὰ πλέων ἐστίν. εἰ μὲν γὰρ κενεὸν ἦν, ὑπεχώρει ἂν εἰς τὸ κενόν· κενοῦ δὲ μὴ ἐόντος οὐκ ἔχει ὅκῃ ὑποχωρήσει. (§ 8) Πυκνὸν δὲ καὶ ἀραιὸν οὐκ ἂν εἴη. τὸ γὰρ ἀραιὸν οὐκ ἀνυστὸν πλέων εἶναι ὁμοίως τῷ πυκνῷ, ἀλλ' ἤδη τὸ ἀραιόν γε κενεώτερον γίνεται τοῦ πυκνοῦ. (§ 9) Κρίσιν δὲ ταύτην χρὴ ποιήσασθαι τοῦ πλέω καὶ τοῦ μὴ πλέω· εἰ μὲν οὖν χωρεῖ τι ἢ εἰσδέχεται, οὐ πλέων· εἰ δὲ μήτε χωρεῖ μήτε εἰσδέχεται, πλέων. (§ 10) Ἀνάγκη τοίνυν πλέων εἶναι, εἰ κενὸν μὴ ἔστιν. εἰ τοίνυν πλέων ἐστιν, οὐ κινεῖται.

the limits of big chains' (ἀκίνητον μεγάλων ἐν πείρασι δεσμῶν, *ibidem*, 26) and, rather abundantly, 'it stays the same in the same and lies at itself, and thus it stays there fixed' (ταὐτόν τ' ἐν ταὐτῷ τε μένον καθ' ἑαυτό τε κεῖται / χοὔτως ἔμπεδον αὖθι μένει·, *ibidem*, 29/30). Finally, in line 37 of the same fr. B 8, Parmenides states that Destiny has bound it (Being) to be one and immovable and that for that reason all things postulated by mortal men, like becoming and perishing, being and not-being and changing place (τόπον ἀλάσσειν), are mere names (ὀνόμασται).

In Parmenides the negation of the existence of place (*topos*) is a consequence of the one-ness of Being: if there were place next to Being, it would *ipso facto* be something other than Being (i.e., presumably, not-Being) and Being itself would no longer be one. However the nature of this *topos* is left unspecified: it is just one of the possible forms which not-Being may take in the minds of mortal men. Melissus transformed the Parmenidean Being. He endowed it with extension, thus making it, so to speak, more concrete. Moreover, he made it *infinite*, both *qua* extension and *qua* duration.[35] Thus it is not surprising that also Melissus' not-Being takes the guise of empty *extension*, i.e. of the void. In the fragment just cited Melissus argues from the non-existence of the void to the non-existence of motion: if there were a void, Being would move into it. It is obvious that *to kenon* here stands for empty *space*, or an empty *place*,[36] i.e. it is what I called a (b) *kenon*. In this respect it is

---

[35] In doing so he probably consciously elaborated and changed what in Parmenides was merely an *image*, viz. the representation of Being as extended, as a finite sphere (fr. B 8, 43 ff. DK). Against a 'spatial' interpretation of Parmenides' Being, cf. e.g. W. J. Verdenius, *Parmenides, Some Comments on his Poem* (Groningen 1942, repr. Amsterdam 1964) 4. See also Tarán (1965) 150–160. A. P. D. Mourelatos, *The Route of Parmenides, a Study of Word, Image and Argument in the Fragments* (New Haven/ London 1970) 123–129 regards the image of the sphere as a 'metaphor in the sense of a model'. Indeed, most scholars now either think that the spherical character of Parmenides' Being is a mere metaphor for its perfection and determination, or prefer to look at it as a sign that Parmenides could not even himself really and successfully break away from the conventions of thought of mortal men. I cannot, of course, go into this here. See for further references Guthrie (1967–1981) II, 34–39 and 43–49.

[36] This has been denied, however, by Sedley (1982) 175–193, esp. 178. He argues that Melissus only thought of a vacuum of the (c) kind. This he infers from his observation that the void mentioned in § 7 is an *internal* void. However, even if it were admitted that we are here dealing exclusively with an internal void (on which see the next note), it does not necessarily follow, otherwise than Sedley seems to suggest, that such an internal void is not an

immaterial whether we interpret the counterfactual motion of Being as a motion of Being as a whole into an *external* void, or as an internal motion of (parts of) Being into an *internal void*.[37]

Also the argument of (8) can be linked up with some passages in Parmenides' poem. Thus in Parmenides fr. 28 B 8, 23–25 DK we read that Being

> [...] is indivisible, since it is equal as a whole (πᾶν ὁμοῖον); nor is it at any place more (scil. Being), which could keep it from being kept together, nor is it (scil. at any place) less (scil. Being), but as a whole it is full of Being. Therefore it is as a whole continuous; for Being borders on Being.

All this seems to suggest that Melissus substituted *to kenon* for Parmenides' more abstract not-Being. Thus § 8 of fr. B 7 may read as an adaptation of these Parmenidean lines: also Melissus' Being does not admit of a more or a less. 'More Being' would amount to 'more full' and 'less Being' to 'more void'. The possibility of the existence of *to kenon* (not-Being) being excluded, the possibility of an internal admixture of void and the concomitant degrees of density is also excluded. In this passage the focus is clearly not on empty *places*, but on more (κενεώτερον) or less empty parts of Being itself. Though in § 9 void is more or less defined as that which can 'receive' something, i.e. which can serve as place, in § 8 the phenomenon of locomotion is not what is at stake. We are therefore here dealing with a concept of *to kenon* of the (c) kind. We may conclude, then, that the philosophy of Melissus shows

---

empty *place*. At any rate Melissus speaks of 'moving *into* the empty' and the essential point seems to be that in order to make it possible for Being to move, there must be some place for Being to move *into* (i.e. a void of the (b) kind). Sedley's argument that such an interpretation would have to ignore § 8 of Melissus' text as 'an irrelevant intrusion' is mistaken. For as I have tried to show, §§ 7 and 8 each elaborate on a different aspect of Parmenides' Being. For a brief but, to my mind, adequate refutation of Sedley's argument in this respect, see now Furley (1987) 112, n. 10.

37 I for my part do not believe, however, that the latter alternative is what Melissus intended: the argument speaks of motion of 'it' (i.e. in the singular), into *the* void (i.e. also in the singular) whereas an *internal* motion of Being would *ipso facto* imply its being plural, which in its turn would imply that it is (plural) not-Being rather than (singular) Being which moves into the internal void spaces. It may be, though, that Melissus' thought was not altogether clear on the subject. G. Reale, *Melisso. Testimonianze e frammenti*, Biblioteca di studi superiori 50 (Firenze 1970) 182 defends the thesis that Melissus argued against both an internal and an external void, because his argument was aimed at the Pythagorean notion of void. For the latter thesis, however, there is no evidence at all.

how the Parmenidean ontology could provide what was, so to speak, the negative archetype of both the (b) and the (c) conception of void.

2.2.3 We shall now turn to the way in which Leucippus and Democritus conceived of *to kenon*. Since it is impossible to determine the individual contribution to atomism made by each of them, I shall as a rule refer to them as 'the early atomists'. As is well known, hardly any *verbatim* fragments of their physical writings survive. Our main witness is Aristotle, the testimonies of later Greek doxographical tradition and of the Aristotelian commentators being in their turn largely dependent on him. As we shall see, our eventual picture of the early atomists' conception of the void is closely connected with the question of the reliability of Aristotle's testimony on the issue.[38]

Leucippus and Democritus are universally agreed to have developed their physical system under the sway of the 'Eleatic challenge'. For what it is worth, the ancient biographical tradition links up Leucippus with Zeno in a teacher-pupil relation,[39] whereas Democritus is said to have referred to the ideas of the 'followers' of Parmenides and Zeno, which in his days (the second half of the fifth century) were very much talked of.[40] The limits of our subject do not allow us to go into the details of the relationship between Eleatic philosophy and early atomism.[41] I shall here mainly argue that the early atomists were committed to all three of the conceptions of *kenon* outlined above and that of these at least their using the (b) and the (a) concept—but possibly also their using the (c) concept—was called forth by what might be called the 'Eleatic challenge'.

---

[38] I cannot here go into the details of the vexed question of Aristotle's reliability as a witness of presocratic philosophy in general. For a successful qualification, at least in so far as Democritus is concerned, of the largely negative judgement of Cherniss (1935), see the discussion by Löbl (1976) 58–116. See also Guthrie (1957). On the 'doxographical' side of the matter see Mansfeld (1986) 7–23.

[39] D. L. IX, 30 (= fr. 67 A 1 DK): οὗτος ἤκουσε Ζήνωνος. Cf. also the outline of the Eleatic succession as it appears in Clem. Al. *Strom.* I, 64 (= fr. 67 A 4 DK). Ps.-Gal. *Phil. Hist.* 3 (*DG* 601), Hippol. *Philosoph.* 12 (*DG* 564), and Tz. *H.* II, 980 (all three in fr. 67 A 5 DK), represented Leucippus as a pupil of Melissus who must have been more or less his contemporary.

[40] D. L. IX, 41.

[41] Cf. e.g. the pertinent observations of Furley (1967), ch. 6 ('The Atomists' Reply to the Eleatics') 79–103.

First, as Aristotle (*GC* 325 a 1 ff. = fr. 68 A 7 DK = fr. 146 Luria) already implied, in arguing that the atoms are *separated* by the void, the early atomists subscribed to the Eleatic thesis that there cannot be many (beings) without something to separate them.[42] This argument probably constituted the negative counterpart of Parmenides' contention that being must be one and continuous (fr. 28 B 8, 23–25 DK, referred to in the previous subsection). It may also have been at the basis of the second lemma of one of Zeno's arguments against plurality (fr. 29 B 3 DK), a *reductio ad absurdum* of the thesis that 'there are many', in which it is argued that two beings need a third item to separate them and that this third item in its turn needs another item to separate it from the first, etc. *ad infinitum.* The atomist conception of void to which this argument gave rise, is a void of the (b) kind outlined above: void as an empty space or place separating the atoms.

In the same context (*GC* 325 a 4) Aristotle refers to yet another Eleatic argument to which Democritus and Leucippus appealed. Motion, the argument says, is impossible if there is no separate void.[43] This in fact amounts to a restatement of one of the premises of Melissus' argument against motion as encountered in the previous subsection. As we saw there, this argument uses a notion of void of the (b) kind, void as an empty place *into* which a thing may move.

It may be added that, on the level of the atoms, such a void of course was thought to be an *external* void (i.e. external to the atom). On the level of phenomenal bodies it was regarded as *internal.* Thus, according to Aristotle, the early atomists not only tried to prove the existence of a void, by appealing to the alleged fact that it was required in order to make motion possible, but they also pointed to the phenomena of compression[44] and growth,[45] as presupposing that the (phenomenal) bodies which are compressed, c.q. which grow, 'contain' void spaces into which particles (of the bodies themselves in the case of compression, of another body in the case of growth) could move. If we may believe Aristotle (*Phys.* Δ 216 b 22 ff. = fr. 256 Luria, not in DK) the early atomists

---

42 *GC* 325 a 5: οὐδ' αὖ πολλὰ εἶναι μὴ ὄντος τοῦ διείργοντος.

43 *GC* 325 a 3: κινηθῆναι δ' οὐκ ἂν δύνασθαι μὴ ὄντος κενοῦ.

44 *Phys.* Δ 213 b 15–18, an argument refuted by Aristotle at 214 a 32–b 1.

45 *Phys.* Δ 213 b 18–22, an argument refuted by Aristotle at 214 b 1–9.

connected these phenomena with the notion of *manon* ('rare') and *puknon* ('dense'), which they may have taken from the Eleatic tradition.[46] As Aristotle makes clear, they defined the 'rare' as 'that which has many separate voids'.[47] If we may trust this testimony—and I do not see any reason why we should not—it appears that also the early atomists could slide into a *façon de parler* which used a concept of void of the (c) kind (void as an empty part of a (phenomenal) body).[48] At the same time the use of the term 'separate (κεχωρισμένα) voids' indicates that for them this could be *only* a *façon de parler*, since in reality a phenomenal body was nothing but a collection of atoms separated by voids of the (b) kind. So even if it may have been convenient for the early atomists to speak in some contexts (e.g. in their proofs from compression or growth) as if phenomenal bodies contained empty parts or pores, i.e. voids of the (c) kind, their elementary physics allowed them always to reduce these phenomenal (c) voids to 'real' (b) voids.

Up to this point our account has, I think, been fairly uncontroversial. However, whether or not the early atomists were also committed to a conception of void of the (a) kind (void as mere space) seems to be a moot point in the scholarly literature. It has been doubted by Jammer and, more recently, by Sedley and Inwood.[49] This doubt seems to have been called forth by scepticism concerning the reliability of Aristotle's account of the atomist *kenon*, in the course of his general discussion of void, which is appended to his famous discussion of *topos* in the fourth book of the *Physics*. I shall here try to show that

(1) there is no reason to subscribe to the strong scepticism of these scholars concerning Aristotle's reliability as a witness of early atomist physics; and that

---

[46] See the remarks on Melissus in the previous subsection.

[47] *Phys.* Δ 216 b 30: εἰ μὲν οὖν τὸ μανὸν λέγουσι τὸ πόλλα κενὰ κεχωρισμένα ἔχον etc.

[48] The fact that the early atomists occasionally chose this perspective, even if only as a manner of speaking, also explains Aristotle's statement, *Metaph.* A 985 b 4 ff., that Leucippus and Democritus called the atoms and the void στοιχεῖα: phenomenal bodies are so to speak made up of atoms and void, which is why Aristotle calls them—in what is this time, without doubt, his own terminology—a 'material cause' (αἴτια ὡς ὕλη).

[49] Jammer (1954) 9; Inwood (1981); see also Solmsen (1960) 140–142. Sedley (1982) 179–183, denies the early atomists a conception of void of the (b) kind as well as of the (a) kind.

(2) also the atomist conception of void of the (a) kind can be argued to have been based on elements of the Eleatic tradition.

Let us first have a look at the testimonies which attribute an (a) concept of *kenon* to the early atomists. Though this view is repeatedly encountered among the later Aristotelian commentators, their testimonies cannot count as independent evidence, since they are of course dependent on Aristotle, while also showing a tendency to identify the views of the early atomists with those of Epicurus.[50] At *Phys.* Δ 213 a 15 ff. Aristotle submits that:

> The proponents of the void (*to kenon*) represent it as a place (space) and as a kind of vessel, and they think it is full when it contains the bulk of which it is capable and that it is void (*kenon*) when it is empty. So the same thing is 'empty', 'full' and 'place', but their being is not the same.

This is a quite general statement, and perhaps for that reason neither Diels/Kranz nor Luria included it among the testimonies concerning the early atomists. However, I do not see which philosophers Aristotle could have had in mind other than the early atomists. Anyway, the view of *kenon* as 'space' is also attributed to the atomists in a fragment, preserved by Simplicius *In Cael.* p. 294, 33 Heiberg (= fr. 68 A 37 DK = fr. 172 Luria), from Aristotle's lost monograph 'On Democritus':

> Democritus is of the opinion that the nature of the eternal things is: small beings which are infinite in number. He assumes that, underlying these, there is another thing, place/space (*topos*) which is infinite in extent. And he calls place/space by the following names: 'void', 'nothing', and 'infinite' (τῷ τε κενῷ, καὶ τῷ οὐδενὶ καὶ τῷ ἀπείρῳ).

The mere fact that Aristotle wrote a monograph on Democritus indicates that he was well acquainted with the early atomist physics. Since, moreover, his judgement of Democritus in general is rather favorable,[51] there seem to be no *prima facie* reasons to

---

[50] Cf. e.g. Simp. *In Ph.* 571, 27 (= fr. 254 Luria, not in DK): τὸ δὲ διάστημα τοῦτο οἱ μὲν περὶ Δημοκρίτου καὶ Ἐπίκουρου κενὸν εἶναι λέγουσιν οὕτως ὥστε ποτὲ μὲν πληροῦσθαι σώματος, ποτὲ δὲ καὶ κενὸν ἀπολείπεσθαι. See further the extracts from the Aristotelian commentators conveniently printed together by Luria as (parts of) frs. 255–260.

[51] Cf. his remark on Democritus' views on coming to be and passing away in *GC* 315 a 34 ff.: ὅλως δὲ παρὰ τὰ ἐπιπολῆς περὶ οὐδένος οὐδεὶς ἐπέστησεν ἔξω Δημοκρίτου. οὗτος δ' ἔοικε μὲν περὶ ἁπάντων φροντίσαι κτλ.

assume that his testimony concerning the early atomists *must* be biased. What then are the reasons for which others reject his testimony concerning the early atomist conception of void as space?

It has been argued that 'die Bezeichnung des *kenon* [...] als Raum oder besser Ort, der einmal mit einem Körper ausgefüllt, dann wieder leer sein kann, zu sehr von der peripatetischen Raumvorstellung als 'Ort' bestimmt ist und deshalb dem demokritischen Leeren nicht gerecht wird'.[52] This contention lacks all ground. For the most typical feature of the Peripatetic, or at any rate of the Aristotelian, concept of *topos* is that it is *not* a 'Raumvorstellung' but a concept of *place*. The atomist void is introduced by Aristotle as a subspecies of a kind of concept of *topos*, viz. *topos* as an independent extension, which has been defended by *others*, but which he rejects himself.[53] If Aristotle's representation of void as *space* is mistaken at all, this is certainly not because he reads his own 'Raumvorstellung' back into atomist physics.

In this connection it has been argued by Sedley that it suits Aristotle to treat void as place, because 'he has already defined place in such a way as to deprive it of independent existence, and he now seizes the opportunity to tar void with the same brush'.[54] Though Sedley's argument purports to show that even a (b) concept of void was originally absent from the early atomists' physics, it is nevertheless apposite to discuss it here, since if it were correct it would *a fortiori* prove that the early atomists were not committed to an (a) concept of void (a concept which Sedley does not even consider in relation to the early atomists).

Sedley's argument is not really plausible. Nothing in fact indicates that Aristotle seizes the opportunity to reject in a few words the existence of a void, now that the arguments for a self-subsistent space have already been refuted. In his own introduction to the subject, at *Phys.* Δ 213 a 12, he motivates his inquiry into the nature and existence of the void by pointing out that the subjects

---

52 Löbl (1976) 221; Löbl, however, is fairly inconsistent in this respect, as is apparent from his statement, *ibidem*, 224: 'Zunächst ist das Leere der Raum in dem sich die Atome bewegen'.

53 On the difference place/space see my remarks in ch. 1; on the dialectical arguments of *Phys.* Δ see below, ch. 4.

54 Sedley (1982) 179.

topos and *kenon* are closely connected.[55] There is no reason to doubt the sincerity of this remark. Apparently *others* were used to treating the two subjects under one heading. In addition, we may observe that if the atomist void really had nothing at all to do with space or place, as Sedley would have us believe, Aristotle could as well have refrained from discussing it in this context altogether.

Moreover, on closer view Aristotle does not appear to tar the void with the same brush with which he tarred place-as-extension. At the beginning of his discussion of the void, at *Phys.* Δ 213 a 20, he declares that he is taking his starting-point from the arguments used by the proponents, as well as from those propounded by the opponents of the void. These arguments are of a quite diverse nature. It is true, the argument that void is a necessary condition for motion is indeed presented as basically an argument for a separate self-subsistent place and it is said that this notion has already been proved to be untenable.[56] However, Aristotle also has a number of arguments in stock which are new and which are primarily directed against the conception of an 'internal void' or void as what Sedley called a 'space filler'.[57] In other words Aristotle was not short of ammunition and there was no need for him to resort to improper arguments.

It seems then that Sedley's attitude towards the reliability of Aristotle's testimony, in the passages at stake here, is unnecessarily sceptical. One rather gets the impression that Aristotle was confronted with a diversity of arguments for the existence of *kenon*, some of which were based on a concept of *kenon* of the (a) kind (void-as-space), others on a concept of the (b) or (c) kind (void as empty space or as an empty part of a phenomenal body).

---

[55] *Phys.* Δ 213 a 12 ff.: Tὸν αὐτὸν δὲ τρόπον ὑπολήπτεον εἶναι τοῦ φυσικοῦ θεωρῆσαι καὶ περὶ κενοῦ, εἰ ἔστιν ἢ μή, [...] ὥσπερ καὶ περὶ τόπου· καὶ γὰρ παραπλησίαν ἔχει τήν τε ἀπιστίαν καὶ τὴν πίστιν διὰ τῶν ὑπολαμβανομένων· οἷον γὰρ τόπον τινὰ καὶ ἀγγεῖον τὸ κενὸν τιθέασιν οἱ λέγοντες etc.

[56] *Phys.* Δ 216 a 24–26: οἱ μὲν οὖν οἴονται τὸ κενὸν εἶναι, εἴπερ ἔσται ἡ κατὰ τόπον κίνησις, ἀποκεκριμένον καθ' αὑτό· τοῦτο δὲ ταὐτόν ἐστι τῷ τὸν τόπον φάναι εἶναι τι κεχωρισμένον· τοῦτο δ' ὅτι ἀδύνατον, εἴρηται πρότερον.

[57] At *Phys.* Δ 216 b 30 ff., e.g., Aristotle counters the argument that the assumption of a void is necessary to account for τὸ μανόν and τὸ πυκνόν by pointing out that (1) if this void is thought of as *separate*, it cannot exist, for the same reason for which a separate place cannot exist; and that (2) if, on the other hand, this void is thought of as *not separable*, i.e. as a part of a particular body, other absurdities follow (e.g. that this void must move along with the body).The latter argument is directed specifically against the conception of void as a space filler (= our void type (c)).

Indeed, at times, he displays a certain skilfulness—despite the in end unsatisfactory character of his arguments against the void in general—in playing off the different uses of *kenon* against each other.[58]

It has become clear, then, that Aristotle's testimony about the early atomists' void serving also as a void of type (a) is not to be eliminated as a mere caricature, serving to facilitate Aristotle's task of disproving the existence of void. Nor, as we have shown, can it be explained away as yet another instance of Aristotle reading his own concepts (this time allegedly: 'space') into the systems of his predecessors. We may therefore now turn to another consideration, which may corroborate our thesis that Leucippus and Democritus were not only committed to a concept of void of the (b) or (c) kind, but also to one of the (a) kind, i.e. a concept of void as an underlying space, partly occupied by the atoms, partly empty.

Here again, I am afraid, our conclusions are directly opposed to those of Sedley. Sedley argues that one would not really expect any serious reflection on space as such on the early atomists' part. 'Questions about the status of place', he says, 'are higher- order metaphysical questions which should be thought to be no more pertinent to the cosmological system of the atomists than to those of Anaxagoras and Empedocles'[59]. However, this argument neglects the fact that questions concerning the status of place were conspicuously raised by Eleatic ontology. Parmenides' Being is not in a place, but *in itself*, presumably because otherwise Being would no longer be one.[60] Zeno reforged this Parmenidean thesis into his so-called paradox of place, which in reality may well be read as a paradox of plurality.[61] According to Zeno, if all being is

---

[58] Thus at *Phys.* Δ 217 a 4 ff. he states that one of the absurd consequences of assuming the existence of a κενὸν which is μὴ χωριστόν is that the void would then turn out to move. But πῶς οἱόν τε φορὰν εἶναι κενοῦ ἢ τόπον κενοῦ; *κενοῦ* γὰρ γίγνεται *κενόν*, εἰς ὃ φέρεται. It is clear that the latter sentence uses the term κενόν in two different senses and it seems legitimate to assume that this ambiguity was part of the position taken by the atomists.

[59] Sedley (1982) 180.

[60] Cf. fr. 28 B 8, 29–30 DK: ταὐτόν τ' ἐν ταὐτῷ τε μένον καθ' ἑαυτό τε κεῖται / χοὔτως ἔμπεδον αὖθι μένει.

[61] At least it is not implausible to regard Zeno's paradox in this way. Saying that a being (x) must be in something would, on Zeno's line of thought, normally imply that there has to be an (y) for (x) to be in. If this (y) *is*, there has to be a (z) for it to be in etc. *ad infinitum*. So, on the

necessarily somewhere, i.e. in something, and if that something exists, then that something will itself have to be in something etc. *ad infinitum*.[62]

Now if we assume that early atomist physics should largely be regarded as an attempt to come to terms with the different challenges concerning not-being, motion and plurality, which were posed by the Eleatic arguments, there seems to be no reason to make an exception for the problems concerning place. Thus the fact that Democritus called the void 'nothing'[63] may be regarded as, among other things, an attempt to forestall the vicious consequences of Zeno's *regressus*. For he regarded the void *qua* place or space in which all beings (i.e. all atoms) are as a nothing which nevertheless exists.[64] This position allowed him to steer clear of the Scylla of the first lemma of Zeno's paradox—by showing that void, in being a nothing, does not itself have to be in something—and of the Charybdis of the second lemma—this time by showing that, since void is not utterly non-existent, we do not have to assume that the beings which are in it are *nowhere*, so that we do

---

assumption that every being is in something else, one ends up with an infinity of beings. That Zeno's paradox itself, which is unfortunately not extant in its original wording, at least used the expression ἐν τίνι εἶναι, or something similar, may be inferred from the fact that Aristotle tried to avert its consequences by pointing out that the word ἐν is among the πολλαχῶς λεγόμενα, and that place itself—when conceived in the Aristotelian fashion—is 'in' something else in a non-local sense (*Phys.* Δ 210 b 23 ff.). We may surmise that Zeno deliberately opted for the vague expression ἐν τίνι εἶναι as the basis of his argument, because he was not so much concerned with refuting any specific conception of place, but rather with attacking one of the possible guises of ontological pluralism. On the possibility of Zeno's argument having originally exhibited a dilemmatic structure, comparable to that of the two so-called arguments against plurality (fr. 29 B 1–3 DK), see Mansfeld (1985) 262– 263. In that case the 'paradox of place' as it became known to us would originally have constituted the first lemma, whereas the second lemma—the existence of which is witnessed by ps.-Arist. *MXG* 979 b 25–26, and Pl. *Tim.* 52b—must have amounted to the statement that what is nowhere, is nothing.

62 Cf. Arist. *Phys.* Δ 210 b 23: ὃ δὲ Ζήνων ἠπόρει, ὅτι εἰ ἔστι τι ὁ τόπος, *ἐν τίνι* ἔσται. At *Phys.* Δ 209 a 23 the presentation is slightly different: εἰ γὰρ πᾶν τὸ ὂν *ἐν τόπῳ*, δῆλον ὅτι καὶ τοῦ τόπου τόπος ἔσται. Eudemus, as quoted by Simp. *In Ph.* 563, 17 (= fr. 42 Wehrli) has: ἀξιοῖ γὰρ πᾶν τὸ ὂν *ποῦ* εἶναι· εἰ δὲ ὁ τόπος τῶν ὄντων, ποῦ ἂν εἴη.

63 Cf. e.g. the fragment from Aristotle's lost 'On Democritus' quoted above, 47.

64 For a review of some modern interpretations of the ontology of the Democritean void see now the otherwise not very illuminating article of Voelke (1990) esp. 343.

not have to assume that they do not exist. In other words, from the point of view of the 'Eleatic challenge' it is only natural to expect that the void also figured as space, i.e. as a void of the (a) kind.

As a final point, I would like to draw attention to the term *apeiron* ('infinite') which, as the fragment from Aristotle's 'On Democritus' cited above shows, was used by Democritus to denote the same 'thing' (or rather: 'nothing') as was denoted by the term *kenon*. The natural connotations of this word, 'unlimited' and 'unbounded', would also seem to point to a conception of void as mere extension or space rather than to the stricter conception of void as unoccupied space.

We may conclude the present subsection by stating that the available evidence indicates that the early atomists endorsed several concepts of void *promiscue*. It is a fair guess, if not more than that, that each of these uses originated in different aspects of the Eleatic challenge. If it is thus established that, in the guise of their (a) concept of *kenon*, the early atomists had a concept of space, it must at the same time be admitted that the two other conceptions of *kenon*, remaining at least as important, probably kept them from arriving at a coherently expressed theory of space. As I already indicated, and as I attempt to show in the next subsection, Aristotle skilfully exploited these ambiguities.

2.2.4 We shall next survey the ways in which Epicurus and the Epicureans made use of the term *kenon*. An excellent discussion, for all its brevity, of the Epicurean use of spatial terms is now to be found in Long/Sedley, *The Hellenistic Philosophers*. Their systematic comments contrast favorably with much of the scholarly literature on the subject which is mainly concerned with arguing for fairly conjectural connections between Epicurus' philosophy of space and that of Aristotle and Plato.[65]

---

[65] Adorno (1980) argues mainly for Platonic influence on Epicurus' concepts of space; Inwood (1981) 273–285 is an attempt to read the Epicurean concept of void as an adaptation of Aristotle's concept of τόπος (!). Solmsen (1977), who starts out by pointing to the 'Aristotelian (or Peripatetic) components in his [=Epicurus'] system' winds up his passage on void in Epicurus by the salutary remark that 'Epicurus appears as an innovator' and that the innovations made by Epicurus in the atomist system 'make us wonder whether for the conception of void as place Epicurus really needed Aristotle's precedent'. This statement, however, is immediately followed by the assertion that we still have to make room for Aristotelian influence since the

I shall not enter into the details of Epicurus' theory of space or its historical setting, but confine myself to a single aspect: the way in which Epicurus, on the one hand, succeeded in refining and clarifying the concept of *kenon*, while, on the other hand, he did not faithfully adhere to his own conceptual distinction.

The way in which Epicurus seems to have 'defined'[66] his spatial terms appears most clearly from the testimony of Sextus *M.* 10, 2 (= Usener 271 = LS 5 d):

> Therefore one must grasp that, according to Epicurus, of 'intangible substance' (*anaphês phusis*), as he calls it, one kind is named 'void' (*kenon*), another 'place' (*topos*) and another 'room' (*chôra*), the names varying according to the different ways of looking at it, since the same substance when empty of all body is called 'void', when occupied by a body is named 'place', and when bodies

'area of agreement or closeness' between Aristotle and Epicurus is considerable. The issue is clinched, if we may believe Solmsen, by the fact that Epicurus called his space ἀναφῆς οὐσία, which, Solmsen implies, cannot but remind us of Aristotle, who, *Phys.* Δ 214 a 6 called void that which 'is not filled by body perceptible to touch (ἁπτόν)'. In general, the possibility of Epicurus having been influenced e.g. by Aristotle is not to be excluded *a priori*. In fact, a very strong case for the thesis that Epicurus' theory of minima was called forth by Aristotle's criticisms of early atomist physics has been made by Furley (1967) 111–130. See in this connection also Huby (1978). However, the attempts made by the scholars mentioned above to establish a direct historical connection between the positive theories of space of Plato or Aristotle on the one hand, and of Epicurus on the other, carry very little conviction. They are on the whole merely based on (1) the assumption that great thinkers like Plato and Aristotle *must* have had an impact on Epicurus, and (2) some verbal correspondences, which on closer scrutiny are utterly trivial. Thus e.g. Adorno's assumption, (1980) *passim*, that Epicurus is citing (!) Plato's *Timaeus* whenever he uses the term χώρα and his remark (*ibidem*, 257), that 'gran confusione si genera se [...] non si tengono distinti i significati dei termini da lui usati in un contesto storico e semantico preciso' betrays a dangerous lack of awareness of the fact that (a) Epicurus was in general not at all interested in taking over specific technical terms from others, and advocated the use of words in their everyday sense instead (on which see the pertinent comments in Long/ Sedley (1987) I, 101) and that (b) the spatial terms applied by him were very common. If one is willing to believe that Epicurus cites Plato when using the term χώρα, one may as well believe that Newton cites Lucretius when using the term *spatium*.

[66] It should be noted, however, that Epicurus never worked with a very strict notion of definition. What we usually call his 'definitions' were arguably only meant as 'outline accounts' (the term he uses for giving such accounts is ὑπογράφειν: cf. *Ep. Men.* 123. The elaborate dialectical technique, leading up to strict technical definitions, which was used by e.g. Plato or the Stoics, was rejected by Epicurus as basically a merely linguistic exercise: cf. D. L. X, 31 (= fr. 257 Usener). Long/ Sedley (1987) I, 97–101, include some important texts—with illuminating comments—in this connection.

> roam through it (χωρούντων δι' αὐτῆς σωμάτων) becomes 'room' (*chôra*). But generally it is called intangible substance in Epicurus' school, since it lacks resistant touch.[67]

The upshot of this account seems to be that according to Epicurus the terms *topos*, *chôra* and *kenon* all *denote the same thing*, viz. space, which in its broadest sense could also be referred to as *anaphès ousia*. The eventual choice of *topos*, *chôra* or *kenon* was thus made to depend on the context. This seems also to be what is implied by the doxographical testimony of Aëtius,[68] which states that 'Epicurus says that the difference between void (*kenon*), place (*topos*) and room (*chôra*) is one of name'.

Nevertheless it is important to note that his introduction of the general concept of *anaphês phusis* in principle allowed Epicurus to anticipate some of the possible objections against the early atomist concept of *kenon*. As we saw, that concept displayed some ambiguities, in that it seemed to refer both to space as such and to unoccupied place or space. Thus Aristotle could point out that the atomists presented the void as something which made motion possible by yielding to the moving body (*Phys*. Δ 215 a 22 ff.). At the same time he observed that the phenomenon which seems to be suggested by the term 'yielding' (*hupeikein*), viz. mutual replacement (*methistasthai*), is impossible in the case of the void because (1) the void is not a body, and (2) if we imagine a wooden cube entering the void, the extension (*diastêma*) which was formerly in the void will interpenetrate the equal dimensionality of the cube.[69] It is clear that Aristotle elaborates on an ambiguity in the early atomist position, viz. the fact that void is, on the one hand, presented as a space occupant which may be *replaced* by a body (as is suggested by the term *hupeikein*) and, on the other hand, as the space itself in which there was formerly emptiness (an empty *diastêma*) and into which a body (atom) may move.

As we indicated in the previous subsection,[70] a similar ambiguity was detected by Aristotle in the early atomist argument that

[67] The translation is that of Long/ Sedley (1987) I, 28.

[68] Aët. I, 20, 2 (= fr. 271 Usener = L/S 5 c): Ἐπικουρος ὀνόμασιν παραλλάττειν κενὸν τόπον χώραν.

[69] *Phys*. Δ 216 a 34 ff.: ἐν δὲ δὴ τῷ κενῷ τοῦτο μὲν ἀδύνατον· οὐδὲ γὰρ σῶμα. διὰ δὲ τοῦ κύβου τὸ ἴσον διάστημα διεληλυθέναι δόξειεν ⟨ἂν⟩ ὅπερ ἦν καὶ πρότερον ἐν τῷ κενῷ (⟨ἂν⟩ suppl. Cornford).

[70] Cf. above, n. 57 and 58.

the existence of rare and dense bodies proves that these bodies must contain void 'parts'. In that case, Aristotle contends, the void parts move along with the body and must themselves be in a place, which—on the atomist assumption—means in a void.

If we may thus safely assume that the fact that the early atomist conception of the void oscillated between what we described as the (a), (b) and (c) types of void called forth criticisms of the sort propounded by Aristotle, we may at the same time add that the only thing really needed to ward off such criticisms was a conceptual clarification which showed what happened to a void, when a body approached it. Here Epicurus provided the right answer: there is really only one kind of spatial extension (*anaphês-phusis*), which is strictly speaking only to be called void, when it is unoccupied. As soon as such a void is occupied by a body, it ceases *ipso facto* to be void in that strict sense and becomes the place (*topos*) of the body.[71] On these lines of thought, the notion of void as a 'space filler', i.e. void as an empty part of a body, only makes sense as a *façon de parler* concerning bodies at the phenomenal level, which has no bearing on the real state of affairs at the atomic level and which, consequently, says nothing about the real nature of space. It may be that for the early atomists the use of a (c) conception of void was also merely a *façon de parler*. The difference, however, is that in speaking of this (c) conception of void they applied the very same term they used to refer to space as such (i.e. the (a) void).

This is not to say that Epicurus remained faithful to his own conceptual distinctions in practice. If we confine ourselves to discussing his use of the term *kenon*, we may begin by citing some passages where he indeed sticks to using *kenon* in the strict sense of 'unoccupied space'. Thus, when at *Ep. Hdt.* 44 the continuous motion of the atoms is explained by reference to the

---

[71] Note that Sextus, *M.* 10, 222 attributes εἶξις to the Epicurean void, which would at first blush seem to contradict our interpretation and make the Epicurean void subject to the same objections as the Democritean void. However, as the context makes clear, εἶξις as an 'essential attribute' of the void is opposed to ἀντιτυπία, the essential characteristic of atoms, which causes them to *rebound* at the occasion of a collision. The opposition, then, seems to be between resistance and non-resistance: the void does not yield in the sense of 'move back', but it lets the atoms go right through it. Cf. Lucr. I, 437–438: sin intactile sit, nulla de parte quod ullam / rem prohibere queat *per se transire* meantem.

specific, unresistent, nature of the void, Epicurus uses the following words:

> For the nature of the void, which separates each atom off by itself (ἡ διορίζουσα ἑκάστην αὐτήν) brings this about since it is unable to lend them any support.

The same strict usage of *kenon* seems to underlie Epicurus' use of the expression 'in a region with many void spaces' (ἐν πολυκένῳ τόπῳ) in *Ep. Pyth.* 89, where he contends that other *kosmoi* can come to be in the intercosmic space (*metakosmion*), as long as that is only 'a region with many void spaces', not a region which is absolutely empty.[72]

However, in another passage, at *Ep. Hdt.* 40, Epicurus seems to be more concerned to stress the interchangeability of the Epicurean spatial terms as all—i.e. including the term *kenon*—denoting space, than to outline their individual ranges of application:

> If there were no place (*topos*),[73] which we call 'void' (*kenon*), 'room' (*chôra*) and 'intangible substance' (*anaphê ousian*), bodies would not have anywhere to be or to move through in the way they are observed to move.[74]

At *Ep. Hdt.* 41–42, moreover, we clearly encounter a conception of *kenon* of the (a) kind. Here Epicurus tries to show that the universe, *to pan*, is infinite, both in respect of the *number* of its atoms and in respect of the *extent* of the *kenon*:

> Indeed, the totality of things is infinite both in the number of the bodies and in the magnitude of the void. For if the void were infinite but the bodies finite, the bodies would not remain anywhere but would be travelling scattered all over the infinite void,

---

72 This use of κενόν in the strict (b) sense seems also attested by two fragments from περὶ φύσεως viz. *PHerc.* 1149, col. 18, ii, 10–11 (= fr. 24, 43, 10–11 Arrighetti) and *PHerc.* 1479/ 1417 col. 12, iii, 10–11 (= fr. 31, 4, 4 Arrighetti).

73 I have adopted—with Long/ Sedley (1987) II, 20—Usener's conjecture τόπος δὲ εἰ μὴ ἦν instead of the harsh τὸ πρόσθεν· εἰ μὴ ἦν of the mss.; Gassendi smoothened the mss. reading into: τὸ πρόσθεν· εἰ ⟨δὲ⟩ μὴ ἦν etc. Cf. on these conjectures, Sedley (1982) 183 ff. It does not really matter much whether we opt for the reading of the mss. or for either of the two proposed conjectures: the upshot seems to be that we shall have to assume the existence of space as a second factor, next to bodies, and that this space may be given different names.

74 Cf. Lucr. I, 425–428: tum porro locus ac spatium, quod inane vocamus / si nullum foret, haud usquam sita corpora possent / esse neque omnino quoquam diversa meare.

for lack of the bodies which support and marshal them by buffering. And if the void were finite, the infinite bodies would not have anywhere to be.[75]

The wording leaves no room for doubt; the conception of void here used is void of the (a) kind: it is that *in which* bodies are and *through which* they move, i.e. space *tout court.* That Epicurus did not mean to argue that there is an infinite amount of *empty* space, is once more confirmed by Lucretius' version of the same argument, which instead of Epicurus' *kenon* has the word *spatium.*[76]

Finally, also the notion of a *kenon* of the (c) kind seems to play a part, though, I admit, probably only in a manner of speaking. Lucretius renders an argument which purports to prove the existence of void on account of the differences in weight exhibited by bodies of the same volume:

> It naturally follows that what is equally large, but seems lighter, reveals that it contains more void, while by contrast what is heavy announces itself as containing more body and much less void. Naturally, then, the thing we call void, which we are seeking with keen reasoning, is mingled with things.[77]

I think we are allowed to conclude that—though compared with the early atomists Epicurus made considerable conceptual progress by isolating, so to speak, a notion of space *as such*—his

---

75 The translation is that of Long/ Sedley (1987) I, 44.

76 Lucr. I, 968–969: praeterea si iam finitum constituatur / omne quod est spatium etc.; *ibidem,* 984–985: praeterea spatium summai totius omne / undique si inclusum certis consisteret oris etc. I would suggest that the *empirical* argument, which infers the existence of void at the atomic level from the fact that on the phenomenal level we see that bodies cannot move without a void (Lucr. I, 334–345 and Phld. *Sign.* 36 (p. 78 De Lacy), slides from a concept of (a) κενόν to a concept of (b) κενόν. For it cannot be implied that on the level of phenomena we see 'void' *qua* empty space. Rather the idea is that when we see phenomenal bodies moving, we perceive this motion as taking place through space and we posit that space as self-subsistent. We may infer that the same holds at the atomic level, so that also there a self-subsistent space has to exist. Given the atomic, discontinuous character of the material particles moving through that space, that space itself has to be partly empty so that there are empty spaces (i.e. 'real' voids of the (b) kind) at the atomic level. For a different explanation see Long/ Sedley (1987) I, 32.

77 Lucr. I, 364–369: ergo quod magnumst aeque leviusque videtur, / nimirum plus esse sibi declarat inanis / at contra gravius plus in se corporis esse / dedicat et multo vacui minus intus habere. / est igitur nimirum id quod ratione sagaci / querimus admixtum rebus, quod inane vocamus (transl. Long/ Sedley (1987) I, 31).

writings and the testimonies concerning his thought betray a less univocal use of the term *kenon* than might perhaps be expected.[78] On the other hand, in view of Epicurus' general lack of interest in definitions and dialectical subtleties this should probably not surprise us too much. The important thing for him was to have shown that there is such a thing as space and that the terms *chôra, topos,* and *kenon* in the end all denote this space.

2.2.5 I shall finally pay some attention to the role played by the void in the physics of Strato of Lampsacus. Strato, who became head of the Peripatetic school in Athens after Theophrastus (*c.* 287 BCE), and who remained so until his death (269 BCE) is one of the most intriguing philosophical personalities of the early Hellenistic period. That he showed serious interest in the problems of space and void is shown by the catalogue of his works, preserved in Diogenes Laërtius V, 59 (= Fr. 18 Wehrli), which informs us of the fact that he wrote a *Peri tou Kenou* ('On the Void'). Before discussing those features of his concept of the void which seem to be of particular interest against the background of the preceding subsections, some remarks concerning the sources for our knowledge of Strato's physics are in order.

The outlines of Strato's theory of the void have to be reconstructed from a few doxographical scraps, which are probably to be traced back to Aëtius, from a number of passages in Simplicius and, according to some scholars, from the preface of the *Pneumatica* of the first century CE physicist Hero of Alexandria. Whether and to what extent the latter work may count as witnessing Strato's thought on the void is a moot point. Diels regarded the preface of the *Pneumatics* as basically nothing less than a summary or extract of Strato's lost *Peri tou Kenou.*[79] He thought the connection between Strato and Hero could be established in the form of a strict ancestry-and-descent relation: Hero was dependent on *De Ingeniis Spiritualibus,* a work—now only extant in an Arabic

---

[78] This is probably what constitutes the background for the quite divergent views on Epicurus' concept of vacuum which are to be found in the scholarly literature. The 'classical' view, represented by Bailey (1929) 294–296, maintains that Epicurus' void is basically space. On the other side there are studies like Rist (1972) 56–57, which exclusively defend the stricter interpretation of void as unoccupied space.

[79] Diels (1893) 101–127.

version and a shortened Latin translation of the Arabic—written by Philo of Byzantium, who probably lived in the second century BCE. Philo's work in its turn was, according to Diels, directly dependent on the work of the rather shadowy figure of Ctesibius, a third century BCE Alexandrian who may well have known Strato personally, since we know that Strato spent some time in Ptolemaic Alexandria as the tutor of the young Ptolemaeus Philadelphus.[80] Apparently by way of 'circumstantial evidence' Diels adduced the fact that some notions in Hero's preface (e.g. the notion of ἀκολουθία πρὸς τὸ κενούμενον) are also known to have been defended by the physician Erasistratus, who lived in the early third century BCE and who is often connected with Peripatetic philosophy.

Gottschalk has recently subscribed to the main features of Diels' thesis.[81] Though he is somewhat more careful than Diels in not immediately assuming a direct *stemmatic* connection between Strato, Ctesibius, Philo and Hero—according to him the preface of Hero's *Pneumatics* and Philo's *De Ingeniis Spiritualibus* 'represent two stages of the same tradition'—he nevertheless regards Hero's preface as containing for the most part Stratonic materials and, accordingly, includes large parts of it in his edition of the fragments of Strato. Still the only *direct* evidence on which all these reconstructions were based was the fact that Hero's preface contains a fragment which also occurs in Simplicius where it is ascribed to Strato.[82] From this Diels and Gottschalk inferred that the whole preface must haven been derived from Strato through one or two intermediary sources.

Doubts concerning the validity of Diels' *Quellenforschung* in this respect were already raised at an early date and other possible sources of the theories of Erasistratus and Hero were pointed out.[83]

---

80 For a convenient survey of these matters, see Gottschalk (1965) 134 ff.

81 Gottschalk (1965) 97–182; see esp. the conclusions at page 141.

82 Simp. *In Ph.* 693, 10; the versions of Simplicius and Hero are printed together as frs. 65 a and 65 b Wehrli.

83 M. Wellmann, 'Zur Geschichte der Medizin im Altertum', VI, *Hermes* 35 (1900) 367–382, tried to show that Erasistratus' suction theory does not derive from Strato but from Erasistratus' teacher Chrysippus (not to be confused with his Stoic namesake); A. Schmekel, *Die positive Philosophie in ihrer geschichtlichen Entwicklung* 1: *Forschungen zur Philosophie des Hellenismus* (Berlin 1938) 110–124, tried to show that Hero's theory is by and large a variant of atomism.

Gottschalk defended Diels' conclusions against these attacks, but unfortunately this defence, for all its brilliance, largely amounted to showing that attempts to localize a *single* alternative source for Erasistratus and Hero were unsuccessful.[84] Thus the problem of the reliability of Hero's preface as a testimony of Strato's physics seemed to be reduced to the question whether it can *as a whole* be traced back to either Strato or a *single* other source. The question of the methodological validity of Diels' procedure—esp. his establishing a *quellengeschichtliche* connection between Hero's preface as a whole and Strato on the basis of the mere fact that one passage in Hero is paralleled, with an ascription to Strato, in Simplicius—has received little attention. Nevertheless, there are serious indications that Hero, or his source, combined materials from quite heterogeneous sources.

For one thing, the text contains an explicit reference to Archimedes, who by the time Strato died was only 18 years old and who cannot therefore have been cited by Strato.[85] Furthermore, Furley has recently shown that Hero's text also contains probable references to Epicurean and even to Stoic theories.[86] A more detailed investigation of the relationship between Hero's preface and Greek theoretical physics surely seems a *desideratum.* In the meantime I think a prudent attitude towards the value of Hero's *Pneumatics* as evidence for Strato's theory of the void is justified. In what follows I shall accordingly confine myself to a discussion of the testimonies in the doxographical tradition and Simplicius.

My aims in the present subsection are limited. I shall only consider the nature of Strato's concept of void.[87] It will be shown that the distinction made above between three different concepts of void may help to clear up some of the difficulties which still surround the interpretation of Strato's concept. My suggestions are these:

---

[84] Gottschalk (1965) 128 ff.

[85] Hero *Spir.* p. 114, 24 ff. Gottschalk; in his commentary, 153, Gottschalk points out that this passage is a digression and a later contribution, perhaps Hero's own.

[86] Furley (1985) 594–610, esp. 606.

[87] For a more thorough discussion of Strato's theory, his arguments, connections with Aristotle, Theophrastus and some contemporary and later theories of *horror vacui* I refer to the studies of Gottschalk and Furley.

(1) Strato's *kenon* is a void of the (c) kind outlined above: an empty part of a body. He rejects the (b) concept (void not as an empty part of a body, but as an empty part of space between bodies) and the (a) concept (void as space).
(2) His rejection of a concept of void of the (b) kind is motivated by his empiricism, as well as by his sticking to the Aristotelian ontology of substances and the Aristotelian notion of the continuum.
(3) His rejection of a concept of void of the (a) kind (void as space) does not imply a rejection of a concept of three-dimensional self-subsistent space. He simply preferred to reserve the term *topos* to denote such a space, probably influenced by Aristotle's criticisms of the use made by the early atomists of the term *kenon* in this connection.

How far are these suggestions actually supported by the evidence? Simplicius *In Ph.* p. 693, 10 ff. Diels (= fr. 65a Wehrli) unambiguously credits Strato with a theory of 'micro-void':

> Strato of Lampsakos tries to show that the void exists interspersed in every body so that it is not continuous. He says that otherwise water and air and other bodies could not be penetrated by light or heat or any other corporeal power (*dunamis sômatikê*). How could the rays of the sun penetrate to the bottom of the jar? If the liquid had no pores in it, and the rays forced a way through, then full jars would overflow, and it would not be the case that some rays would be reflected upwards and others would go downwards through the liquid.[88]

Though this testimony would seem to leave little room for doubt, the attribution of a micro-vacuum to Strato has been challenged by Gatzemeier.[89] Gatzemeier's arguments against this attribution are conveniently summarized by Furley:[90]

(a) it is inconsistent with two other testimonies of Simplicius (*In Ph.* p. 601, 19 and 618, 20 Diels) which state that void is *de facto* always filled with body;
(b) it is inconsistent with Strato's criticism of the atomists' arguments for the existence of void;
(c) it is inconsistent with Strato's view of the continuity of body;

[88] I have taken over Furley's translation of the passage ((1985) 597).
[89] Gatzemeier (1970) 93–97.
[90] Furley (1985) 597–598.

(d) Strato's theory of matter would make sense of a theory of potential void (rather than of actual micro-vacua), which is not inconsistent with any of these.

As to (a), Furley has argued, with reference to Gottschalk, that the micro-void theory is not inconsistent with the view that 'the void' (*sic*) is isometric with the cosmic body and is always filled with body.[91] I think he goes a long way in the right direction, in so far as he shows how the fact that the cosmic *space* is always *de facto* filled up could be squared with the assumption of the existence of micro-vacua. Obviously Strato's sticking to the Aristotelian substance ontology—in spite of the changes he introduced into the Aristotelian system as a whole—and to the Aristotelian view of the continuum allowed him to regard an internal pocket void *not* as something having an existence *sui generis*, but as part and parcel of a specific body (substance).

What apparently neither Furley nor Gottschalk recognized, is the fact that this theoretical distinction between overall space on the one hand, and internal pocket voids on the other, was probably backed up by a terminological distinction between *topos* and *kenon*. Thus Simplicius *In Ph.* 601, 19, attributes to Strato the view that *topos*, i.e. *space*, is isometric with the cosmic body. Admittedly, at p. 618, 20 ff. he substitutes *kenon* for *topos*, but he makes clear that this is his own interpretation (καὶ Στράτωνα *οἶμαι* [...] ταύτης γενέσθαι τῆς δόξης) and a quick glance at the context in which this reference occurs, may show what lies at the roots of this interpretation.

The reference to Strato occurs in the course of a list of different concepts of *kenon*. The context as a whole is Simplicius' *Corollarium de Loco*, a kind of appendix to Simplicius' running commentary on Aristotle's discussion of *topos* in *Phys.* Δ, which is—despite Simplicius' critical remarks concerning certain aspects of Aristotle's discussion—strongly influenced by Aristotle's arrangements of the topics at issue. Now, as is well known, Aristotle

---

91 However, Gottschalk does not say what Furley makes him say: he, rightly, argues that it is Strato's theory of *place* which would seem to exclude the possibility that any empty space should actually exist (Gottschalk (1965) 131: 'Presumably Strato regarded the interstices as so intimately bound up with the bodies to which they belong that he ignored them in his definition'); Furley (1985) 598, substitutes 'void' for Gottschalk's 'place', a substitution which was also made by Simplicius, but which, as my remarks in the text above show, I regard as unwarranted.

regarded the discussion of the void as closely connected with the discussion of space and place, because he regarded all theories of a void as basically theories of a self-subsistent space and described at least some theories of self-subsistent space as theories of void.[92]

The view ascribed to Strato by Simplicius in fact amounts to the view which Aristotle ascribed to Plato (void as a space always filled up *de facto*).[93] Simplicius makes clear that he thinks it was the view of a number of Platonists. This use of *kenon* as an equivalent for 'space' (i.e. our (a) *kenon*) was, as I have shown, quite common in Greek philosophy and will in the present context have been induced by the fact that Aristotle used the term in this way. Thus Simplicius' remark may be read as implying that Strato held a theory of three-dimensional self-subsistent space, but it may not be read as positively implying that Strato himself used the term *kenon* in this connection.

Furthermore, that Strato did not use *topos* and *kenon* synonymously seems at least to be implied by the definitions preserved by Aëtius. (I, 18, 4 and I, 19, 3, printed together as fr. 55 Wehrli):[94]

> Strato said that there was no void (*kenon*) outside the cosmos but that it was possible for there to be one inside the cosmos;[95]

and:

> Strato said that space/place (*topos*) was the extension between container and contained.[96]

---

[92] Cf. *Phys.* Δ 208 b 27, 213 a 12 ff., 214 a 13 ff.

[93] Cf. *Phys.* Δ 214 a 13 ff, on which see my remarks below, ch. 3, 112–113.

[94] Note that Wehrli wrongly prints the testimony of Theodoretus as independent evidence (fr. 54), whereas it is really just another witness for Aëtius and thus should be printed—if at all—together with the version of Stobaeus (= fr. 55 Wehrli).

[95] We cannot be sure about what is meant by the somewhat enigmatic 'it was possible for there to be one' (δυνατὸν εἶναι in the version of Theodoretus; δυνατὸν γενέσθαι in the version of Stobaeus). It may be a reference to the possibility of a larger void (i.e. a void of the (b) kind) occurring *unnaturally*, i.e. a void created *between substances*, by artificial means. Such a view might be at the basis of a theory of *horror vacui*. However, we do not know—if, that is, we are skeptical about the value of Hero's *Pneumatics* (which does propound such a theory) as a testimony—whether Strato held such a theory. Anyway I would suggest that we should not attach too much value to the wording of this doxographical testimony. The important fact is not so much *how* Strato defined place and void, but rather that he defined them *separately*.

[96] For the right interpretation of this curious, and distorted, 'definition' see Furley (1985) 595 and Gottschalk (1965) App. 2, 169. It ultimately goes back to one of the possible candidates for being τόπος which was rejected by

These few scraps of evidence allow us to infer tentatively that Strato made a distinction between *topos* and *kenon*, *topos* being the self-subsistent three-dimensional extension underlying each and every body, whereas the term *kenon* was reserved to refer to internal micro-voids occurring within otherwise continuous substances.

As additional evidence we may adduce a text which has hitherto been generally neglected by those who studied Strato's theory of space and void, probably because it originally did not belong to any of the purely physical treatises (it is not cited by Simplicius in a physical context) and because Wehrli, rightly, printed it among the logical fragments. It is a fragment from Strato's book *Peri tou Proterou kai Husterou* ('On prior and Posterior') preserved by Simplicius *In Cat.* 432, 1 ff. (= fr. 27 Wehrli).

Among the different ways in which one thing can be prior to another is 'being prior by nature':

> As prior by nature, as not admitting of a reversal of the order of being/of the ontological priority (ὡς μὴ ἀντιστρέφον κατὰ τὴν τοῦ εἶναι ἀκολούθησιν) [we may regard] that which is capable of existing when the other [scil.: in respect of which it is prior] does not exist, like place (*topos*) in relation to body and body to colour.

Strato here clearly refers to his concept of a three-dimensional self-subsistent space or place. It may be that, in presenting *topos* the way he does here, he is consciously taking over one of the descriptions of such a space provided by Aristotle—who of course rejects this conception of *topos*—at *Phys.* Δ 208 b 34–209 a 3.[97] Our passage in Simplicius, which is probably a *verbatim* quotation, shows that Strato used the term *topos* to refer to space as such.

We may conclude that, though the evidence is scanty, it indicates that Strato differentiated between *topos* (space/place) and *kenon* (void as it naturally occurs, i.e. as empty parts of bodies). This *kenon* was regarded as intimately bound up with the substance in which it occurred, so that this conception of micro-

---

Aristotle: τὸ μεταξὺ τῶν ἐσχάτων (cf. *Phys.* Δ 211 b 8). Wehrli's interpretation (*Straton von Lampsakos,* 57) 'einen supponierten Hohlraum zwischen Gefäss und Gefässinhalt' is nonsensical.

[97] *Phys.* Δ 208 b 33–209 a 2: 'If τόπος were such, its power would be astonishing, and prior to all other things (πρότερα πάντων). For that without which nothing else is, but which can itself exist without all other things, is necessarily prior to everything else. For τόπος is not destroyed when the things in it are destroyed'.

void is seen not to be inconsistent with the thesis that the cosmic space is in fact always full of body (where body should be taken to mean 'substances').

We now come to Gatzemeier's second argument against the ascription of the micro-void theory to Strato: its alleged inconsistency with Strato's criticisms of the early atomists. Simplicius *In Ph.* 652, 18 (= fr. 61 Wehrli), informs us that Strato reduced the reasons for assuming the existence of void as, according to Aristotle, these were advanced by the early atomists,[98] to two: spatial movement and compressibility. Moreover, Simplicius states that Strato added one other argument to the four listed by Aristotle, viz. the argument from magnetism.[99] Thus Strato would seem to have recognized basically three (kinds of) reasons for assuming the existence of void: (1) spatial movement, (2) compressibility, (3) magnetism.

Of these we know that he rejected (1) (Simplicius *In Ph.* 659, 20 = fr. 63 Wehrli) and (3). His rejection of the argument from spatial movement seems to have taken the form of the experimental proof that we can see that movement takes place in conditions where we cannot see that there is a void (which in practice means: where we do not notice compression). His experiment is described by Simplicius in the following terms (*In Ph.* 659, 20 ff. = fr. 63 Wehrli):

> If you drop a pebble into a vessel filled with water and invert the vessel, blocking the exit at the mouth, the pebble moves to the mouth of the vessel as water moves around into the place of the pebble. The same thing happens with swimmers, and fish, and so on.[100]

Furley argued that 'the most the example can prove is that the pebble does not need a *perceptible* void space to move into'. I think, however, that this was precisely what Strato regarded as sufficient proof: the existence of void spaces had to be admitted only in so far as the phenomena unambiguously pointed in that direction. The phenomenon of motion, of bodies replacing each other, could excellently be explained without the assumption of an empty

---

[98] Viz. (1) motion, (2) compression and expansion, and (3) growth. Cf. Arist. *Phys.* Δ 216 b 22 ff.

[99] On the atomist explanation of magnetism see fr. 68 A 165 DK (Democritus) and fr. 64 A 33 DK (Diogenes of Apollonia).

[100] Furley's translation ((1985) 601).

place into which the bodies move. One only had to posit a three-dimensional self-subsistent underlying space and assume the infinite divisibility of matter.

Strato's rejection of (3), the argument from magnetism, seems to have developed along similar lines: the phenomenon of magnetism does not *unambiguously* point to the existence of small empty spaces. Simplicius (*In Ph.* 663, 2 ff. = fr. 62 Wehrli) cites what appear to be Strato's own words:

> Strato refutes also the argument from magnetism: 'Neither does magnetism', he contends, 'force us to posit the void. For whether magnetic attraction itself exists is not altogether clear [...] nor is it clear, if there is magnetic attraction, whether the stone attracts other things through the void and not by other means. For those who say so do not *prove* (ἀποδεικνύουσι) the existence of void but merely *assume* (ὑποτίθενται) it'.

The upshot of Strato's criticism seems clear: proving the existence of void by referring to its role in magnetic attraction is an instance of *obscurum per obscurius*. We know too little about the causes of magnetism to warrant such an inference. Interestingly, the hypothetical and merely speculative explanation of magnetism that was provided by the early atomists should not, according to Strato, be called a *proof*; it is merely a supposition.[101] It seems that he was committed to a strictly empiricist 'philosophy of science'. To him it was not enough for a hypothesis to adequately account for the phenomena. It could apparently be accepted as valid if and only if the phenomena themselves left no room for an alternative explanation.

He probably thought the phenomenon of compression provided such an unambiguous indication. At any rate, our sources do not inform us that he also rejected this third argument. The only positive proof for the existence of micro-void attested for Strato by Simplicius (*In Ph.* p. 693, 10 ff. = fr. 65 a Wehrli, quoted above)—the argument of the 'percolation' of sunrays and of heat through other bodies—again shows his strong empiricist tendencies.

---

[101] That this was the general tenor of Strato's criticism of the early atomist physical theories appears also from a passage in Cicero who, *Acad. Pr.* II, 121 (= fr. 32 Wehrli) tells us that Strato spoke of the early atomists' theory of the atoms and the void as: somnia haec [...] esse Democriti, non *docentis*, sed *optantis*, ipse autem singulas mundi partes persequens quicquid aut sit aut fiat naturalibus fieri aut factum esse *docet* ponderibus et motibus.

We may conclude, then, that Strato's criticisms of the early atomist doctrines were primarily of a *methodological* nature. Strato rejected the physical principles of the atomists as metaphysical 'dreams', which cannot be positively confirmed by actual phenomena: he inherited Aristotle's common-sense view of substances[102] and the Aristotelian notion of the infinitely divisible material continuum. Individual entities in this system, substances instead of atoms, did not need independent void spaces to move into: they could just move into each other's places. He tried to corroborate this view by showing 'experimentally' that we do not actually perceive void spaces when ordinary things move. Nevertheless, some phenomena—compression and extraction of one and the same body, as well as the interpenetration of heat and light and (other) bodies—necessitated the assumption of void interstices. These interstices, however, were conceived of as being a kind of pores, or cavities, in the individual substances, which could also move together with those substances. They clearly belong to what we above called a (c) *kenon*. So we are not allowed to conclude that Strato's positing the existence of micro-voids is inconsistent with his rejection of atomism.

The third objection (c) against the ascription of a theory of micro-vacuum to Strato, viz. that it is inconsistent with Strato's statement that matter is continuous and infinitely divisible, has in fact already been refuted: Simplicius' statement that according to Strato void was interspersed in every body, so that it is not continuous' should not be read as implying that Strato subscribed to the atomist thesis that there is no continuous body. He was only committed to the view that continuous stretches of matter are interrupted by void interstices.

---

[102] Diels (1893) 112, argued that 'Straton stellte sich im Wesentlichen auf den Standpunkt der Abderiten'. I would rather agree with Furley (1985) 608, who called him a 'reforming Aristotelian' or with A. Stückelberger, 'Vestigia Democritea: Die Rezeption der Lehre von den Atomisten in der antiken Naturwissenschaft und Medizin', *Schweizerische Beiträge zur Altertumswissenschaft* 17 (Basel 1984) 139: 'Straton hat sich durchaus als Peripatetiker gefühlt'. Peripatetic elements in Strato's thought are e.g. his use of the terms οὐσία and ποιόν and πόσον in the fragment from περὶ τοῦ προτέρου καὶ ὑστέρου by Simp. *In Cat.* 423, 1–34 (= fr. 27 Wehrli), his finite and geocentric universe, his theory of infinite divisibility of matter (fr. 82 Wehrli) and places, and his view that qualities such as heat and cold are irreducible principles (fr. 42–48 Wehrli).

This leaves us with the final objection, (d), Gatzemeier's claim that his ascription to Strato of a theory of 'potential void' makes better sense of the evidence. Here we should note, in the first place, that Gatzemeier's whole case mainly rests on what is arguably a wrong interpretation of Simplicius *In Ph.* 618, 20 ff. As was pointed out above, Simplicius there interprets Strato's space as an (a) void. In other words, although he uses the term κενόν, what he is actually referring to is Strato's concept of cosmic *space.* His additional statement that this cosmic space is only τῇ ἐπινοίᾳ empty should therefore be taken to mean that space is always occupied and is only empty when regarded conceptually ('in itself'). This has escaped Gatzemeier, who takes κενόν here to refer to the Stratonic void in the strict sense and who accordingly interprets the words τῇ ἐπινοίᾳ as implying that the Stratonic micro-voids should only be assumed 'in Gedanken'. But there are also other objections against Gatzemeier's interpretation of the Stratonic void as a merely 'potential' void. Furley has rightly pointed out, first, that the evidence (particularly the argument from percolation in fr. 65a Wehrli, quoted above) rather suggests that what Strato had in mind was something more than merely potential channels in the structure of penetrable matter, and, secondly, that in the end the whole idea of potentially void interstices cannot be made coherent sense of.[103] It seems then that we are justified in rejecting Gatzemeier's interpretation of Strato's conception of the void.

We may conclude the present subsection by summarizing our findings. Strato's theory of the void was elicited by the need to account for some phenomena which apparently could not be explained otherwise: compression and the 'interpenetration' of substances taking place in the case of the propagation of heat or light. This only led him to endorse the minimalistic (c) concept of void. His Aristotelian substance ontology and his views on continuity would not naturally allow him to assume that individual substances are separated by void spaces (= (b) concept of *kenon*). In line with his general empiricism, moreover, he rejected the atomist views on this matter as not being supported by observation and as mere unfounded 'assumptions'.

Though he opted for an unaristotelian concept of *topos* as a three-dimensional self-subsistent extension, he does not seem to

---

103 Furley (1985) 598–600.

have used the term *kenon* (as what we called the (a) kind of *kenon*) to denote this 'space'. Our sources indicate that in this connection he stuck to using the term *topos*. He could thus maintain that when a body containing empty pores moves, the pores move along with it, without being *ipso facto* committed to the view that space or place itself moves, a view which, as we noticed, Aristotle put forward as a *reductio ad absurdum* of the early atomist notion of void. It seems then that Strato's restriction of the use of the term *kenon* was a response to the same arguments that probably induced Epicurus to refine the early atomist terminology in this respect. Like Epicurus, and like the early Stoics, Strato succeeded in isolating a concept of self-subsistent space and separating it from *to kenon* in the strict sense. Strato called it *topos*, Epicurus opted for *anaphês ousia*, and the Stoics for the periphrastic 'that which can be occupied by being' (τὸ οἷόν τε κατέχεσθαι ὑπὸ σώματος). For Strato *kenon* in the strict sense was a micro-void internal to substances, for Epicurus it was primarily the empty space between the atoms, and for the Stoics the infinite space extending outside the cosmos.

2.2.6 It appears, to conclude, that the different conceptions of *kenon* outlined in the present section were probably in the end all rooted in the ways in which a substantivated term like *to kenon* could be applied. This explains why even philosophers who had a clearly defined *concept* of void were nevertheless sometimes seen to use the *term* in different, even inconsistent, ways. As an example we may adduce the Stoic astronomer Cleomedes (who probably lived at the end of the third century CE).[104] As is known, the Stoics by definition restricted the use of the term *kenon* to cases where strictly unoccupied place—which in their system meant in practice: the extracosmic space—was referred to.[105] Nevertheless Cleomedes states in one passage (*Cael.* I, 1, 17–19 Todd) that

> The void outside it [sc.: the cosmos] stretches to infinity in all directions. Of this [sc.: of the void!] the part which is occupied by body is called *topos*, and that which is not occupied will be *kenon*.

So even at this late date an adherent of a school which invested the term *kenon* with a quite specific meaning, and who, to judge

---

[104] See my remarks on the date of Cleomedes below, ch. 6, n. 22.
[105] See below, ch. 6, 264–266.

from the last part of the citation, was well aware of this specific meaning, could not resist the temptation to slide into a different application of the term and to use it in addition as an equivalent of 'space'.[106]

As we saw, in the case of early atomist physics the juxtaposition of different conceptions of *to kenon*—such as it can be inferred from Aristotle's testimony—can be given a more specific explanation. There it was probably caused by the need to respond to *different* aspects of what has been termed the Eleatic challenge. In its turn, this peaceful jumble of different conceptions of void called forward criticisms of the kind propounded by Aristotle. Both the Epicurean and the Stratonic, and possibly also the early Stoic 'refined definitions' of *to kenon* should be regarded as attempt to ward off these or similar criticisms.

Conclusions

In this chapter we attempted to show to what extent the philosophical discussion about space, place and void was, in the period with which we are dealing, dominated by the ambiguity of the Greek spatial terms. It was shown that, otherwise than is often suggested in the scholarly literature, the Greek language did not have a one-word equivalent of our 'space'. The first philosopher to isolate a notion of space in an explicit way was probably Epicurus. Arguably also previous thinkers had a more or less clear-cut conception of space. However, they referred to it as *chôra* or *topos* (both of which could also mean 'place'), or as *kenon* (a denomination which presented problems of its own).

Next, we concluded that the ways in which substantivated adjectives like *to kenon* could be used in ordinary Greek provided the basis for the different applications of the term *kenon* in early atomist physics. It was shown that these different applications

[106] Cf. also the argument at *Cael.* I, 1, 20–24 Todd: Πᾶν σῶμα ἔν τινι εἶναι ἀναγκαῖον. Τοῦτο δὲ ἐν ᾧ ἐστι, τοῦ κατέχοντος αὐτὸ καὶ πεπληρωκότος ἕτερον εἶναι δεῖ, ἀσώματον ὂν καὶ οἷον ἀναφές. Τὴν οὖν τοιαύτην ὑπόστασιν, οἵαν τε οὖσαν δέχεσθαι σῶμα καὶ κατέχεσθαι, *κενὸν* εἶναι φάμεν. As this passage suggests, the temptation to use *kenon* in the sense of 'space' may have been the stronger because of the fact that the Stoics lacked a neat one-word equivalent of our 'space': instead they used periphrastic expressions like τὸ οἷόν τε κατέχεσθαι ὑπὸ ὄντος (an expression which is here clearly echoed by Cleomedes); for further details see below, ch. 6, section 6.1.

could be regarded as answers to different questions raised by the Eleatics, whereas they were skilfully played off against each other in Aristotle's critique of the Abderites. Strato, Epicurus and the Stoics tried to evade, each in their own way, these and similar problems by restricting the application of the term *to kenon* to specific cases.

CHAPTER THREE

CONCEPTS OF SPACE IN PLATO'S *TIMAEUS*

Introduction

In his 1929 introduction to the ninth volume of the Loeb translation of the works of Plato, R. G. Bury wrote: ' ... there is no point more obscure or more vehemently disputed than the nature of the so-called 'platonic matter''. And indeed, from antiquity onwards the problem of the identification of the *hupodochê*, the third factor introduced in the course of Plato's famous cosmological account, has kept scholars busy trying to fit the relevant passages into a coherent framework. The quite depressing *diaphonia* which has resulted from these attempts should warn us that perhaps there is something wrong with the very idea underlying most of them, the idea—that is—that Plato's thoughts on the matter at issue can be made really coherent sense of anyway. And, to declare my hand right away, I happily enroll myself among those who experience a feeling of alienation when reading Archer-Hind's exulting statement that Plato's presentation of the receptacle is that 'than which there is no more masterly piece of analysis in ancient philosophy'.[1]

The present chapter does not contain the umpteenth attempt to read the *Timaeus* account of the receptacle as a coherent theory of space. What I attempt to establish is rather that the very different and often incompatible descriptions provided by different historians are in most cases to be regarded as overcharitable interpretations, each of them singling out and working out what is in fact only one among several coexisting characterizations applied to the receptacle by Plato himself.

Traditionally the discussion about the 'real nature' of the Platonic receptacle has centered around the question whether it should be described as space or as matter, or perhaps as both.[2] Concerning

---

[1] Archer-Hind (1888) 44.

[2] A rather idiosyncratic interpretation of the receptacle passages is now available in the form of Ashbaugh (1988). I am afraid this book can hardly be regarded as a useful contribution to the interpretation of Plato's cosmo-

this issue the present chapter claims two things. First, that the *exclusive* identification of the receptacle with either space or matter is often based on what I previously called unwarranted essentialist preconceptions concerning 'the real nature' of space or matter. It will be argued that these appellations are not *necessarily* incompatible and that the text of the *Timaeus* itself contains some clues indicating in what sense the receptacle can be regarded as space and in what sense as matter.

Second, as to the sense in which the receptacle may be called space, I shall try to show that in the *Timaeus* as a whole and more especially at *Tim.* 48e–52d Plato seems to be committed to different and sometimes incompatible concepts of space. It will be pointed out, moreover, that the occurrence of these inconsistencies should be connected with the ambiguous way in which Plato sketches the relationship between the sensible world and the realm of intelligible Ideas. In addition, Plato's tendency towards presenting a *Nebeneinander* of different and partly incompatible concepts of space may have been enhanced by the various senses—partly outlined in the previous chapter—in which common parlance used expressions like 'to occupy a place', 'to be in a place' and 'to be in something'. Here Plato's natural philosophical tendency seems rather to have been towards taking for granted such equivocations and obscurities as an inevitable part of any attempt to explain the phenomena of the sensible world, rather than towards actively trying to get rid of them. In this connection we should realize of course that the account of the receptacle is part of Plato's 'likely account': in *Tim.* 29c–d Plato has Timaeus announce that 'if in many points concerning many things, concerning the gods and the genesis of the universe we prove unable to give accounts that are always in all respects self-consistent and perfectly exact,

---

logical dialogue. Its philosophical approach is anachronistic on the whole (on which see below, n. 19) and, even worse, often unclear (on which see e.g. below, n. 11). In addition, it appears to me to be flawed by its cavalier dismissal of Aristotle's testimony, as well as by its failure to take account of some, admittedly traditional but nonetheless fundamental, problems concerning the receptacle account—the very problems, in fact, which are central to the present chapter. But also at the level of details the book contains a number of odd claims and factual mistakes (as an example one may adduce the claim concerning the *hupodochê*, *op. cit.* 100, that 'the noun *dochê* (from *dokeô*) means *vision* and is related to *dokêma* meaning *opinion* or *expectation*').

be not surprised'[3] and he adds that concerning these matters it becomes one to accept τὸν εἰκότα μῦθον. Moreover, the receptacle itself is introduced at 49a as 'a difficult and obscure kind' (χαλεπὸν καὶ ἀμυδρὸν εἶδος).

In so far as such an approach does not leave Plato with a 'masterly piece of analysis' but, on the contrary, rather seems to highlight his weaknesses, it may seem of little interest to those favouring the historical method of what might be labelled 'rational reconstruction'[4]: the attempt to interpret charitably the theories of philosophers of the past in such a way that they are coherent and basically interesting. I believe, however, that the interest of the present analysis, if correct, is rather that it provides Plato's attempts to cope with the problems of space with a historical and philosophical context in linking them up both with his own metaphysical system and with the ordinary usage of spatial terms in his own time. Not only does this do more justice to Plato than merely admiring him for qualities which were not really his, it also puts us in a position to estimate adequately the contributions of Aristotle to the philosophical debate concerning the issue.

Two additional remarks will be in order. First, throughout the present chapter the discussion will concentrate on *Tim.* 48e–52d. This is the part of the *Timaeus* in which the receptacle is first introduced. I shall take it for granted that this part can be studied more or less by itself.[5] It is immediately followed by the famous description of the precosmic chaos and the ordering process by which the Demiurge creates elementary geometrical particles of some kind (the four regular solids) out of elementary triangles. It is a moot point whether this 'elementary chemistry' should be read back into *Tim.* 48e–52d, more particularly whether the 'things entering and leaving the receptacle' in the latter passage should be taken to be geometrical forms introduced in the passages which follow.[6] Though I tend to subscribe to the view

---

3 *Tim.* 29c: Ἐὰν οὖν ... πολλὰ πολλῶν πέρι, θεῶν καὶ τῆς τοῦ παντὸς γενέσεως, μὴ δυνατοὶ γιγνώμεθα πάντῃ πάντως αὐτοὺς ἑαυτοῖς ὁμολογουμένους λόγους καὶ ἀπηκριβωμένους ἀποδοῦναι, μὴ θαυμάσῃς.

4 Cf. R. Rorty, 'The Historiography of Philosophy: Four Genres', in: R. Rorty, J. B. Schneewind, Q. Skinner (edd.), *Philosophy in History: Essays on the Historiography of Philosophy* (Cambridge, etc. 1984) 49–75, esp. 49 ff.

5 On the more or less self-contained character of *Tim.* 48e–52d see Lee (1966) *passim.*

6 Such was the opinion of Baeumker (1890) 131, Schulz (1966) 108–110,

that *Tim* 48e–52d does not yet presuppose this theory, and that it simply tries to depict the relationship between intelligible Ideas and sensible things in general, I do not think the problem as such has much relevance for the present discussion. We shall here be primarily concerned with the status of the receptacle, in particular the question whether and in what sense it serves as space. In this connection it is immaterial whether that which it receives is all kinds of 'qualities' or immanent characteristics, or geometrical planes and figures. As will become clear, the main problems (among which the question of the separability versus the inseparability of the receptacle is the most prominent) remain the same.[7]

Second, the present chapter only deals with the views on space as they are found in the *Timaeus*. It is well known that in some ways also the testimonia preserving information concerning Plato's so-called 'unwritten doctrines' contain information that might be of interest to a historian writing about concepts of space.[8] Still, I shall only refer to these *agrapha dogmata* in passing, in connection with Aristotle's criticisms of the *Timaeus*. The reasons for doing so are the following.

In the first place the problem of the interpretation of the *agrapha dogmata*—as well as the concomitant problem of the reliability of the sources—is itself the subject of enduring controversy. In this area there is very little firm ground to stand on, so that I feel that extensive references to these unwritten doctrines would in general be of little value as additional evidence. Second, though the Platonic dialogues are of course in many interesting ways interconnected, they were conceived as self-contained philosophical treatises.[9] Accordingly, it should be possible to study the

---

and Ross (1924) I, 168; against this view see Guthrie (1967–81) V, 270, n.1, and Cornford (1937) 182–183.

[7] Cf. Sorabji (1988a) 35: 'There remains the complication of how vestigial fire in the period of chaos relates to fire as we know it, which has been packaged into shaped corpuscules by the Creator-God. But it is natural to assume that space is neither more nor less a subject of properties in the orderly world produced by God than in the chaotic world that precedes it'.

[8] The same goes, *mutatis mutandis*, for the *Philebus*. See, however, the qualifying remarks on the relationship between *Tim.* and *Phil.* made by Guthrie (1967–81) V, 270, n. 1. See also the convenient survey of opinions on this issue as provided by M. Isnardi-Parente in an additional note to Zeller/ Mondolfo (1974) 14 ff.

[9] Cf. Guthrie (1967–81) IV, 324: 'Each dialogue of Plato is a self-contained,

Timaeus, strictly speaking Plato's only cosmological dialogue, on its own.

The present chapter will contain the following sections. In section 3.1 the arguments of those who have proffered an exclusive interpretation of the receptacle as either space or matter will be scrutinized and criticized. It will appear that the text of *Tim.* 48e–52d suggests at least one sense in which Plato could coherently maintain that the receptacle is both matter and space at the same time. In section 3.2 we shall try to show that, though in *Tim.* 48e–52d one particular view of the receptacle seems to be predominant, the presentation as a whole defies attempts at a consistent interpretation. However, the occurrence of these inconsistencies can at least partly be explained by reference to the problems Plato usually experiences in explaining the phenomenon of *methexis.* Furthermore, and most importantly for the purpose of the present study, I think it can be plausibly shown that in so far as Plato thinks of the receptacle as 'space' he is committed to several rather inarticulate common conceptions of space. Against this background it will be shown, in section 3.3, that Aristotle's criticisms of the receptacle in the *Timaeus* are, on closer scrutiny, contrary to what is often asserted, very much to the point, and that there is no good reason for accusing him of being either ignorant or unfair.

3.1 *The receptacle as matter and space*

3.1.1 In *Tim.* 48e–52d the receptacle is introduced as a third factor needed in the account of the genesis and structure of the cosmos—next, that is, to the eternal and unchangeable Ideas on the one hand, and the floating unstable realm of the sensible world on the other hand. In the course of the passage the nature of the receptacle is illustrated by means of a number of similes.[10] Problems arise in so far as it is, at least at first blush, not clear how the various descriptions applied by Plato are to be squared with one another. On the one hand some of the denominations (e.g. the use

---

organic unity, yet each is bound to its fellows by a subtle web of interconnecting threads'. The interconnection of course springs forth from the fact that some main elements of Plato's thought recur in various contexts in various dialogues. For that matter, it will be shown in the present context that the way Plato tries to cope with the phenomenon of *methexis* in the *Timaeus* can be paralleled from other dialogues.

[10] Conveniently listed by Guthrie (1967–1981) V, 263–264.

of the terms *hedra* and *chôra* in 52a) rather suggest that Plato was thinking of a *spatial* factor, a something-in-which.[11] On the other hand there are others (e.g. the use of the term *ekmageion* in 50c and the gold analogy of 50a)[12] which at first sight rather seem to point to a something-out-of-which, a kind of matter. Moreover, there is also the testimony of Aristotle who contended that in the *Timaeus* Plato so much as identified space (or place) and matter.

It is not surprising, then, that already at an early date questions arose as to the right interpretation of the receptacle passages. The issue seems to have divided scholarship ever since the days of Aristotle.[13] Most interpreters have been of the opinion that the two different descriptions were incompatible and that consequently only one of them conveyed Plato's *real* thoughts. In the present section I intend to show that their attempts to end up with a coherent theory of the receptacle-as-matter or the receptacle-as-space are in the end based on (1) a selective or wrong reading of Plato's actual text and (2) wrongheaded (essentialist) presuppositions concerning the 'real nature' of space and matter with, as a corollary, premature conclusions about the incompatibility of space and matter as labels of one and the same entity.

I want to stress, however, that I do not intend to replace these traditional interpretations by a new interpretation of Plato's views of the receptacle as a coherent theory, this time a coherent theory

---

11 For the translation of *chôra* by 'space' see the previous chapter. Ashbaugh (1988) *passim*, appears to opt for the translation 'spatiality'. This rather abstract term is defined as follows, *op. cit.* 114: 'spatiality is neither a body nor a place. It is containment itself, the idea (*sic*) of physical limit'. Elsewhere, *op. cit.* 114, we are told that 'spatiality is the ability of bodies to be besouled'. For the two latter claims concerning the nature of 'spatiality' I can find no basis at all in Plato's text. Nor can I find any passage where Plato opposes *chôra* ('spatiality') to *topos* ('place') in the way presupposed by Ashbaugh although it is not entirely clear what her own views on this issue are. At *op. cit.* 133 place is said to be 'the visible individual *analogue* of space' (my italics), and the two terms are said to relate 'as the different items that Plato usually juxtaposes to each other in a dialectical inquiry: soul and city, philosopher and sophist, knowledge and opinion, original and image, to cite a few'. At p. 134, however, place is said to be a *part* of spatiality. Perhaps we had best leave these modern dialectical musings for what they are and turn to Plato's own text where the terms *topos* and *chôra* appear to be used synonymously at *Tim.* 52b, a passage quoted in the text below, 106.

12 *Tim.* 50a: εἰ γὰρ πάντα τις σχήματα πλάσας ἐκ χρυσοῦ ... etc.

13 See Baeumker (1890) 151 ff. where a review of opinions from Aristotle down to the 19th century is given; for a survey of more recent positions on the issue see Guthrie (1967–81) V, 265 n. 3 and 5; see also Schulz (1966) 9–10.

of the receptacle-as-both-space-and-matter. It is true, the present section will be concluded by an outline of what I shall call Plato's predominant mode of presentation of the receptacle, but these conclusions will be immediately qualified in section 3.2. I think it can be shown that Plato's presentation of the receptacle is in the end fundamentally incoherent.[14] The aim of the present chapter as a whole is to set out these inconsistencies and—partly with the help of the results of the preceding chapters—to explain their origin rather than to explain them away.

3.1.2 A number of students, both ancient and more recent, of Plato's *Timaeus* have held that the receptacle can best and most adequately be described (or even that it should exclusively be described) as matter.[15] This could be done by playing down, or just ignoring, the 'spatial' characterizations of the receptacle and by stressing the alleged incompatibility of the labels 'space' and 'matter'.

To begin with the former point, one of the obvious problems which these scholars would seem to run into is the plain fact that Plato quite unambiguously calls his receptacle *chôra*. Two ways are apparently open to counter this objection. One could either silently pass over Plato's use of the term *chôra* in 52d,[16] or point out that the word *chôra* does not so much describe the *essence* or *nature* of the receptacle but rather its *function*: 'Nicht als den Raum als solchen wolle Plato damit die Materie bezeichnen; es sei ihm vielmehr die Materie der besondere Stoff, welcher den in ihn eintretenden Ideen, resp. deren Nachbildern Platz und Raum gewähre'.[17]

It goes without saying that simply ignoring the fact that Plato

---

[14] This conclusion is apparently shared by Sorabji, who concludes his discussion of the receptacle passages, (1988a) 32–35, by remarking that 'Plato does not, then, keep to a single account'.

[15] Among these are, according to Simp. *In Ph.* 227, 23 ff., the Stoics and the neoplatonist Pericles of Lydia (a pupil of Proclus). Of the so-called Middle Platonists we may in this connection mention Atticus and Plutarch. In the 19th century this position was defended among others by Hegel and Ueberweg. For a more extensive discussion and further references see Baeumker (1890) 152 ff. In the 20th century the receptacle has been most conspicuously identified with matter by Eva Sachs (1917) 223–232.

[16] Thus Sachs (1917). On this see Schulz (1966) 8.

[17] Baeumker (1890) 180, referring to the position taken e.g. by Ueberweg and Bassfreund.

calls the receptacle *chôra* is an obvious *pis aller*, the more so since the term *chôra* is in fact one of the most explicit denominations used by Plato in this connection. This leaves us with the only serious argument against the identification of the receptacle with space (*chôra*), viz. that Plato's using the latter term should *not* be taken to imply that he thinks the receptacle is *essentially* space, but rather that to be that in which the ideas are reflected is, according to Plato, merely one of the *functions* of the receptacle (which, on this line of thought would be *essentially* matter).

A first objection to this view is that there is nothing in Plato's text which suggests that he thus invested the 'spatial' aspect of the receptacle with an inferior, merely *functional*, status. The several descriptions applied are in the end *all* functional characterizations. In fact, the whole distinction between 'function' and 'essence' of the receptacle seems to play no role at all. On the contrary, though Plato's wording is not always explicit, he clearly seems to suggest that the nature and the function of the receptacle are essentially bound up with one another. At *Tim.* 49a, for instance, it is asked 'What function are we to conceive it to fulfil by its own nature?' (τίν' οὖν ἔχον δύναμιν κατὰ φύσιν αὐτὸ ὑποληπτέον;).[18] The answer is that it is 'the receptacle, and as it were the nurse, of all becoming' (πάσης εἶναι γενέσεως ὑποδοχὴν αὐτ ἠν, οἷον τιθήνην). These *functions* apparently belong to the *nature* of the receptacle, they are κατὰ φύσιν. A similar linking of the function (*dunamis*) and the 'nature' (*phusis*) of the receptacle is found at *Tim.* 50b where the receptacle, here apparently in its spatial role, is called 'the *nature* (thing) which receives all bodies' (τῆς τὰ πάντα δεχομένης σώματα φύσεως). Furthermore it is said that the receptacle must always be called by the same name because it never departs from its function (ἐκ γὰρ τῆς ἑαυτῆς τὸ παράπαν οὐκ ἐξίσταται δυνάμεως), which is further specified by the statements that it always receives all things (δέχεταί τε γὰρ ἀεὶ τὰ πάντα) and that it lies by its own nature as a kind of moulding stuff (*ekmageion*) for everything (ἐκμαγεῖον γὰρ φύσει παντὶ κεῖται).

These passages contain important information in that they show that (1) the receptacle is represented as a 'thing' (*phusis*)

---

[18] I have taken over the reading of A; following FY, with the OCT edition, we get δύναμιν καὶ φύσιν ('what function and nature should we conceive it to have'). For our present purpose it is immaterial which of these *variae lectiones* is preferred.

characterized at different times by different functions (either 'space' or 'matter')[19] but that (2) there is at first blush no indication that Plato here regarded the one function as more 'essential' than the other. The receptacle is by virtue of its own nature both an out-of-which (it is an *ekmageion*) and an in-which (for it 'receives all things': δέχεται [...] τὰ πάντα). In other words, there is no attempt on Plato's part to specify the thing-like character of the receptacle independently of the functional characterizations.[20]

---

[19] In this connection it may be noted that Plato's own wording provides us with an argument against a subjective-idealist interpretation of the receptacle as well as against its modern pendant, viz. the view that Plato treated the ἐν ᾧ and the ἐξ οὗ as—to borrow Wieland's well-known characterization of the Aristotelian causes—as mere *Funktionalbegriffe*. For, quite in line with the persistent realism pervading Greek thought, the receptacle—both when described as an ἐν ᾧ and when described as an ἐξ οὗ—is presented as something more than merely a theoretical concept: it is reified, it is a 'thing' (φύσις). Subjective-idealist interpretations of the receptacle were quite common in the nineteenth century (with scholars like Ritter) and of course among the Neokantians of the early twentieth (Cohen, Natorp). A convenient survey of these interpretations can be found in Zeller/ Mondolfo (1974) 55–57. Ashbaugh (1988) professes to take a middle position between the traditional 'realist' interpretatation of the *Timaeus* and a subjective-idealist one. According to Ashbaugh, *op. cit.* 1, the *Timaeus* constitutes an answer to the question 'if reason could persuade nature to show itself, what would the soul see?', and she takes it that Plato here 'explains not what an objective cosmos is, but how it comes to be known by the soul'. Unfortunately the question from which she starts out (realist vs. subjective-idealist interpretatation) is anachronistic and, hence, misdirected. In addition her account is at times blurred and unclear in so far as it does not seem to distinguish between the respective roles of *cosmic* (i.e. 'objective') and *human* (i.e. 'subjective') soul or intelligence (thus it is unclear *whose* thinking is at issue at p. 75, where we are told that thinking 'constitutes the cosmos', or *whose* intelligence at. p. 113–114, where we find that 'space results from the demands of intelligence upon being').

[20] It is perhaps tempting to assume that at *Tim.* 48e–52d the receptacle may be regarded as 'being essentially' mere extension, this extension serving as a space (an 'in which' in a not strictly physical sense) for the immanent forms and at the same time as a material constituent for physical bodies. But—apart from the fact that, as I shall try to show, even *Tim.* 48e–52d is in this respect not consistent—it may be observed that this characterization of the receptacle as mere extension would certainly not seem to apply to the way it is presented in the passages dealing with the precosmic chaos (52d ff.). There it seems to be already endowed with elementary qualities (Plato speaks of ἴχνη, traces of fire etc.) before the Demiurge starts working on it. So when taken *ad litteram* this receptacle, which may be called 'matter', but which, since it is said to move, would hardly seem to be space, is not the same as that which in the passage immediately preceding (48e–52d) was described as absolutely without qualities in the primordial state (and which could be described as 'space'). However, the fact that in both cases the receptacle may conveniently be described as that *in which* form is

If the distinction between *functional* and *essential* characteristics has thus been shown to be inapplicable, we may try to scrutinize the other arguments which have been adduced in order to prove that the receptacle is merely, or only essentially, matter, and which thus forced scholars to ignore, or to explain away, Plato's explicit use of the labels *chôra* and *en hôi.*

Eva Sachs seems to have been committed to the following argument:

(1) the receptacle is represented as matter (i.e. as a constituent of physical reality); if the receptacle is matter,[21] then:

(2) the receptacle is corporeal;[22] if the receptacle is corporeal, then:

(3) the receptacle cannot be space.[23]

Though (1) has been denied by quite a few interpreters of the

---

imposited and *out of which* phenomenal bodies are thus created (which means that its *functional characteristics* are in both cases similar) suggests, I presume, that Plato tended towards using his 'third kind' as a *Funktionalbegriff,* but that the fact that he nevertheless stuck to presenting it in a reified manner resulted in what appear to be different and incompatible presentations. Such an interpretation would be preferable, I think, to that which tries to explain away the differences between the chaotically moving receptacle of 52d ff. and that of 48e ff. by e.g. stressing that a moving space (52d ff) is an admirably modern notion (Schulz (1966) 88). Alternatively one might take 52d ff. as a merely mythical excursus (G. Vlastos, 'The Disorderly Motion in the *Timaeus*', in Allen (1965) 379–399; see also, by the same author, 'Creation in the *Timaeus*: Is it a Fiction?' in Allen (1965) 401–419 and Vlastos (1975) 69).

An interesting implication of the above interpretation, if correct, would be that the problems concerning the Platonic receptacle would turn out to be in some respects comparable to the problems concerning the way in which we should view the juxtaposition within the Aristotelian physical system of primary matter and secondary matter. For Aristotle's realism with regard to his physical principles see Wagner (1967) 2, 337 ff., a critical discussion of Wieland (1962), a study which rather one-sidedly focuses on the functional aspect ('Funktionalbegriffe') of Aristotle's main physical concepts.

[21] This is inferred (1) from the analogies used in *Tim.* 48e–52d (on which see below, esp. n. 25) and (2) from *Tim.* 52d ff. (i.e. the passages on the 'precosmic chaos', on which, however, see above, 74–75.

[22] The corporeality of the receptacle was, by the way, also implied by F. Ueberweg, 'Ueber die platonische Weltseele', *RhM*, N.F. 9 (1854): 'Platon konnte eine die Figuren in sich aufnehmende *Masse* ... unbeschadet ihrer *Materialität,* da sie auch das Prinzip der Räumlichkeit bildete, sehr wohl auch ... den Sitz und Ort jener Figurenbildung nennen'[my italics].

[23] The inference from (2) to (3) seems implicit in the argument of Sachs (1917) 224–225, that the receptacle as τιθήνη appears 'an Stelle der Elemente aus denen Früheren die Welt bildeten ... diese Kritik der Elementenlehre ... setzt einen *Stoff* voraus, nicht den bloszen *Raum*'.

Timaeus, it will, as we shall see, have to be agreed on by anyone who takes the text of the *Timaeus* seriously. However, the inference from (1) to (2) is probably based on the fact that *our* common parlance term 'matter' apparently bears connotations of corporeality according well with the thing-like presentation of the receptacle. However, though it may be part of the modern stereotype of matter that matter is a corporeal stuff, we are of course not allowed to assume that all historical concepts of matter answered this stereotype. Whether Plato's alleged concept of matter involved the connotation of corporeality is to be decided on the basis of an investigation of the contexts in which he seems to have used it.

At this point Sachs claims that her view finds support in the text of the *Timaeus.* She observes that the basic elements in a number of metaphorical descriptions of the receptacle (gold, a mother, a wax tablet) are themselves material in the sense of *corporeal.*[24] But nothing in the text suggests that it is this *corporeal* character of gold or wax which furnishes the *tertium comparationis.*[25] This is perhaps most clear at *Tim.* 50d where the receptacle is compared to a mother (which I take to be a reference to its material aspect) whereas the Ideas are compared to a father. In the latter case it is clear that it is not the corporeality of a father which we are supposed to transfer to the Ideas. The upshot of the comparison is of course that the Ideas, like a father, are made to play a formative role in the creation of the cosmos. Similarly, the receptacle is

---

[24] Sachs (1917) 224: 'Dass hier ein realer Stoff gemeint ist, geht schon aus der Entwicklungsreihe, in die diese Betrachtung Platons Neuschöpfung einordnet, hervor ... es wird aber noch klarer an dem Bilde, dessen Platon sich bedient, um den neuen Begriff deutlich zu machen ... das schon erwähnte Beispiel vom Goldarbeiter .... wenn irgendwo, so ist hier der Begriff des Stoffes als eines 'Arbeitsmateriales', das Verhältniss des formenden Künstlers zur Idee und Stoff so klar dargestellt, dass nur ein im voraus befangenes Urteil die Beziehung leugnen könnte.' etc.

[25] However, Sachs (1917) 224 concludes: 'Man wird dem Sinne dieses anschaulichen Gleichnisses so wenig wie dem der Homerischen Vergleiche gerecht, wenn man nur nach dem tertium comparationis sucht. Es ist doch kein Zufall, dass dieser Goldarbeiter gerade Dreiecke formt und dass nach Anwendung dieses Bildes mit den Worten ὁ αὐτὸς δὴ λόγος καὶ περὶ τῆς τὰ πάντα δεχομένης φύσεως die drei Prizipien, die Materie, die Form und der geformte Stoff eingeführt werden'. I think, however, that the use of the word λόγος in this connection ('the same story can be told about ...') suggests that it is the way gold stands to the forms in the gold analogy, rather than the specific corporeal character of the gold, which is transferred to the receptacle (and the way it is related to the forms).

compared to a mother, because of her alleged receptive *role*, not because of her corporeality.

Even if (2) were accepted, however, the inference from (2) to (3) would not necessarily be correct. This would seem to depend on the specific sense in which the term 'space' is used in (3). For even if we avoid lapsing into talk about functional versus essential characteristics—a distinction which we already showed to be not really applicable in the case of Plato's receptacle—it might well be that Plato only called the receptacle *chôra* because it is that 'in which' forms or qualities inhere. In that case the receptacle, even if corporeal, could serve as the *ex hou* for sensible bodies and as the *en hôi* for the inherent qualities. In other words, the receptacle might be at the same time matter and space, though not with respect to the same things. In fact, as I shall try to show below, this was indeed how Plato in the main presented his receptacle.

It seems, then, that Sachs based her exclusive identification of the receptacle with matter—and her concomitant neglect of the denominations *chôra* and *en hôi*—on an inconclusive argument which was, moreover, based on mistaken *a priori* assumptions. Our provisional conclusion must therefore be that attempts to justify our description of the receptacle as *merely* or *essentially* matter have been unsuccessful and that its 'spatial' descriptions, *chôra* and *en hôi*, are not explained away very easily. For the moment, therefore, we may posit that the receptacle is presented as a something which Plato further characterizes by endowing it with two functions. If we are to take his wording seriously, these functions are (1) to be an *en hôi* and (2) to be an *ex hou*.

3.1.3 On the other side of the debate there are those who favoured an exclusive identification of the receptacle as space. As the 'matter party' had to explain away the explicit denomination *chôra*, so the 'space party' has to explain away those features in the text which seem to point to an identification of the receptacle with matter. In addition, also on this side of the debate much stress has been laid on the alleged incompatibility of the labels 'space' and 'matter' or *en hôi* and *ex hou*.[26]

---

[26] Cf. e.g. Keyt (1961) 297: ' ... since the third factor is space it cannot also be matter, unless Plato's conception is inconsistent'.

As to the latter argument, some additional comments may be in order here. The thesis of the incompatibility of the labels 'space' and 'matter' is usually motivated by an explicit or implicit commitment to at least one of the following two essentialist presuppositions:

(1) matter is corporeal, and

(2) space is absolute.

In the previous paragraph we already remarked that (1) is mistaken and, moreover, probably in itself insufficient to establish the alleged incompatibility. For a corporeal matter may well still in some sense also be space. On the other hand, (2) would seem to be sufficient to establish the required incompatibility. For an absolute space can *by definition*, i.e. in virtue of its very absoluteness, not be a constituent factor of the bodies which it contains. However, (2) no less than (1) belongs with what in chapter 1 were called unwarranted 'initial generalizations'.

Many different concepts of matter and many different concepts of space have been defended throughout the history of philosophy and science. Though among the concepts of matter that which takes matter as corporeal and among the concepts of space that which takes space as absolute may be the ones that first come to mind (e.g. because they found their way into common parlance) we are not allowed to assume in advance that *if* Plato had a concept of space and a concept of matter he *must ipso facto* have been committed to these two specific concepts. If we define 'matter' quite generally as 'the underlying constituent of physical reality', we may recognize that within different physical systems different 'things' answer this description. There is no reason to assume *a priori* that it is out of the question that any of these 'things', whether corporeal or not, served at the same time as space—as long, that is, as space is not taken as an absolute container. The final answer, then, to the question whether the receptacle can serve both as space and as matter cannot be given before we have established *in what sense* the receptacle is presented as matter and *in what sense* as space.

A first attempt to establish how Plato used the terms *en hôi* and *ex hou* will be made in the next paragraph. We shall now first examine the attempts made by the 'space party' to explain away those features in Plato's own text which seem to point to an identification of the receptacle with matter. In this respect the champions of the exclusive identification of the receptacle with space have two arguments to offer:

(1) Plato nowhere explicitly called his receptacle 'matter',[27] the use of the expression *ex hou* in the gold analogy being allegedly nothing but an exigency of the metaphor.[28]

(2) It was Aristotle who first mistakenly read a concept of matter, presumably even his own concept of *hulê*, into the *Timaeus* account which in reality leaves no room for such a concept.[29]

On closer scrutiny these arguments may be countered in the following way:

Ad (1). As to Plato's not having a technical term for matter, it is one thing to notice that throughout his writings the word *hulê* is only used in its original sense of 'wood'.[30] It is quite another thing, however, to maintain that this shows that Plato did not have anything resembling what we might call a *concept* of matter. Anyway, it would certainly need some arguing to prove that this is not what he tried to convey by metaphorical descriptions like *ekmageion* ('moulding stuff' or 'wax-tablet'), or *mêtêr* ('mother'), or the periphrastic *ex hou* occurring in the gold analogy. Especially the latter expression can hardly be taken otherwise than as referring to a material constituent.

---

27 Thus Cornford, according to whom bodies are nothing but bundles of properties without a substratum remarks, (1937) 181: 'There is no justification for calling the Receptacle 'matter'—a term not used by Plato. The Receptacle is not that 'out of which' (ἐξ οὗ) things are made; it is that 'in which' (ἐν ᾧ) qualities appear, as fleeting images are seen *in* a mirror'; also H. Happ (1971) 128 stresses the alleged distinction between the ἐν ᾧ and the ἐξ οὗ. Cf. also Cherniss (1944) 121. Also Ashbaugh (1988) 119 rejects the identification of the receptacle and matter. One of her reasons for doing so is that Plato does not use the term *hulê* in the receptacle account, and that where he does use it elsewhere it does not mean 'matter'.

28 Cf. e.g. Mohr (1985) 94: 'I suggest that Plato's description of the gold as that out of which shapes are formed (50a6) is an exigency of the metaphor and is not the relevant aspect of the gold that is being compared to space'.

29 Cf. Ross (1951) 222: 'But ... Aristotle is wrong in identifying this [i.e. the receptacle] with the 'matter' which plays so large a part in his own philosophy'. See also Cherniss (1944) 171: 'Aristotle's inability to understand that the ὑποδοχή is, as Plato says it is, χώρα and not his own prime matter *manquée* accounts for much of his infelicitous criticism of Platonic physics'.

30 Cf. *Tim.* 69a: οἷα τέκτοσιν ... ὕλη παράκειται τὰ τῶν αἰτίων γένη. Cf. Claghorn (1954) 6, and Ast (1835–38) III, 432–433. There is no reason at all to assume with Skemp (1960) 209, that Aristotle's remarks are meant to imply that Plato *himself* actually called the 'participant' ὕλη. The same misleading suggestion is to be found in Cherniss (1944) 121.

This takes us to the argument that the out-of-which character of the gold in the analogy is not the aspect of the gold which is relevant for the comparison (the *tertium comparationis*). Even if this argument—which of course has much of an *ignoratio elenchi*—were accepted, it would not prove enough. For in order to prove that the receptacle cannot be called matter it is not sufficient to point out that it is not explicitly represented as an *ex hou*. There are a number of other features of Plato's description of the receptacle which on this line of thought would also have to be either ignored or explained away. We already called attention to its being called *mêtêr* and *ekmageion*. Why, moreover, does Plato say that the receptacle is the only thing which, when pointing at it, we may call a 'this', whereas the qualities entering and leaving it are merely 'suchlike' (49e–50a)?[31] Why not just call it 'here', if—that is—it is merely space in an absolute sense? Furthermore, how can we make sense of the assertion that (phenomenal) fire is in reality the 'ignified part' of the receptacle (τὸ πεπυρωμένον μέρος 51b) otherwise than by regarding the receptacle itself as a kind of material constituent?

We may conclude, then, that Plato's text presents a number of indications which point to his taking the receptacle as a constitutive factor of sensible reality. At any rate, strong arguments will be needed to persuade us to neglect these indications. As yet the scholarly literature has not succeeded in providing such strong arguments.

Ad (2). Next, there is the important question of the reliability of Aristotle's testimony. As a preliminary remark we may note that the question whether Aristotle, when calling the receptacle of the *Timaeus* 'matter', was implying that Plato was committed to his (i.e. Aristotle's) own concept of *hulê*, is in practice often confused with a different one, viz. whether we ourselves are, when speaking of a concept of matter on Plato's part, *ipso facto* investing it with all features of the *Aristotelian* or of any other specific concept of matter.[32] However, after what has been said—in the above and in

---

[31] On this passage see the extensive and detailed discussion of Cherniss (1954); see also the survey of comparatively recent literature on this passage by M. Isnardi-Parente in a supplementary note to Zeller/ Mondolfo (1974) 19 ff.

[32] Compare Mohr (1985) 93, who objects to those who made 'Plato's space look much like Aristotle's matter' that 'if we distinguish the philosophical

chapter 1—about how these kinds of 'labelling generalizations' should be applied by the historian, the second question does not need a separate discussion here. What, then, about Aristotle? Was he silently reading *his own* specific concept of matter into the *Timaeus* ?

In the first place we may note that the fact that Aristotle does not even himself apply his own terminology in any rigidly defined way—the specific sense in which he uses his basic terms often differing according to the context—lends little *prima facie* plausibility to the thesis that he is reading back very specific concepts of his own devising into his predecessors.[33]

Secondly, as a survey of the cases in which Aristotle uses the term *hulê* in his descriptions of the systems of his predecessors shows, the negative judgements of Cherniss and others about his reliability as a witness are based on a wrong conception of the nature of the Aristotelian descriptive labels.[34] Aristotle is quite

---

functions of the two sorts of entities in each philosopher's system, we shall see that they have little in common'. This statement is followed by a survey of the 'five basic functions which matter serves in Aristotle's metaphysics' and a series of arguments purporting to show that most of these functions have no connection with Plato's space. A similar position has been defended by Kung (1988) 175. I cannot here go into the details of the arguments adduced by these scholars. It is enough for the present purpose to point out that they appear to be simply wrong in presupposing that any identification of the receptacle with matter or any comparison between the receptacle and Aristotelian matter would only be correct if the receptacle served *all* functions of the *Aristotelian* matter. A much more sensible position is taken by Baeumker, who correctly argues (1890) 185) that 'Der Begriff der Materie wird ein verschiedener sein, je nach den Voraussetzungen, die zu ihn führen. Bei Aristoteles entspringt derselbe einer Analyse des Werdens. ... Anders bei Plato.... Nicht wie aus einem Sinnendinge ein anderes entstehe, ist hier Problem, sondern wie überhaupt Sinnendinge neben den Ideen sich begründen lassen'. I think Aristotle was sensitive to the latter difference, witness his remarks about the Platonic position identifying matter and space in *Phys.* Δ 211 b 29 ff., quoted in the text below, 113.

33 The acknowledgement that Aristotle applied his own terminology in a fairly equivocal way is what led to Wieland's rather extreme view of the Aristotelian causes as *Funktionalbegriffe* (Wieland (1962) *passim*). Cf. further Owens (1951) 107–135 ('The Aristotelian Equivocals') and, on the concept of ὕλη in particular, Happ's conclusions (1971) 804–806.

34 The only author of whom I know who seems to be aware of this fact is Happ (1971) 129, who stresses that Aristotle can ... 'mit einem sehr allgemeinen Begriff von ὕλη arbeiten ... ohne auf den spezifischen Gehalt seiner eigenen Lehren über πρώτη ὕλη und ὕλη νοητὴ Rücksicht nehmen zu müssen'. Others, probably still in the aftermath of Cherniss' famous studies on the attitude of Aristotle towards his predecessors (see e.g. (1935) 355, 359 ff.) usually just *assume* that Aristotle is misrepresenting Plato when he

aware of the fact that in his historical descriptions the label *hulê* refers to a variety of concepts which may often enough be very different from his own concept of *hulê*.[35] In this respect he is a

credits him with a concept of matter. The underlying idea is that Aristotle *must* be identifying the receptacle of the *Timaeus* with *his own* concept of matter. A fairly eccentric version of this view is defended by Claghorn (1954) 15–18. In his zeal to prove that Aristotle's account of the *Timaeus*' receptacle is not in the least misleading, he tries to show—after having had to admit that Aristotle's identification of Platonic space and matter 'cannot be verbally true' because Plato 'never uses the term 'matter'' (on which see my remarks above)—that 'Aristotle was accurate in spririt ... in the sense that his view of matter corresponds to the intent of Plato's receptacle'. Thus he turns matters upside down: since it cannot be proved that Plato's receptacle was meant to be actually the same thing as Aristotle's matter, it has to be shown that Aristotle's matter was really intended to be the same thing as Plato's receptacle. That the receptacle also has a 'spatial' aspect does not appear to constitute a problem: Claghorn tries to show that also for Aristotle matter and space are in a sense to be identified, since things are *in* matter 'in the sense that matter is a necessary constituent for surrounding bodies which made place possible'. This, however, is not based on any evidence in any of Aristotle's texts. On the contrary, it completely ignores the fact that the introduction of a sharp distinction between place (τόπος) and matter is one of the main elements of Aristotle's theory of τόπος as it is propounded in the *Physics*.

35 The so-called doxographical passages occuring e.g. in the first books of the *Physics* and the *Metaphysics* make this abundantly clear. There Aristotle investigates to what extent his own scheme of four causes was endorsed by his predecessors. In the course of these investigations it becomes clear to him that the earliest thinkers (viz. Thales, Anaximander, Anaximenes, and to some extent also Anaxagoras and even the Pythagoreans) only conceived of a *material* cause. In itself this already shows that in speaking of the first principle of these philosophers as 'matter' Aristotle did not at all mean to imply that they conceived of matter in exactly the same way as he did. For in his own system the concept of matter functions in close connection with a concept of form which, as we saw, he thought to be absent as such from the systems of the early monists. Moreover his own four causes, including the material cause, are primarily meant to provide a conceptual framework for the explanation of physical phenomena like locomotion, change and the coming into being and passing away of individual things. Still this does not preclude his attributing a material principle to the Eleatic Melissus whose one Being, admitting of no kind of change or motion whatsoever, does not, according to Aristotle, belong to the domain of physics at all. Cf. *Phys.* A 184 b 25 ff.: Τὸ μὲν οὖν εἰ ἓν καὶ ἀκίνητον τὸ ὂν σκοπεῖν οὐ περὶ φύσεώς ἐστι σκοπεῖν and *ibidem* 185 a 12: ἡμῖν δ' ὑποκείσθω τὰ φύσει ἢ πάντα ἢ ἔνια κινούμενα εἶναι. See also *Metaph.* A 986 b 12 ff. Without being concerned here with the question whether Aristotle does or does not misrepresent Melissus' position in the passage at issue, we may safely conclude that where he calls Melissus' Being 'One in matter' (κατὰ τὴν ὕλην, *Metaph.* A 986 b 20) he is using the term ὕλη in a non-specific sense and that he certainly does not imply that this ὕλη fulfils any of the physical functions served by his own concept of matter. On Aristotle's general procedure in these passages and on his sources see now Mansfeld (1986) 7 ff.

more conscientious historian than those historians who were criticized in chapter 1 for having endorsed unwarranted essentialist presuppositions about what a concept of space is.

This, of course, does not mean to imply that Aristotle's descriptions are never biased. It should be stressed, however, that if he may sometimes be guilty of misrepresenting the thought of his predecessors in so far as he credits them with a concept of *hulê*, this is not so much because he is reading *his own concept* of matter back into their systems, but rather because he is wrongly ascribing to them *any concept of matter whatsoever.* Whether or not we judge these Aristotelian labellings wrong or misleading will in each individual case depend on an analysis of the physical systems at issue. At any rate, merely pointing out that the philosopher at issue was not committed to *Aristotle's* concept of matter will not do.

When seen from this perspective the question whether Aristotle misrepresents Plato's intentions when he describes the receptacle as *hulê* boils down to the question whether Plato's receptacle can or cannot, by any reasonable stretching of the meaning of the term, be called 'matter'. In section 3.3, after we have studied the ways in which Plato actually presents his receptacle, we shall revert to this question concerning Aristotle's reliability. In the meantime we may conclude the present subsection by stating that the matter-like characterizations of the receptacle in Plato's text cannot easily be ignored or explained away, and that *a priori* arguments purporting to prove the incompatibility of the labels 'space' and 'matter' are inconclusive.

3.1.4 So far it has been argued that there are no *a priori* reasons to deny that the receptacle can be consistently called space and matter at the same time. Moreover, we have seen that the text of the *Timaeus* seems to support both identifications. We may now proceed to investigate in some more detail in which particular way Plato himself juxtaposed the labels *en hôi* and *ex hou*, and in particular whether and how he thought these descriptions could be squared with one another. The label *en hôi* is applied in a number of passages throughout *Tim.* 48e–52d. The label *ex hou* occurs only in the context of the famous gold analogy (*Tim.* 50a). What we shall try to find out is whether the wording and presentation of these passages contain any indications as to how Plato thought these two characterizations of the receptacle could be compatible.

Guthrie maintains that in the passages at issue, *Tim.* 50a and 50e, the phrases *en hôi* and *ex hou* are used synonymously; more particularly, that in the phrase ἔν τισιν τῶν μαλακῶν σχήματα ἀπομάττειν (50e) the word *ἐν* is used in 'exactly the same way' as the *ἐκ* in the phrase εἰ γὰρ πάντα τις σχήματα πλάσας ἐκ χρυσοῦ etc. (50a).[36] Moreover, according to Guthrie, Plato uses the two terms *promiscue* in other contexts. In this connection he argues that the phrase 'in which' is ambiguous in Greek as in English: 'A clay bust is in space, but also modelled in clay'.

However, the only parallels adduced by Guthrie (and, as far as I know the only evidence to be found) for the alleged synonymy of the phrases *ex hou* and *en hôi* (in at least one of its senses) are two other passages from Plato[37] which do not really furnish independent evidence concerning the use of ἐν and ἐξ. They too closely resemble the examples from *Tim.* 50a and *Tim.* 50e in that they all appear in a context where what is described is the activity of manufacturing an object (δημιουργεῖν) *out of* a pre-existing material. It may be true that in these cases both English and Greek allow one to speak of the object being made *in* that material. It should be borne in mind, however, that in the specific Greek examples here adduced the use of ἐν, though it might at first sight seem to be synonymous to ἐξ, is called forward by the idea that it is the *form* of the final product which is worked out *in* clay whereas that which is said to be made *out of* clay is not the form but the final product.

That these are the senses in which the words 'in' and 'out' should be taken in the quotations just referred to (*Tim.* 50a and 50e), becomes particularly apparent when we consider the actual meaning of the two verbs used. The object of the verb πλάττω, that which you mould out of gold, is the result of the act of modelling. The verb ἀπομάττω, however, means 'to copy, to imitate, to take impressions from (x) into (y)' (on which see LSJ, *s.v.*, III). In other words, it implies the act of copying a model into something else, in this case into gold. This entails that when Plato speaks of

[36] Cf. Guthrie (1967–81) V 265 n. 3; for a similar view see Schulz (1966) 108, n. 338.

[37] *Phil.* 59e: ἐξ ὧν ἢ ἐν οἷς δεῖ δημιουργεῖν; *Pol.* 288d: σώματα ... ἐξ ὧν καὶ ἐν οἷς δημιουργοῦσιν.

σχήματα πλάσας ἐκ χρυσοῦ the σχήματα he has in mind are the finished products, whereas when he speaks of ἔν τισιν τῶν μαλακῶν σχήματα ἀπομάττειν the σχήματα are rather the forms which are copied from a model into the gold. So although the gold is both the 'in which' and the 'out of which' it does not have these qualifications *with respect to the same things.* We may also put it otherwise: the identity between the *en hôi* and the *ex hou* is material, not formal.

If the image is transferred to the receptacle itself, it appears that calling the receptacle an *en hôi* is not incompatible with also treating it as an *ex hou.* For the same thing which provides room for—and thus in a sense constitutes the space of—the (images of the) Ideas (i.e. of what we might call 'instantiations' of the Ideas) may at the same time conveniently be described as a substratum for the eventual phenomenal bodies (i.e. 'instances' of Ideas) and, as such, as an *ex hou.*[38]

It is important to realize how, following this line of thought, we should imagine the interrelation of the three 'factors' (Ideas, sensible flux, receptacle) introduced at the beginning of 48e. On the interpretation I propose Plato regards the objects in the phenomenal world as the *result* of the entering of the forms (or images of Ideas) into the receptacle. Here, in other words, the receptacle is presented as a *constituent* of the phenomenal world. In what sense, then, can Plato also call it *chôra* and in what sense may we call it 'space'?

First, Plato most probably just called his third factor *chôra* because it was an 'in which', viz. that 'in which' the immanent forms are reflected.[39] The overall perspective, in that case, is metaphysical rather than physical.[40] Second, I would suggest that also on a more purely physical level Plato had a motive for calling the

---

[38] For the use of 'instances' *versus* 'instantiations' in this connection see below n. 65 and 70.

[39] That χώρα apparently is the χώρα *of something* also appears from the fact that the term is used as a synonym for τόπος and ἕδρα, which are usually used in such a relational setting.

[40] Solmsen (1960) 43, expresses the same idea in other words: 'Plato's own approach remains metaphysical. The fire in this world is an imprint which the form of Fire makes on the receptacle. The nature of the receptacle itself is not affected by these imprints; all that happens is that for the time being a part of it takes on a fiery appearance. These ... are the points on which Plato dwells...'

receptacle *chôra*, viz. the fact that for individual physical bodies the receptacle serves as a kind of principle of extension. We noted in chapter 2 that both *chôra* and *topos* may mean 'region' or 'extension'. We also noted, in chapter 1, that when space is in this way regarded as a three-dimensional extension, it is not always implied that the extension is separate and absolute. The expression κατέχον χώραν τινά, to which Plato refers at *Tim.* 52b, might simply mean 'having a certain extension'. Though it has to be admitted that in this respect Plato's text does not provide unambiguous clues, we may note that such a background for Plato's calling the receptacle by the denomination *topos* was at least suggested by Aristotle.[41]

When Plato's text is interpreted along these lines, the receptacle may be fitted into the history of concepts of space as an example of space of what in chapter 1 was called the (a) kind.[42] *Chôra* is a principle of extension of phenomenal bodies and a receptive medium or substrate for the immanent characters, *not* a kind of absolute extension underlying physical change and motion. To use comparisons which are familiar in this connection, it is like a mirror receiving images rather than a box into which things can be put and inside which they can be moved around.[43] In so far as on this line of thought matter and space are materially identical, it will be impossible to use this concept of space in the explanation of the motion or the location of physical bodies.

---

41 *Phys.* Δ 209 b 6 ff., on which see below, 114.

42 For comparisons between the Platonic and the Cartesian concepts of space see Guthrie (1967–81) IV, 265 and Baeumker (1890) 187.

43 Cf. L. Robin, *Platon* (Paris 1938) 168: 'Son troisième terme n'est pas un tel contenant: il est ce dans quoi se produisent la génération et la corruption mais en ce sens qu'il est *le sujet* du Devenir'. It is to be noted that Plato himself never compares the receptacle to a mirror. Kung (1988) argues that on the principles of Plato's own 'katoptrics' (as set out at *Tim.* 46a6–b7) such a comparison would be inappropriate, because (1) a mirror is being acted upon (while the receptacle is not), (2) a mirror distorts that which it reflects (the receptacle, according to Kung, does not), (3) a mirror neither contains nor constitutes images. I connot go into these points here, but I wish to point out that if *we* compare the receptacle to a mirror, we are not *ipso facto* committed to regarding it as *in all respects* resembling a mirror, nor to regarding it as in all respects resembling a mirror *as Plato himself would conceive of it.*

3.2 *The receptacle as space: incompatible perspectives*

3.2.1 So far our interpretation has only been based on observations concerning the way in which the phrases *en hôi* and *ex hou* were used in the immediate context of the gold analogy (*Tim.* 50a and 50e). It may be that in doing so we have given a hyperinterpretation of these passages. We shall therefore proceed to investigate whether other passages dealing with the receptacle (particularly in *Tim.* 48e–52d) confirm our hypothetical interpretation.

Generally speaking, the overall scheme of *Tim.* 48e–52d at first sight certainly seems to corroborate the interpretation just outlined. As has been shown by Lee,[44] *Tim.* 48e–52d on the whole seems to be naïve as to motion of phenomenal objects, in the sense that this is not explicitly raised or solved as a problem (exceptions will be noted presently). Stress is laid on the different ontological strata, and on their connection by what may be called the 'metaphysics of the image'. What is more, even the phenomenal flux itself—'that which we perceive to be constantly changing from one state to the other' (ἀεὶ ὃ καθορῶμεν ἄλλοτε ἄλλῃ γιγνόμενον), as Plato at 49d describes it—is here represented as being very much concerned with *change* (either coming into being and passing away, or qualitative change), rather than with locomotion in the ordinary sense.[45] On this interpretation, when Plato speaks of the receptacle as 'that in which they are always appearing and coming into being and out of which, in turn, they perish',[46] the 'they' he is referring to are not phenomenal sensible bodies but qualities, images of Ideas entering and leaving the receptacle, or, as some would prefer to say, the Ideas in their immanent status. The receptacle is that which, so to speak, allows the Ideas to be

---

44 Lee (1966).

45 Cf. also Cherniss' remark (1944) 130: 'Of the phenomenal flux itself nothing more can be said than that it is the resultant of these entrances and exits in the receptacle'; cf. also Driscoll (1979) 256: '...Plato's use of the words τόπος and χώρα notwithstanding, it would be a mistake to identify the Receptacle with any conception of place or space that allows for an identity of objects through change'. For a critical discussion of other aspects of Driscoll's paper, see chapter 4. The only description of the receptacle in *Tim.* 48e–52d which seems to speak out against this interpretation is to be found at 50b where Plato speaks of τῆς τὰ πάντα δεχομένης σώματα φύσεως. But see my remarks in the next paragraph.

46 *Tim.* 49e: ἐν ᾧ δὲ ἐγγιγνόμενα ἀεὶ ἕκαστα αὐτῶν φαντάζεται καὶ πάλιν ἐκεῖθεν ἀπόλλυται.

mirrored in it and which is thus provided with forms, not that through which things move.

Apart from the famous gold analogy already mentioned this approach seems to be predominant in the following passages:

– 49e–50a: A distinction is here made between the receptacle which we may describe by employing the terms 'this' and 'that' on the one hand and the entities which appear in it (φαντάζεται) which are to be called 'suchlike' (τοιοῦτον). In describing the phenomenal world the only correct way of distinguishing 'things' is by calling them e.g. '*this* fire-like thing', and we should bear in mind that in doing so we actually only refer to the receptacle. The ontological implication of this semantical thesis is that for the phenomenal things as they appear to us the receptacle is an indispensable constituent.[47]

– 50c: The receptacle is like a moulding-stuff (ἐκμαγεῖον) which is changed[48] and marked (κινούμενόν τε καὶ διασχηματιζόμενον) by the entering figures (ὑπὸ τῶν εἰσιόντων).

– 50d: The Ideas are compared to a father, the receptacle to a mother and that which is engendered in their midst (which should be interpreted as: that which has both of them as constitutive factors)[49] is likened to the offspring (τὴν δὲ μεταξὺ τούτων φύσιν ἐκγόνῳ).

– 50d–e: The receptacle is called that which will receive all kinds (τὸ τὰ πάντα ἐκδεξόμενον ἐν αὑτῷ γένη); consequently it has to be

---

[47] The passage at issue has been extensively analyzed and discussed in Cherniss (1954); for a discussion of the semantical aspects see De Rijk (1986) 267–271 and 274–275.

[48] I translate κινούμενον with 'changed', for which translation see Cornford (1937) 186, n.1; it may also be that Plato means nothing more than that the receptacle 'moves' in so far as what really happens to it, viz. the entrances and exits of qualities, appears *to us* as motion; on this see Cherniss' remark quoted above, n. 45.

[49] Some commentators who want to stress what is to their mind the strictly spatial character of the receptacle point out that Plato refers to the 'current view' that the father is the sole cause of creation, the mother only furnishing nourishment and room (χώρα) for the infant (cf. Cornford (1937) 187). However it is doubtful whether the 'current view' is as well-articulated as is suggested. (At any rate it was certainly not universally accepted among philosophers: cf. those listed in Aëtius V, 5 (p. 418 Diels) for having held that women also produce semen). Moreover Plato's choice of words here does not really suggest this view. In particular the words τὴν μεταξὺ φύσιν rather suggest that the sensible world shares in both the Ideas and the receptacle. I am afraid that Cornford's references to views held by the natives of S.E. Australia or to Boswell's *Life of Johnson* are not sufficient to decide the issue. Incidentally, I fail to see any justification for Ashbaugh's claim, (1988) 99, that the 'recipient [...] is *mêtêr* in relation to both the offspring and the intelligible model'.

itself devoid of all these forms (ἄμορφον ὂν ἐκείνων ἁπασῶν τῶν ἰδεῶν ὅσας μέλλοι δέχεσθαι πόθεν). The process of the receptacle receiving these characteristics is compared to the fabrication of fragrant ointments when odourless liquids are taken as a kind of raw material in order not to affect the nature of the odours which are added subsequently. The suggestion clearly is that the receptacle has to be entirely without quality, precisely because it is in some way a constituent factor of the sensible world, just as the liquid is a constituent factor for the ointment. It is reasonable to take the terms ἰδέα and γένος to refer to qualities, or immanent characteristics, not to sensible bodies.

– 51b: Here the question of the nature of the receptacle is taken up again and it is stated that what appears to us as fire is really the fiery part of the receptacle, what appears as water the liquefied part, etc. (πῦρ μὲν ἑκάστοτε αὐτοῦ τὸ πεπυρωμένον μέρος φαίνεσθαι κτλ.). This, again, would seem to point to the receptacle serving as a constituent factor of phenomenal bodies, receiving qualities rather than receiving the bodies themselves.[50]

As we noticed, this way of describing the receptacle nicely fits in with the overall perspective of *Tim.* 48e–52d, which is metaphysical rather than physical, and makes sense both of the denominations *chôra* and *en hôi* and of the denomination *ex hou* (as well as of the other characterizations of the receptacle as matter). It is for this reason that perhaps we may regard it as the predominant mode of presentation in *Tim.* 48e–52d. However, Plato apparently does not always stick to this mode of presentation.

To begin with, there are a number of passages in which he at least leaves room for doubt. Though these passages in principle allow of an interpretation along the lines just sketched they may also be read as suggesting that the receptacle is the substrate in which *phenomenal things* appear. On this line of thought the receptacle is not so much a *constituent factor* but rather a *container* of the second factor, the phenomenal flux:

– 50c: Here the things entering and leaving the receptacle are called copies of the eternal Beings (τῶν ὄντων ἀεὶ μιμήματα),

---

[50] Notice, however, that directly after the sentence quoted the receptacle is said to become the two remaining phenomenal elements, air and earth, in so far as it receives their μιμήματα. Here the use of the latter term seems to suggest that the receptacle receives rather than constitutes phenomenal bodies (which when they were introduced as the second factor at the outset of 48e were called copies, μιμήματα, of the eternal forms). This shows the extent to which the two approaches distinguished here were really intertwined in Plato's thought.

which copies are stamped from them (τυπωθέντα ἀπ' αὐτῶν). In itself this might be merely a way of describing the immanent qualities (or instantiations of the Ideas). On the other hand the wording reminds us of 48e–49a where it is the second factor itself, the world of phenomenal appearances, which is called a copy of the eternal Forms, subject to becoming, and visible (μίμημα δὲ παραδείγματος ... γένεσιν ἔχον καὶ ὁρατόν). So it is not at first sight clear whether the things entering and leaving the receptacle are just qualities (instantiations) or rather phenomenal bodies (instances). The same goes for 51b, where the receptacle is said to receive μιμήματα of earth and air.

– 52a: the sensible world is contrasted with the realm of the transcendental Ideas and the individual sensible 'thing' is described as 'named after and similar to the [transcendental] Idea' (τὸ δ' ὁμώνυμον ὅμοιόν τε ἐκείνῳ ... αἰσθητὸν) and 'ever carried about' (πεφορημένον ἀεὶ) because it is ever 'coming to be in a place and again perishing from it' (... γιγνόμενόν τε ἔν τινι τόπῳ καὶ πάλιν ἐκεῖθεν ἀπολλύμενον). If we may identify *topos* here with the receptacle, as is plausible in view of the use of the term *topos* directly afterwards in 52b, it is to be concluded that here again *phenomenal bodies* are said to be what enters and leaves the receptacle.[51] On the other hand the wording is rather similar to that of 49e–50a, where what enters and leaves the receptacle is not phenomenal bodies but characters or qualities.

– 52b: The receptacle is said to be that which provides a seat to the things that come to be (ἕδραν δὲ παρέχον ὅσα ἔχει γένεσιν πᾶσιν). Though this might still be a reference to the qualities entering and leaving the receptacle, the wording at the same time seems to suggest that Plato means that the receptacle provides room for phenomenal entities, the second factor, which was, as we recall, earlier (50c) called 'subject to becoming' (γένεσιν ἔχον).

– 52c: Though the meaning here is rather obscure, it seems that it is said either of the immanent qualities or of the sensible phenomena that 'they fleet (φέρεται) ever as a phantom of something else' and that for that reason they must 'come to be in something else' (ἐν ἑτέρῳ τινὶ γίγνεσθαι).[52]

---

[51] Still, it is significant that the 'motion' (πεφορημένον) of this phenomenal body is apparently nothing but its (?) coming to be in a certain place and its again perishing from it. As I take it, this presentation is ambiguous too. The implication may be that what we perceive as the motion of an instance of a certain Form is really the process of the instantiations of this Form (the qualities) entering and leaving different parts of the receptacle. For if it is the phenomenal body itself which is entering and leaving the receptacle, why should it not just move *through* the receptacle instead of perpetually entering and leaving it?

[52] On the possible connection between motion (φέρεται) and coming to be (γίγνεται) see above n. 45 and n. 48.

Finally, there is one case which seems to leave no room for doubt. For here it is obviously suggested that the receptacle receives phenomenal bodies, i.e. that it serves as a kind of absolute space:

> – 50b: The receptacle is called that which (or: the nature which) receives all bodies (τῆς τὰ πάντα δεχομένης σώματα φύσεως).[53]

We may add that this is also the predominant way of using the spatial terms *topos*, *chôra* and *hedra* in the final, 'cosmological' part of the *Timaeus*. In fact in the course of the accounts of the elements and their motion and of the respiration of animals the terms *topos*, *chôra*, and *hedra*—which do all also occur in the crucial section 48e–52d—are unambiguously used to denote the place *into which* and *out of which* a *phenomenal body* may move.[54]

Along with each of the two modes of presentation outlined here goes a different view on the role of the receptacle and on phenomenal motion. On the former approach, the receptacle is the *ex hou* of the phenomenal world and only an *en hôi* in the sense of 'that in which qualities, or immanent Ideas, inhere'. In these cases there is no indication at all that the receptacle figures as an absolute space underlying the motion of physical bodies. In fact, this approach is naïve as to the explanation of physical motion and it would indeed have difficulties in providing such an explanation.

On the second approach, however, the receptacle is presented as a kind of box in which the things of the phenomenal world are

---

53 Against the view of those scholars who take the context of this quotation to be concerned with the geometrical shapes of the primary corpuscles described later in the *Timaeus*, see Cornford (1937) 182, n. 4. I side with Taylor (quoted by Cornford, *loc. cit.*) in explaining the 'transient characters as the characteristics of different sensible bodies ... various sounds, colours, scents etc. revealed to us in different regions'. Taylor and Cornford seem unaware, however, of the fact that Plato's use of the word σῶμα in the present quotation cannot be easily fitted into their interpretation. We must assume, then, that—as I am arguing here—Plato wavered between two modes of presentation. Sometimes the receptacle is said to receive just images, at other times it is presented as receiving sensible bodies.

54 Cf. *Tim.* 60b–c: γενόμενος δὲ ἀὴρ *εἰς* τὸν ἑαυτοῦ *τόπον* ἀναθεῖ; 60 c: συνέωσέν τε αὐτὸν *εἰς* τὰς *ἕδρας ὅθεν* ἀνῄει ὁ νέος ἀήρ; 79d: τὸ θερμὸν [...] *εἰς* τὴν αὑτοῦ *χώραν* [...] ἰέναι. Ashbaugh's translation of *chôra* in the receptacle passages as 'spatiality', (1988) *passim*, is not only too abstract, but it also makes nonsense of her claim, (1988) 101, that there is 'a consistent use of the term chôra throughout the dialogues', since elsewhere chôra usually and obviously has the rather more concrete meaning of 'land' or 'place' or 'space'.

put and through which they move. On this approach there is no sign of the receptacle functioning as an *ex hou*, or as matter.

The *Nebeneinander* of these two perspectives may be due to the fact that Plato was on the one hand concerned to represent the things of the phenomenal world as a 'projection' of the Ideas into the receptacle, whereas on the other hand he stuck to describing the phenomenal bodies *as they appear to us*, i.e. as objects which keep their identity through motion. Thus the transition from the one perspective to the other may be analysed into the following 'phases':

phase (1): the predominant mode of presentation; phenomenal bodies are nothing but the images of Ideas being projected in the receptacle, or rather, parts of the receptacle which are 'formed' by images of Ideas. The images 'appear' in and 'vanish' out of the receptacle. This does not mean that they go elsewhere but just that they stop being projected into space. As we already noted, this approach is naïve as to motion of phenomenal bodies. Presumably the idea is that motion of phenomenal objects as it appears to us (with the concomitant idea of identity of objects through motion) is unreal: there is only the continuing process of images being projected in ever different parts of the receptacle. (Cf. the talk of ἐν ᾧ [...] φαντάζεται and ἐκεῖθεν ἀπόλλυται in 49e–50a, the use of εἰσιόντα καὶ ἐξιόντα [...] μιμήματα in 50c).

phase (2): the images are now not said to enter and leave the receptacle but to fleet through it (cf. 49e where the image is called περιφερόμενον).[55] Though here this may still only be a way of speaking, the consequences of this 'way of speaking' get serious outlines in:

---

[55] The difficult passage in full reads (in Cherniss' translation): 'But <it is safest> not to speak of these as severally distinct but so to call the such and such *that always recurs alike* (περιφερόμενον) in each and all cases together, for example <to call> that which is always such and such fire and so with everything that comes to be'. (Cherniss (1954) 114; see also Cherniss' concluding remarks, (1954) 128: 'The distinctive names, naïvely and improperly applied to phenomena, denominate in each case 'the such and such, whatever the correct formula may be, that is always identical in each and all of its occurrences''). Note that Cherniss—rightly, as I see it—makes a distinction between the *phenomena*, the transient alterations of the receptacle, and the 'such and such' which is denominated when we apply any distinctive name (see also Cherniss (1954) 129–130).

phase (3): the phenomenal bodies, i.e. the bodies as they appear to us are said to move through the receptacle (cf. 52a which speaks of the second factor, the sensible world, as αἰσθητόν, γεννητόν, πεφορημένον ἀεί).

As we noted, the presentation of phase (1) fits what seems to be the overall metaphysical perspective of *Tim.* 48e–52d, whereas the presentation of phase (3) seems to fit the way the receptacle is introduced at 52b in order to account for ordinary conceptions of space and place as well as the way in which the terms referring to the receptacle (*chôra, topos, hedra*) are used elsewhere in the *Timaeus* (on which see above).

3.2.2 It might be argued that there is no real difference between the two perspectives or modes of presentation when one considers the eventual sensible bodies as *nothing but* aggregates of qualities in space. In that case it makes no difference whether one says that the 'things' in the receptacle are immanent qualities or, alternatively, sensible bodies (which are themselves nothing but bundles of qualities). Such an interpretation, along with the description of the receptacle as a *medium* (as distinct from both *space* and *matter*), has been advanced by scholars who wanted to stress that what they called the Platonic form-ontology (as distinct from the Aristotelian substance-ontology) could well do without a concept of matter[56] or, alternatively, that the Platonic ontology could well do without a concept of space-as-a-container (space type (c)).[57] On this interpretation, it was suggested, the matter-or-space question was basically misdirected.

It has to be admitted that on this line of thought the status of the *objects contained* by the receptacle does not differ on either mode of presentation, which means that we may elegantly preserve the triad Ideas—sensible world—receptacle, as presented by Plato. Nevertheless, we are still left with the fact that the status of the *receptacle itself* (more particularly its separability or inseparability) is *not* the same on either mode of presentation. In this respect the differences of perspective in the text of the *Tim.* cannot be that easily ironed out.

---

[56] Cf. Cornford (1937) 180–181; and Keyt (1961) 298, who maintains that Aristotle did not see, or did not want to see, that whereas for himself 'a body is a compound ... of form and matter, for Plato a body is a form'.

[57] Cf. e.g. Mohr (1985) 92 ff.

In the first place there are the passages, subsumed above under what we called the first mode of presentation, where a constitutive role of the receptacle seems to be suggested in terms which are too strong to be ignored, e.g. when it is said that fire is 'the ignified part' (τὸ πεπυρωμένον μέρος) of the receptacle. Now also in these cases one may well choose to label the receptacle 'medium' instead of 'matter' as long as it is realized that this 'medium', in so far as it is *extension,* is the extension of the 'images = phenomenal bodies'. In other words, the 'medium' will here be a kind of *Prinzip der Räumlichkeit* for the phenomenal bodies.

In the second place there are those passages, subsumed under what above we called the second mode of presentation, where the receptacle is presented as the extension *through which* the 'images = phenomenal bodies' move, as they by Plato are said to move (φέρεται, περιφερόμενον). Now we may well choose also here to label the receptacle 'medium' instead of 'space' as long as we realize that, if we may take for granted that the phenomenal bodies are extended, the receptacle will have to constitute a *separate* extension for them to move through. So the receptacle-as-medium according to the second mode of presentation would still turn out to be different from the receptacle-as-medium according to the first mode of presentation.

In fact the only way to make coherent sense of the receptacle being a medium for aggregates of qualities constituting phenomenal bodies would seem to be to ignore Plato's own presentation. Thus we might suggest as a possible escape that the extension we perceive (i.e. the extension of phenomenal bodies) is not provided by the receptacle but that it is itself just one of the 'qualities' (or properties), present in each bundle. However, though in Neoplatonist and Patristic philosophy such a view is not entirely unheard of, there is not the slightest indication in the text of the *Timaeus* which shows that this was what Plato intended.[58]

---

[58] Cf. Plot. *Enn.* II, 4, 11: οὐ τοίνυν ὄγκον δεῖ εἶναι τὸν δεξόμενον τὸ εἶδος, ἀλλ' ὁμοῦ τῷ γενέσθαι ὄγκον καὶ τὴν ἄλλην ποιότητα δέχεσθαι. What might be regarded as a patristic radicalization of this Platonist 'devaluation' of the role of matter, viz. the elimination of receptive matter as a factor in the creation of the sensible world and the concomitant description of this sensible world as nothing but a 'configuration of qualities' (συνδρομὴ τῶν ποιοτήτων) is found e.g. in Gregory of Nyssa, *In Hexaemeron*, Migne, *PG* 44, 69 c, where God's thinking ὁμοῦ τὰ πάντα δι' ὧν ἡ ὕλη συνίσταται τῷ σοφῷ τε καὶ δυνατῷ θελήματι κατεβάλετο ... τὸ ξηρὸν, τὸ ψυχρὸν, τὸ θερμὸν, τὸ χρῶμα, τὸ

Alternatively, it may be assumed that Plato's idea was that 'real bodies', i.e. bundles of qualities, have no extension and that consequently our phenomenal bodies, which *do* appear as extended, are not the 'real bodies'. What we call the same *extended* body moving through space would in that case really be nothing but a bundle of qualities having no extension of itself but getting its extension from always different parts of the receptacle. Thus Plato would be able to rescue both the separability of the receptacle (viz. from what it contains) and the fact that the receptacle nevertheless constitutes the extension of what we perceive as bodies. It should be objected, however, that on this line of thought the images or qualities would no longer be identical to the phenomenal bodies, the bodies as they appear to us. Moreover, this view would involve a quite radical break with common conceptions of what bodies actually are (more precisely, that bodies of themselves possess an extension and that this extension moves along when the bodies themselves are moving), with common conceptions of what space is (viz. something connected with bodies as they appear to us, i.e. *extended* bodies), and also with common conceptions of what exactly *motion* amounts to (the idea of *motion* of unextended incorporeal qualities being hard to grasp).[59] It is indeed not uncommon for a philosopher to invest some quite common terms with an uncommon technical meaning, and thus it would *in principle* be conceivable that in the receptacle passages Plato was trying to convey such a novel metaphysics. In that case, however, one might expect this unfamiliar position to be explained in a more unambiguous manner. At the very least one might expect that the ordinary meanings of the terms used are not confused with their

σχῆμα, τὴν περιγραφὴν, *τὸ διάστημα*. For a full discussion of this text (and for parallels and further references) see M. Alexandre, *L'exégèse de* Gen. *1, 1–2a dans l'* In Hexaemeron *de Grégoire de Nysse: deux approches du problème de la matière* in : H. Dörrie/ M. Altenburger/ U. Schramm (edd.), *Gregor von Nyssa und die Philosophie* (Leiden 1976) 168 ff. An interesting discussion (with more examples) of this Neoplatonist tendency to devaluate the role of matter is now to be found in ch. 4 ('Bodies as Bundles of Properties') in Sorabji (1988a) 44–59.

[59] It is true that Plato occasionally speaks of the immanent characteristics *entering and leaving* the receptacle (instead of moving through the receptacle), a way of presentation which would accord quite well with the interpretation here discussed. However the fact that he also speaks of these characteristics as *moving* through the receptacle (φέρεται) shows that he is not consistently committed to a single particular view.

new specific connotations. What we see, however, is that throughout the *Timaeus* Plato uses words like 'moves' (φέρεται) and body (σῶμα)[60] in what appear to be their colloquial senses, without any further qualification.

It appears, then, that an unprejudiced reading of Plato's own text shows that the alleged form-ontology cannot be held to be present in a pure and consistent form in the *Tim.* The use of the label 'medium' as distinct from both 'space' and 'matter' turns out to be only a *quasi*-solution in that it still leaves open the important question whether the receptacle is or is not a constitutive factor of the phenomenal bodies. In other words, using a different label to denote the receptacle does not alter the fact that the text of *Tim.* 48e–52d, employs different and partly incompatible perspectives.

We may accordingly conclude that, although Plato's account starts out from *three* explanatory factors (Ideas, sensible phenomena, receptacle, as suggested at *Tim.* 48e and 50d), we end up with *four* items playing a role in the metaphysics of *Tim.* 48e–52d, viz.:

(a) the transcendental Ideas;

(b) the self-identical characters, images, or qualities appearing in the receptacle;

(c) the phenomenal bodies;

(d) the receptacle.

It was seen that the *en hôi*-character of the receptacle operates in different ways with respect to (b) and (c) respectively. As an *en hôi* of (b) the receptacle is what might be called a *metaphysical* space, which may be identified with matter. As an *en hôi* of (c) it is a *physical* space which may on no account be identified with matter.

Two additional remarks. First, if we may thus conclude that the presentation of the receptacle as a constitutive principle cannot be explained away, we are provided with an answer to an important question raised in the previous section.[61] There we concluded that the question whether Aristotle misrepresented Plato's views

---

[60] Cf. e.g. *Tim.* 31b where the created world (τὸ γενόμενον) is called σωματοειδὲς and where it is implied that a σῶμα must consist of elements (fire and earth). See also the way in which spatial terms like τόπος and χώρα, and verbs like ἰέναι are used in *Tim.* 60b, 60c, 79d (the passages quoted above n. 54).

[61] See above, 89.

when he described the receptacle as *hulê* could not be decided by just pointing out that Plato was not committed to *Aristotle's* concept of *hulê*. Rather we suggested that an inquiry into the context in which the receptacle was actually presented would have to establish whether it could, by any reasonable stretching of the term, be called 'matter' (or *hulê*). Now we see that the answer to the latter question should be affirmative: the receptacle is at least in some important sense a constituent factor and as such might indeed be described as matter.

Second it may be remarked that the scholarly literature at times shows a surprising lack of awareness of the inconsistencies which, as we saw, inevitably result from the juxtaposition of the two different modes of presentation.[62] On the other hand, in those cases where the inconsistencies *are* noticed, it is usually assumed that one of the two inconsistent perspectives has to be neglected or explained away.[63] Thus those who hold that the receptacle is just matter highlight the first mode of presentation, whereas the champions of the identification of the receptacle as space, in spite of the awareness they usually show of the restricted sense in which the term space is used according to the first mode of presentation (viz. space in a non-absolute sense), sometimes show themselves blind to the full implications of this first mode.[64]

In the next two subsections I shall venture to show that a partial explanation for the occurrence of the inconsistencies we noticed can be found if we pay attention to (1) the metaphysical context of the receptacle passages in the Timaeus, and (2) the ambiguity of the spatial vocabulary applied by Plato.

---

62 Baeumker, for instance, seems to be basically correct in holding the view that the receptacle is both space and matter. Apparently, however, it did not occur to him that this position could not unqualifiedly be squared with the (first part of the) statement ((1890) 177) that 'die platonische Materie ist der leere Raum, d.h. die blosse Ausdehnung', nor with the statement that Platonic matter is not just the 'Ort der Formen' but that 'ohne Beschränkung, die dritte Gattung sei der Raum' (*op. cit.* 183).

63 Cf. Mohr (1985) 92: 'So Plato, while setting forth the philosophical functions of the third thing uses metaphors which, if read literally, would, when taken together, assign incompatible properties to the contents of the receptacle. *Which set of properties should we choose?*' [my italics].

64 Cf. Cornford (1937) 185: 'Plato's space is not a void. It is a recipient which affords a basis for images reflected in it as in a mirror'. Still he thinks the difference between the out-of-which and the in-which important enough to justify the conclusion (*op. cit.* 181) that 'there is no justification for calling the receptacle 'matter''.

3.2.3 As has been indicated, the difference between the two approaches can be quite adequately illustrated by the application of the conceptual distinction between *instantiations* and *instances*.[65] The first mode of presentation outlined above may be characterized as being concerned primarily with instantiations of the Ideas appearing in and disappearing from the receptacle (no matter whether one interprets these instantiations as being just immanent qualities[66] or as the Ideas themselves, viewed in their immanent status).[67] The second mode on the other hand focuses on the receptacle as containing instances of the forms (particular sensible bodies 'sharing in' certain Ideas).

Now admittedly there are contexts in which Plato draws a clear boundary between sensible instances of the Forms and the Forms themselves or their instantiations. Thus when at *Phaedo* 74c—in a context concerned with *anamnêsis*—a distinction is made between the Form of Likeness and the individual things which, in being 'alike', partake of that Form of Likeness, the instantiations of Likeness are treated on a par with the transcendental Form and opposed to instances of Likeness.[68] Here the perspective is epistemological and when seen against this background it is not surprising that Plato distinguishes between the ever fleeting phenomena on the one hand and the self-identical characteristics (instantiations) on the other. It may be interesting to note that the only more or less semantical or epistemological passage in *Tim.* 48e–52d seems to make the same distinction between phenomena (instances) and self-identical characteristics (instantiations). At 49e Plato maintains, to quote Cherniss, that the distinctive names naïvely and improperly applied to phenomena, denominate in each case 'the such and such, whatever the correct formula may

---

[65] Cf. the explanation of the difference as given by de Rijk (1986) 42, n. 28: 'An instance (x) of a property *F* is a suppositum 'having' *F*; an instantiation is the particular case of 'being *F*' occurring in the instance (x). The expression 'a case of holiness' etc. is ambiguous in that it may stand for an instance as well as for an instantiation of holiness'.

[66] This amounts to the viewpoint taken by Cornford (1937) *passim*.

[67] This is the view of de Rijk (1986) 61, who follows Proclus in identifying what enters and leaves the receptacle as 'enmattered forms'.

[68] *Phaedo* 74b–c:—ἆρ' οὐ λίθοι μὲν ἴσοι καὶ ξύλα [instances] ... τῷ μὲν ἴσα φαίνεται, τῷ δ' οὔ;—Πάνυ μὲν οὖν.—Τί δέ; αὐτὰ τὰ ἴσα [instantiations] ἔστιν ὅτε ἄνισά σοι ἐφάνη, ἢ ἡ ἰσότης [the Form considered in and by itself] ἀνισότης;—Οὐδεπώποτέ γε, ὦ Σώκρατες.

be, that is always identical in each and all of its occurrences.'[69]

On other occasions, on the other hand, where Plato's approach is rather ontological and where he focuses, so to speak, on the hierarchy of reality—as in the *Symposium*'s speech of Diotima—the distinction between instances and instantiations is blurred and the big gap arises between instances and instantiations on the one hand and the Form in itself on the other.[70]

With this in mind we need perhaps not be too much surprised by the fact that, although *Tim.* 49e might seem to suggest a clear distinction between instantiations and instances on Plato's part, the way in which he at *Tim.* 48e ff. presents that which enters and leaves the receptacle remains rather ambiguous. The perspective here, as in the *Symposium*, is ontological and metaphysical and what Plato is actually doing may perhaps best be described as trying to explain in cosmological and cosmogonical terms how we ought to conceive of the phenomenon of *methexis*. In this connection we may recall that in the eyes of his most famous pupil he never really succeeded in illuminating the way in which he thus apparently thought to bridge the gulf between his 'two worlds'.[71] We may also recall that at *Tim.* 49a Plato himself spoke of the receptacle as a third kind which was 'difficult and obscure' (χαλεπὸν καὶ ἀμυδρὸν εἶδος), the difficulty of the subject being, in his view, due among other things (ἄλλως τε καὶ) to the fact that it was necessary first to discuss the preliminary problems (προαπορηθῆναι) concerning the real nature of fire and its fellow elements, i.e., presumably, the problems concerning the real

---

69 Cherniss (1954) 128. Cherniss' interpretation of this difficult passage is on the whole very convincing.

70 Cf. *Symp.* 210b: ... πολλὴ ἄνοια μὴ οὐχ ἕν τε καὶ ταὐτὸν ἡγεῖσθαι τὸ ἐπὶ πᾶσιν τοῖς σώμασι κάλλος [instantiation]· τοῦτο δ' ἐννοήσαντα καταστῆναι πάντων τῶν καλῶν σωμάτων [instances] ἐραστὴν. And at 210d–211b it is said that he who is thus rightly initiated in and taught τὰ ἐρωτικά will finally ἐξαίφνης κατόψεταί τι θαυμαστὸν τὴν φύσιν καλόν [...] οὐδέ που ὂν ἐν ἑτέρῳ τινι [instantiation] [...] ἀλλ' αὐτὸ καθ' αὐτὸ μεθ' αὐτοῦ μονοειδὲς ἀεὶ ὄν [transcendental Form]. Similarly, at *Soph.* 255e, where Plato discusses the interrelations between the five 'Most Important Kinds' (μέγιστα γενῆ) at times the terms 'same', 'other' etc. (the names of the five Kinds) refer to the particular instantiations of the Forms occurring in particular things, whereas at other times the particular instances themselves seem to be meant. See the lucid comments made on this passage by de Rijk (1986) 161.

71 Cf. *Metaph.* A 991 a 20 on the alleged relationship between the Forms as paradigms and sensible reality: τὸ δὲ λέγειν παραδείγματα αὐτὰ εἶναι καὶ μετέχειν αὐτῶν τἆλλα κενολογεῖν ἐστὶ καὶ μεταφορὰς λέγειν ποιητικάς.

nature of the objects in the phenomenal world. This may well be read as a modest *captatio benevolentiae*: Plato will try his best but the subject is so difficult as not to allow of an explanation in clear and unambiguous terms.

3.2.4 There is a second factor which may have furthered Plato's ambiguous presentation of the receptacle. The receptacle is explicitly introduced to account for, among other things, our common sense convictions about space and place. As we saw in chapter 1 and 2, these were in Plato's days (as they are now) hardly particularly articulated. A closer look at the passage in which Plato explicitly refers to the common sense views on *topos* and *chôra* may once more confirm this. At *Tim.* 52a–b Plato speaks of the receptacle as:

> The third kind [...] [which] is itself apprehensible by a kind of bastard-reasoning by the aid of non-sensation, barely an object of belief, with respect to which we are dreaming while watching, and affirm that it is somehow necessary that all that exists should exist *in* a place and should occupy some space and that that which is neither on earth nor in the heavens is nothing.[72]

In the first place it should be noticed that the two examples adduced at the end of the quotation are particularly inappropriate. For saying that something is 'on' earth or 'in' the heavens in fact amounts to determining its localization *in relation to its direct surroundings.* To be sure, as we saw in chapter 1, this is one of the possible ways in which one can say that something is 'in' something else (as a fish is *in* the water). However there is no way in which the receptacle—as represented in *Tim.* 48e–52d—can be used to account for this specific sense of the phrase 'to be in something'.

This already suggests that Plato was not really interested in differentiating between the several ways in which spatial terms, in this case the term *topos*, could be used. It appears, furthermore,

---

[72] *Tim.* 52a: τρίτον δὲ αὖ γένος ὂν [...] αὐτὸ δὲ μετ' ἀναισθησίας ἁπτὸν λογισμῷ τινι νόθῳ, μόγις πιστόν, πρὸς ὃ δὴ καὶ ὀνειροπολοῦμεν βλέποντες καί φαμεν ἀναγκαῖον εἶναί που τὸ ὂν ἅπαν ἔν τινι τόπῳ καὶ κατέχον χώραν τινά, τὸ δὲ μήτ' ἐν γῇ μήτε που κατ' οὐρανὸν οὐδὲν εἶναι. For the common intuition that to exist implies to be 'somewhere' see Kahn (1973) 313 ff., 317 ff., and 387 ff. On the possible connection of the argument that what is not 'somewhere' is nothing with Zeno's so-called paradox of place cf. Mansfeld (1985) 262–263. See also above, chapter 2, n. 61.

that he just assumed that to be 'that in which', the denomination applied to the receptacle, was *ipso facto* to be a '*topos* in which' or a '*chôra* which is occupied'. In itself such an identification is not *necessarily* wrong or misleading, as long as one invests the expressions 'in which' and 'in a place' with a sufficiently specified sense. Now this is precisely what Plato does not do. Though the *en hôi* of his predominant mode of presentation also has its roots in common parlance,[73] it is not what common parlance generally means by 'a place in which', and is certainly not connected with the use of 'in' in expressions like 'in the heavens'. The spatial concepts here adduced, in other words, are vague, and Plato seems to slip from one use of *en hôi* into another.

Moreover the wording of the section just quoted seems to suggest that Plato acquiesces in this vagueness: in his view of things, the whole conception of the 'third kind' is only reached by a kind of 'bastard reasoning'. In order to corroborate his account of it he sticks to referring to some common conceptions concerning space. It even seems as if Plato implies that these common conceptions *cannot* be further clarified or refined: when using the expressions like 'being in a place' people are only 'dreaming' and a philosophical inquiry into the nature of space cannot really get much further. The nature of the subject matter will not allow of a real and accurate *logismos*.

[73] Also in common parlance the phrase εἶναι ἐν may also denote what we would call the 'inherence' of qualities or characteristics in a thing. It is the occurrence of this sense of εἶναι ἐν ... in ordinary Greek usage which could lead to Aristotle's more or less technical use in the *Categories* of ἐν ὑποκειμένῳ εἶναι ('to be in a subject')—as distinct from καθ' ὑποκειμένου λέγεσθαι ('being said of ...') to denote the way in which non-substantial characteristics presumably inhere in substances. Aristotle himself explains this technical use as follows: 'By 'in a subject' I mean what is in something, not as a part, and cannot exist separately from what it is in' (*Cat.* 1 a 24–25); Ackrill (1963) 74, rightly comments that 'the 'in' which occurs twice in this definition cannot be the technical 'in' of the definiendum. It must be a non-technical 'in' which one who is not yet familiar with the technical sense can be expected to understand. Presumably Aristotle had in mind the occurrence in ordinary Greek of locutions like 'heat in the water', 'courage in Sokrates''. In such expressions the 'that-in-which', the ἐν ᾧ, is in a way the place of the inherent qualities, but only *in a way*, i.e. not in a strictly local sense. In this way the phrase ἐν ᾧ can even be used in a metaphorical sense: τοῖς λόγοις ἔνεστι ἀμφοῖν κέρδος ('there is advantage in the words of both', Sophocles *El.* 369–370, and οὐ πόλλ' ἔνεστι δεινα τῷ γήρᾳ κακά; ('are there not many disadvantages in old age?', Aristophanes *V.* 441).

It is illuminating to contrast the text just quoted with a passage from Aristotle's *Physics*, where at the outset of the treatment of *topos* the following observation is made:

> That place exists seems clear when we consider the phenomenon of replacement. For where there is now water, there will be—when the water has gone out as if from a vessel—air, and at another time still another body will occupy that same place. That [place] seems to be different from the things entering it and replacing each other. For in that in which (*en hôi*) there is now air, there was formerly water, so that it is clear that place (*topos*) and room (*chôra*) was something different from both, into which and out of which each of the substances went.[74]

Aristotle, like Plato in the passage just quoted, refers to common conventional parlance, but his attitude towards such conventions is totally different. In general he is quite well aware of the fact that the concepts of *topos* which are commonly held may be ambiguous and even fully wrongheaded. This is why, as will be shown in chapter 4, he takes them, in *Phys.* Δ, as the starting point of a careful dialectical procedure which may be described as a gradual weeding out of those connotations of *topos* that can no longer be maintained when more closely investigated, and that prove to be of no use for the purpose of physical science (i.e. which prove to be useless if what we want is to give an account of the location and locomotion of physical bodies).

In the above passage a first selection is already made. For though Aristotle here assumes the equivalence of the expressions 'where', 'in which' and 'in the place in which', he makes it clear that all of them are used in the particular sense relevant to the phenomenon of replacement. In other words, they all point to a 'something' apart from, i.e. separable from, the located body, receiving it as a kind of vessel (ἀγγεῖον). This leaves no room for putting *topos* in this sense on the same level as an *en hôi* in the inherence sense. That Aristotle was serious about making the right conceptual distinction between the two senses of *en hôi* which were apparently applied *promiscue* in the *Timaeus* appears from *Phys.* 210 a 14 ff., a passage occurring in his discussion of *topos*; he distinguishes various ways in which a thing is said to be *in* something else. One among them is precisely the 'in' of

---

74 *Phys.* Δ 208 b 1–8.

inherence: 'In another as health is in hot and cold things and, generally, as form is in matter'.[75] This sense of 'in' is clearly distinguished from 'the most basic way of all, as in a vessel, and, generally, in a place'.[76]

As we shall see in the next section, Aristotle's awareness of these ambiguities in the common usage of spatial terms allowed him to come up with a fairly pertinent critique of the receptacle passages in Plato's *Timaeus*. Though he states that Plato is to be praised for being the first to try to state what space actually was,[77] others apparently merely *using* the concept, he also explicitly remarks that one of the difficulties facing the philosopher who wants to investigate the nature of *topos* is the fact that as yet nobody has accurately formulated the problems (οὐδὲν προηπορημένον) or made any useful suggestion.[78]

As far as Plato is concerned, this conclusion seems to be corroborated by our findings concerning the 'concept of space' of Plato's *Timaeus*. The only passage where Plato makes something clear about the nature of the problems concerning the receptacle is *Tim.* 49a, where he states that the subject is particularly difficult because it is necessary first to discuss the problems concerning the real nature of phenomenal objects (note that he also uses the term προαπορηθῆναι in this connection). I pointed out in the previous paragraph that this should probably be interpreted as a reference to the difficulties inherent, on Plato's line of thought, in *any* account of the relationship between the sensible and the intelligible worlds. But Plato rests content with remarking that the whole subject is obscure. Consequently, there is no attempt to disentangle the imprecise notions of *chôra*, *topos*, and *en hôi*. Plato was obviously convinced that the 'third factor' was—and apparently would always remain—a difficult and obscure thing (χαλεπὸν καὶ ἀμυδρὸν εἶδος, *Tim.* 49a). As a result, his metaphysical

---

75 *Phys.* Δ 210 a 20: ἔτι ὡς ἡ ὑγίεια ἐν θερμοῖς καὶ ψυχροῖς καὶ ὅλως τὸ εἶδος ἐν τῇ ὕλῃ.

76 *Phys.* Δ 210 a 24: πάντων δὲ κυριώτατον τὸ ὡς ἐν ἀγγείῳ καὶ ὅλως ὡς ἐν τόπῳ.

77 *Phys.* Δ 209 b 17. According to Aristotle the basic mode of presentation in the *Timaeus* implies that the receptacle is space and matter. For indications that he was nevertheless aware of the ambiguity apparent in Plato's presentation as a whole, see my remarks in the next section.

78 *Phys.* Δ 208 a 34–36: ἔτι δ' οὐδ' ἔχομεν οὐδὲν παρὰ τῶν ἄλλων οὔτε προηπορημένον οὔτε προηυπορημένον περὶ αὐτοῦ.

statements about the receptacle do not allow us to read them as a coherent theory of physical space.

3.3 *Aristotle's criticisms of Plato's receptacle*

3.3.1 More then once Aristotle has been charged with gross ignorance or unfairness in his criticisms of the Platonic receptacle. It is the aim of the present section to show that when judged against the results of the preceding inquiries Aristotle's remarks can be better understood and that—at least in as far as his critique of the *Timaeus* is concerned—his position appears to be much more fair and to the point than is often assumed.

It is well known that Aristotle connected the receptacle of the *Timaeus* with the second principle, the great-and-small, of the so-called unwritten doctrines.[79] Now the reliability of Aristotle's testimonia concerning the unwritten doctrines of Plato has been the subject of much controversy and has elicited much scholarly activity. I shall here not enter into that discussion but stick as much as possible to the limits set by the present inquiry. Consequently Aristotle's criticism of the second principle of the unwritten doctrines will only be studied in so far as is relevant to our understanding of his position *vis-à-vis* the *Timaeus*. I shall first discuss Aristotle's references to the receptacle in *De Generatione et Corruptione*, and next those found in the *Physics*.

3.3.2 At the outset of the second book of *GC* Aristotle discusses several views that were held by his predecessors on the material principle underlying coming into being and passing away. After briefly mentioning and criticizing the material monists, he refers to the receptacle of the *Timaeus* in the following words:

> What has been written in the *Timaeus* lacks precision (οὐδένα ἔχει διορισμόν). For he did not state clearly whether the 'omnirecipient' is separated from the elements. Nor does he make any use of it after having said that it is a substratum prior to the so-called elements, just as gold is the substratum of objects made of gold.[80]

---

[79] Cf. *Phys.* Δ 209 b 11 ff., 209 b 33 ff.

[80] *GC* 329 a 14–17; the passage is followed by an additional criticism of the gold analogy, Aristotle's point being that the image is inappropriate, since its main point—the fact that all figures moulded out of the gold are still most accurately called 'gold'—does not apply to all kinds of physical

Here the general reproach that Plato's talk about the receptacle is unclear is specified by the further remark that Plato leaves us with doubts concerning the independent status or the absoluteness of the receptacle.[81] Though, as we know, Aristotle thought that the receptacle was primarily intended to serve as a constituent of (and thus to be inseparable from) phenomenal bodies, he here implies that Plato's presentation does not always keep up with this original perspective. This criticism is quite in line with our findings in the previous sections, where we were able to conclude that occasionally the receptacle is said to receive only *qualities*, whereas at other times Plato speaks as if the receptacle receives the phenomenal bodies themselves. In the former view the receptacle is an inseparable space of the (a) kind outlined in chapter 1; in the latter, an absolute space of the (c) kind.

Though the second point raised in the above quotation is not directly relevant to the subject of our present inquiry, it is still interesting enough to be briefly commented on. According to Aristotle the Platonic 'matter' does not reappear in the same form in the rest of the *Timaeus*. Indeed, as we noticed earlier, the 'matter' which the Demiurge puts into order is not the original unqualified receptacle. The Demiurge, according to *Tim.* 52d ff., imposes order on a pre-existent chaos consisting of the receptacle that already contains 'traces' of the Ideas. It seems therefore that Aristotle was also well aware of the relatively isolated status of the passages in which the receptacle was first introduced (*Tim.* 48e–52d) and that he anticipated the distinction between a *matière nue* and a *matière déjà déterminée* made by later exegetes of the *Timaeus*.[82]

3.3.3 We may now turn to Aristotle's references to the *Timaeus*' doctrines concerning the receptacle as they are to be found in the *Physics*. As we already noted, Aristotle realized that *topos* was generally identified with the *en hôi*. For that reason he included

change, but only to qualitative change (ἀλλοίωσις). However this is not a fundamental criticism, but rather a criticism of Plato's presentation (cf. 329 a 17: καίτοι καὶ τοῦτο οὐ καλῶς λέγεται τοῦτον τὸν τρόπον λεγόμενον). It has no real bearing on our present issue.

81 Claghorn (1954) 8, wrongly argues that we have two independent arguments here.

82 Cf. e.g. L. Robin, *Les rapports de l' être et de la connaissance d' après Platon* (Paris 1957) 59; see also Baeumker (1890) 179; see on these problems also Happ (1971) 105, and Claghorn (1954) 9.

(*Phys.* Δ 210 a 14 ff.) a discussion of the various ways in which things can be said to be 'in something else'. In the course of this discussion an *en hôi* in the sense of 'that in which a quality inheres' is clearly distinguished from the sense of *en hôi* which is relevant in the context of a discussion of space or place ('in which' in the sense of in a vessel, and, generally, in a place).[83] Though at this point Aristotle does not explicitly refer to the views of Plato or of any other philosopher, there is at least one other passage which indicates that it was Plato's receptacle in particular which he was thinking of in this connection.

At 214 a 7 ff., in the course of his discussion of *to kenon*, two possible definitions of *kenon* are discussed. The one defines void as that which is not full of tangible perceptible body (αἰσθητοῦ σώματος κατὰ τὴν ἁφήν). The other takes void to be 'that in which there is no *tode ti* or essence of *sôma*', which in this connection should be taken to mean 'that which is formless'.[84] The concept of *kenon* used here is, of course, the type (a) *kenon* discussed in the previous chapter: *kenon* as 'space'.[85] Aristotle identifies the latter of these two views with the view which takes place as matter, so we may plausibly infer that he was thinking of Plato's receptacle, Plato being the only philosopher who is adduced in *Phys.* Δ as having endorsed such a view. Aristotle objects that 'those' who subscribe to this view take the void to be the 'matter of a body', and that in doing so they are not expressing themselves correctly (λέγοντες οὐ καλῶς) because matter is inseparable from the body

---

[83] See above, n. 76.

[84] For the meaning of τόδε τι and οὐσία in this connection ('essence' or 'infima species', instead of the usual 'substance') see *Metaph.* Δ 1017 b 25–26 and Θ 1049 a 35. See, furthermore, Guthrie (1967–81) VI, 215 ff. and, more extensively, Owens (1951) 386–395. This peculiar use of these two terms seems to have been missed by Hussey (1983) 126 (who thinks that what is meant is a body as 'individual' and a 'corporeal substance') and Wagner (1967) 554 (who speaks of 'Gegenstände' and 'in körperhafte Existenz eingegangene Wesenheiten').

[85] The (b) concept of κενόν, void as strictly *unoccupied* space, was rejected by Plato. Cf. *Tim.* 60c, 79b, 80c. Aristotle probably recognized this, if we may take the passage at *Cael.* 309 a 21 ff. as a reference to the *Timaeus*, as is plausible in view of what precedes (cf. 308 b 4 ff.). One will have to agree with Cherniss (1944) that the identification of the receptacle as κενὸν is a deduction made by Plato. However, Cherniss' reconstruction of the deduction hardly makes sense: 'since void ... must be place deprived of matter .... and for Plato τόπος has been identified with matter, Aristotle concludes that the material of body is for Plato the void'. One had better assume that Aristotle here simply uses void in one of its colloquial senses, viz. as 'space'.

of which it is the matter, whereas void—i.e. in this case self-subsistent space—is assumed to be separable (τὸ δὲ κενὸν ζητοῦσιν ὡς χωριστόν). In other words, they are guilty of inconsistency.

The gist of Aristotle's criticism seems to be that you may well take *topos* as matter, and as an *en hôi* in the inherence sense ('that in which there is no *tode ti*'), as long as you do not at the same time confuse such a concept of *topos* with an independent self-subsistent space (and, concomitantly, as long as you do not confuse the *en hôi* in the inherence sense with the *en hôi* in the strictly locative sense). It seems that, here again, it is the tension between what we have called the two modes of presentation of the Platonic receptacle which called forth Aristotle's criticism.

When Aristotle elsewhere criticizes what is according to him the basic mode of presentation of the receptacle in the *Timaeus*, viz. the identification of space and matter, he seems well aware that Plato is using the words 'in' and 'where' in one of their generally accepted meanings (viz. the 'in' of inherence), but he is at the same time quick to point out that such an approach only makes sense if we abstract from phenomenal motion and if we abstract from the existence of individual substances (which might change place):

> One might well think that matter is place, when one considers it in the case of something that is not moving and that is not individuated but continuous. For just as, when there is a qualitative change, there is a something which is now white and which before was black, or now hard while it was formerly soft—which is why we say that matter is something—so too place seems to exist because of a similar phenomenon, except that the former concept (matter) arises because *that which* is now air was formerly water, whereas the concept of place arises because *where* there was air *there* is now water.[86]

So when things are viewed not as individuated and not as moving one is likely to confuse the 'that in which an instance of (x) can come to be' with the 'that *in which* there is an instantiation of (x)', and, accordingly, to confuse matter and space/place.

---

86 *Phys.* Δ 211 b 29 ff.: Καὶ ἡ ὕλη δὲ δοξεῖεν ἂν εἶναι τόπος, εἴ γε ἐν ἠρεμοῦντί τις σκοποίη καὶ μὴ κεχωρισμένῳ ἀλλὰ συνεχεῖ. ὥσπερ γὰρ εἰ ἀλλοιοῦται, ἔστι τι ὃ νῦν λευκὸν πάλαι δὲ μέλαν, καὶ νῦν μὲν σκλῆρον πάλαι δὲ μαλακὸν (διό φαμεν εἶναί τι τὴν ὕλην), οὕτω καὶ ὁ τόπος διὰ τοιαύτης τινὸς εἶναι δοκεῖ φαντασίας, πλὴν ἐκεῖνο μὲν διότι ὃ ἦν ἀήρ, τοῦτο νῦν ὕδωρ, ὁ δὲ τόπος ὅτι οὗ ἦν ἀήρ, ἐνταῦθ' ἔστι νῦν ὕδωρ.

Whereas in these passages Aristotle's criticisms seem to concentrate on the different senses of *en hôi* and the concomitant different ontological perspectives, he elsewhere gives a slightly different account of the way which led Plato to identify place and matter. Thus in *Phys.* Δ 209 b 6–17 Aristotle gives the following possible reason for the identification of space and matter:

> In so far as the place/space (*topos*) of a thing seems to be the extension of a magnitude (τὸ διάστημα τοῦ μεγέθους), it may appear to be its matter. For that is distinguishable from the thing itself. That dimensionality is what is bounded and determined by the form, as, for instance, by a surface or a limit. And such is also matter and the undetermined. For when the limit and the qualities of a sphere are stripped away, nothing is left but its matter.
>
> It is for this reason that Plato, in the *Timaeus*, said that matter and space are the same thing. For the receptacle and space are one and the same thing. And though he there [i.e. in the *Tim.*] did not describe the receptacle (*to metalêptikon*) in the same way as in the so-called 'unwritten doctrines' he still in both cases said that place and space are identical with it.[87] For, whereas everyone says that place is something, only Plato made an attempt to tell what it is.

Here Aristotle concentrates on what I have called the first mode of presentation. In so far as matter, as the underlying constituent of physical reality, is taken to be mere extension, it might be tempting, Aristotle contends, to identify it with space.[88] So whereas in the passages previously quoted he seems to imply that the receptacle was identified with space to the extent that it was an *en hôi* for a *tode ti* or *sômatikê ousia*—in other words, for an instantiation of an Idea—Aristotle now suggests another reason why Plato might have regarded the receptacle as *chôra*, viz. the fact that it serves as a kind of *Prinzip der Räumlichkeit* of the phenomenal bodies (the instances of the Ideas). One is reminded of Descartes for whom matter (or substance) was essentially extension (*extensio*).[89]

---

[87] I have opted for this translation of τὸν τόπον καὶ τὴν χώραν τὸ αὐτὸ ἀπεφήνατο—taking the words τὸ αὐτὸ as elliptical for τὸ αὐτὸ τούτῳ—instead of the traditional 'that place and space are identical' or something of the sort. It has been suggested as an alternative translation by Hussey (1983) 105, and Solmsen (1960) 123, n. 23. The statement that 'place and space' are identical would in the present context be pointless.

[88] Note that the other reason for calling the receptacle space, viz. the fact that it is the ἐν ᾧ of the inherent qualities, seems to be suggested by Aristotle at 211 b 29 ff., quoted above, n. 86.

[89] Although Aristotle himself is of course not committed to the equation he presents here, Simplicius did interpret the passage we are dealing with

Some further remarks on this passage. First, it clearly shows that Aristotle did not ascribe a concept of *corporeal* matter to Plato.[90] Second, though the last sentence praises Plato for his unique attempt to explain the 'nature' of space—others merely using spatial concepts without further explaining them—this praise is not accorded without qualification: Plato got no further than an attempt (ἐπεχείρησεν εἰπεῖν). Apart from the fact that according to Aristotle Plato of course—even in his first mode of presentation—ended up with the *wrong* concept of space and place, we may surmise that Aristotle qualified his praise because, as the above quoted passage from *GC* shows, he thought Plato was also *inconsistent* in *at times slipping* from this first and predominant mode of presentation into another which posited space as a kind of absolute container.

There is one curious reference to Plato's views on the receptacle which still has to be mentioned. At *Phys.* Δ 209 b 33 ff. Aristotle makes the following objection to what somewhat earlier he has described as Plato's views on *topos*:

> Plato, if we must insert an excursus (παρεκβάντας εἰπεῖν) on his thought, will have to tell us why the Forms and the numbers are not in a place, if place is the 'receptive factor' (*to methektikon*), no matter whether the receptive factor is constituted by the great-and-small or, as he wrote in the *Timaeus*, by matter (*hulê*).

It is usually assumed that the question 'why are the Forms not in a place' implies that Aristotle wrongly thought that according to Plato the things entering and leaving the receptacle were the Ideas themselves instead of their images or instantiations.[91] On

---

as an indication that Aristotle himself identified (prime) matter and extension. Cf. Simp. *In Ph.* 229, 6; 230, 19–31; 232, 21–22; 232, 24; 537, 13; 623, 18–19. For a detailed discussion of Simplicius' interpretation see now Sorabji (1988a) 7–22.

90 That Aristotle—contrary to what is sometimes suggested—thought that a material principle could well be incorporeal is also shown by *Metaph.* A 988 a 24 ff.: οἱ μὲν γὰρ ὡς ὕλην τὴν ἀρχὴν λέγουσιν, ... ἐάν τε σῶμα ἐάν τε ἀσώματον τοῦτο τιθῶσιν. Though the context is not too explicit it seems to be suggested that Plato (οἷον Πλάτων μὲν τὸ μέγα καὶ τὸ μικρὸν λέγων) belongs to those who thought of matter as incorporeal. One may also compare Aristotle's speaking of the atoms *and the void* of the early atomists as αἴτια δὲ τῶν ὄντων ταῦτα ὡς ὕλην (*Metaph.* A 985 b 10).

91 This was already the opinion of Simp. *In Ph.* 542, 11; Simplicius' judgement was taken over by Cherniss (1944) 118, Ross (1936) 567, and Wagner (1967) 539, who, however, regards the passage as a *Randnotiz* which was added later.

this line of thought Aristotle mistook the words 'partaking in a most baffling way of the intelligible' (μεταλαμβάνον ἀπορώτατά πῃ τοῦ νοητοῦ), used by Plato in connection with the receptacle, as meaning that the Ideas themselves are in the receptacle.

I think, however, that this is not a really plausible interpretation. First, both the reference to the numbers (*arithmoi*), which do not occur in the receptacle passages of the *Tim.*, and the formula παρεκβάντας εἰπεῖν (make a digression) indicate that Aristotle's attack is here *not* specifically aimed at the presentation of the receptacle *in the Timaeus*; so there is no reason to assume that he specifically took the expression 'partaking in a most baffling way of the intelligible' (μεταλαμβάνον ἀπορώτατά πῃ τοῦ νοητοῦ) as his starting-point. Secondly, if Aristotle had been stimulated by this expression to think that Plato presented the receptacle as containing the Ideas themselves, it would have been more logical for him to ask why the Ideas *are* in space instead of asking, as he does, why the Ideas *are not* in space.

Finally, the wording suggests that the argument centres on the identification of *to methektikon* and *topos*, no matter whether the *methektikon* is regarded in the fashion of the *Timaeus* or in that of the unwritten doctrines. It is, therefore, much more logical to assume that we are dealing with an *ad hominem* argument. This argument starts from the identification of *topos* and *methektikon under all circumstances*. It then infers that if in the case of the *methektikon* of the *Tim.* the things *made out of* the *methektikon* are also *in it*, the Ideas—which are, according to the *agrapha dogmata* as they are interpreted by Aristotle, in a way *constituted by* the *methektikon* (i.e. the great-and-small)[92]—are *in* the *methektikon* and, therefore, in a *place*.

So the digression seems to consist in a shifting of focus of the criticism from the specific text of the *Timaeus* to the Platonic metaphysical system in general, and the argument may be read as an application to Plato's metaphysics in general of the *Timaeus'* identification of the constituent factor of (x) with the place of (x).[93] On

---

[92] Aristotle describes the great-and-small as the ὕλη of the Forms (or ideal numbers): cf. e.g. *Metaph.* A 987 b 19 ff.

[93] I think a criticism of a similar kind lurks behind the brief statement at *Phys.* Γ 203 a 8 ff. where Aristotle says that ἔξω μὲν οὐδὲν εἶναι σῶμα, οὐδὲ τὰς ἰδέας, διὰ τὸ μηδὲ ποῦ εἶναι αὐτάς, τὸ μέντοι ἄπειρον καὶ ἐν τοῖς αἰσθητοῖς καὶ ἐν ἐκείναις εἶναι; the whole *ad hominem* procedure can be parallelled

this interpretation our passage fits the other critical passages we have been discussing, in that it shows how the unwarranted identification of material principle and space, which was the main target of Aristotle's criticism, also turns out to have odd consequences within Plato's own metaphysical system.

3.3.4 We may conclude that Aristotle's criticisms of the receptacle of the *Timaeus* may be reduced to two main points:
(1) Plato is unclear about whether the receptacle is a separable self-subsistent space in which phenomenal bodies are in the strict local sense of 'in which', or whether it is an inseparable constituent factor in which the immanent qualities (or, in Aristotelian vocabulary, the *tode ti*)) are in the non-strict inherence sense of 'in which'.
(2) The latter view only makes sense in a context which disregards locomotion and for that reason it cannot offer a suitable *physical* concept of space or place, i.e. a concept of space or place which can be used in the explanation of phenomenal motion.

On the basis of the investigations of the previous sections we may conclude that both points seem legitimate. There are no reasons to assume that Aristotle 'quietly neglects the possibility that anyone who conceived *topos* in the way described was using the word in a sense totally different from that which he assigns to it himself'.[94] He is not so much criticizing Plato for having defended a concept of space which differs from that which he would choose himself, but rather for having presented an obscure and at times incoherent theory which, though not wholly abstruse, is—contrary to what Plato's own talk about 'things' moving through the receptacle would suggest—unable to account in any satisfactory manner for the phenomenon of locomotion of physical bodies.

---

from *Metaph.* A 992 b 8–10 where it is argued that if the second principle, the great-and-small, is the principle of motion, the Ideas (presumably in so far as they depend on the second principle) will have to be moving: εἰ μὲν ἔσται ταῦτα (sc.: τὸ μέγα καὶ τὸ μικρόν) κίνησις, δῆλον ὅτι κινήσεται τὰ εἴδη.

[94] Cherniss (1944) 115.

Conclusions

In this chapter we tried to establish how the receptacle of Plato's *Timaeus* could be fitted into the history of concepts of space. It was first established that attempts to identify it exclusively with either space or matter were bound to fail. In consequence, we tried to discern how Plato himself thought these two functions could be combined. Thus we established what we called Plato's basic mode of presentation: the receptacle as an *en hôi* for the immanent forms or qualities and as an *ex hou* for the phenomenal bodies. The concept of space connected with this mode of presentation was seen to be that which in chapter 1 has been called the (a) kind.

Next, in section 2 it was observed that Plato did not remain faithful to this basic 'metaphysical' mode of presentation. At times—both at *Tim.* 48e–52d and elsewhere in the same work—he seemed to envisage the receptacle as a receptive *en hôi* for phenomenal bodies, i.e. as a space of what we called the (c) kind. Attempts to explain away the differences were seen to be unsuccessful.

Two possible explanations were given for this shift of emphasis. First, in *Tim.* 48e–52d Plato was primarily concerned with outlining the relationship between the sensible and the intelligible world and in doing so he was experiencing the usual difficulties concerning *methexis*. Thus, when describing what is *in* the receptacle, he was seen to vacillate between two candidates: what might be called the *instances* of the Forms on the one hand, and their *instantiations* on the other. The receptacle was seen to be represented accordingly as either space of the (c) kind or space of the (a) kind.

Secondly, we saw that Plato explicitly introduced his theory of the receptacle as being in conformity with—and accounting for—the rather inarticulate common views about space. Thus, he could, without further qualification, identify his *en hôi* with a *topos en hôi* or a *chôra* which is occupied. His professed pessimism about the possibility of getting any satisfactory clarification in these matters—the nature of the receptacle being according to him obscure and difficult to grasp—may have blinded him to the recognition that he used e.g. the expression *en hôi* in different senses in the same context.

Finally, we concluded that Aristotle's criticisms of the Platonic conception of the receptacle were largely justified. There seems to be no reason not to believe Aristotle when he implicitly presents himself as the first Greek philosopher to notice the real nature of the problems of space and place. Though Aristotle's own final concept of place is of course notorious for the many problems with which it saddled the later Aristotelian tradition, his famous discussion of *topos* in *Phys.* Δ at least had the merit of pointing out, for the first time, a number of conceptual difficulties inherent in the subject.

Plato's concept of space as it appears in the predominant mode of presentation, has more then once been compared to that of Descartes.[95] In this connection we may here note that the same reasons which led Aristotle to criticize the Platonic Receptacle (as it is represented according to the first mode of presentation), to wit its inability to play any role in the account of physical motion, forced Descartes to maintain, next to his quite specific concept of space (*spatium*), a more or less Aristotelian concept of place (*locus*, more precisely: *locus externus*) according to which the location as well as the motion of a particular body is defined in terms of its direct surroundings.[96] Plato's 'theory of space' would arguably have been less inconsistent if he had managed to distinguish in a similar fashion between *physical* and *metaphysical* space or place. It is interesting to note that a tendency to make such a distinction can be discerned among some later Platonists, largely, it would seem, under the impact of Aristotle's critique. Thus Plotinus, *Enn.* II, 4, 12, separates (1) the *matter* of the things in the physical world, which is a kind of mere potentiality (in fact, this is an

---

[95] Cf. above, n. 42.

[96] Cf. R. Descartes, *Principia Philosophiae* II, 14 (ed. Adam/ Tannery, vol. 8, 47): 'Differunt autem nomina loci & spatii, quia locus magis expresse designat situm quam magnitudinem aut figuram, & e contra magis ad has attendimus cum loquimur de spatio. Dicimus enim frequenter unam rem in locum alterius succedere, quamvis non sit accurate eiusdem magnitudinis, nec figurae; sed tunc negamus illam idem spatium occupare; ac semper, cum ille situs mutatur, dicimus locum mutari, quamvis eadem magnitudo ac figura permaneat. Cumque dicimus rem esse in hoc loco, nihil aliud intelligimus, quam illam obtinere hunc situm inter alias res; & cum addimus ipsam implere hoc spatium vel hunc locum, intelligimus praeterea ipsam esse huius determinatae magnitudinis ac figurae'. Cf. also the definition of motion, *ibidem* II, 25 (ed. Adam/ Tannery, vol. 8, 53): '[motum] ... dicere possumus esse translationem unius partis materiae, sive unius corporis, ex vicinia eorum corporum, quae illud immediate contingunt & tanquam quiescentia spectantur, in viciniam aliorum'.

aristotelizing interpretation of the *Timaeus*' receptacle),[97] (2) their extension, which is, next to the qualities, conferred to this matter in the process of emanation; and (3) their *place* (*topos*) which is posterior to both the existence of matter and the existence of the physical bodies themselves.

As another example we may adduce the fifth century CE Athenian Syrianus who preferred to solve the problems just referred to by introducing an elaborate stratification of hypostases. In his commentary on Aristotle's *Metaphysics* he went so far as to distinguish four kinds of *topos*: (1) a place of physical bodies (τόπος σωμάτων φυσικῶν), i.e. physical space or place, (2) a place of 'enmattered forms' (τόπος ἐνύλων εἰδῶν), i.e. matter, (3) a place of mathematical bodies (τόπος μαθηματικῶν σωμάτων), viz. imagination, and (4) a place of the transcendent Ideas (ἀΰλων λόγων), viz. the intellect, or νοῦς.[98] The first two kinds correspond with the two concepts of space or place which, as I have tried to show, appear *promiscue* in the *Timaeus*. A sufficient awareness of the ambiguities exhibited by the *Timaeus* in this particular respect may help to put such developments in later Platonism into the right historical perspective.

---

97 Cf. H.-R. Schwyzer, 'Zu Plotins Deutung der sogenannten platonischen Materie', in: *Zetesis*, Festschrift E. de Strijcker (Antwerpen/ Utrecht 1973) 266–280.

98 Syrian. *In Metaph.* 186, 15–28.

CHAPTER FOUR

CONCEPTIONS OF *TOPOS* IN ARISTOTLE

Introduction

The fourth book of Aristotle's *Physics* contains what has come to be known as his 'classical' discussion of the problems concerning physical place. The definition of *topos* ultimately issuing from this discussion, viz. *topos* as the first unmoved boundary of the containing body, became with some exceptions canonical in the later Aristotelian tradition. However, not very surprisingly, spatial concepts also crop up in a number of other works: the *Categories* (*Cat.*), *De Generatione et Corruptione* (*GC*), and *De Caelo* (*Cael.*). A survey of the ways in which these works use spatial terms, in particular the term *topos*, will reveal important *prima facie* inconsistencies. The present chapter will deal with these various uses of *topos* and will try to fit them all into a more or less coherent picture.

The chapter is divided into two parts. In the first part (sections 4.1–4.3) I shall consider the much-discussed presentation of *topos* in the part of the *Categories* dealing with quantity. Being conspicuously at odds with the 'classical' discussion of the *Physics*, it has troubled Aristotelian commentators from antiquity onwards. For, in presenting *topos* in the *Cat.* as a three dimensional continuous extension, Aristotle seems to be committed to one of the concepts of place he emphatically rejects in the fourth book of the *Phys.*

In section 4.1 the passages in the *Cat.* which speak about *topos* will be briefly discussed, and it will be shown why they are at first sight at odds with the 'classical' *topos* passages in *Phys.* Δ. Moreover, it will be pointed out that, in addition, the presentation of *topos* in the *Cat.* presents problems of its own.

Sections 4.2 and 4.3 will deal with two ways in which scholars and commentators have tried to get rid of the apparent differences between the presentations of *topos* in the *Cat.* and in the *Phys.* One of these ways is to attempt to iron out the inconsistencies: a device as old as Aristotelian exegesis itself. Another is to assume that Aristotle's views on the matter at issue underwent serious changes in the course of his philosophical career. Such an approach to Aristotle's philosophy became particularly fashionable in the aftermath of Jaeger's epoch-making studies on the unity and relative chronology of the Aristotelian works. In our case this approach is exemplified by the thesis that the—early—*Cat.* view on *topos* is still basically Platonic whereas the *Phys.* shows the more mature Aristotelian position.[1]

Section 4.2 deals with some examples of the former ('synthesizing') approach, whereas what is probably the most appealing instance of the latter ('developmental') approach will be evaluated in section 4.3. It will be argued that, contrary to what these developmental or synthesizing approaches seem to presuppose, the *Cat.* account of *topos* should not be read as a consciously and consistently held physical theory.

The second part of this chapter (sections 4.4 and 4.5) will survey the different concepts of place or space to which Aristotle seems to have been committed in his physical writings. Section 4.4 will investigate the dialectical character of Aristotle's school writings in general and look at his inquiry into the nature of *topos* in *Phys.* Δ in particular. It will not contain a full-blown discussion of the many problematic features of Aristotle's final and classical concept of *topos* (some of these will be studied in the next chapter). Rather, it will try to make clear how he arrived at this particular concept and for which reasons he rejected the concept of *topos*-as-extension.

It will be shown that in *Phys.* Δ Aristotle took the conventions of ordinary language as his starting point. From among the several more or less clear-cut concepts of place or space used *promiscue* in ordinary thought and common parlance he selected the concept

[1] A third way to explain away the inconsistencies between the *Cat.* and the *Phys.*, which will be left unconsidered in the rest of this chapter, has here to be indicated. It has been suggested by Jaeger, as well as by quite a number of scholars during the 19th century, that the *Categories* are not a genuine Aristotelian work. However, all arguments to this purpose have been proved inconclusive and the *Categories* are now generally accepted as genuine. Therefore, since there are no reasons to doubt the authenticity of the other works which are relevant to our present investigations (*Phys.*, *Cael.*, *GC*), our point of departure must be the conviction that all passages on τόπος to be discussed in the present chapter are genuinely Aristotelian. Cf. Ross (1924) I, lxxxii; L. M. De Rijk, 'The Authenticity of Aristotle's Categories', *Mnemosyne* IV, 4 (1951) 129–159; Owen (1966) 136 and 139 n.1, repr. in Nussbaum (1986b) 209 and 211 n. 11 and Frede (1987c).

of place as a 'first surrounder'—i.e. the (b) concept of place of our chapter 1—as the only concept which was defensible in the context of physical theory. Moreover I shall argue that the rejected candidates, in particular the concept of *topos* as a three-dimensional extension, represented such inveterate ways of speaking, or of conceiving of the world, that Aristotle could in other contexts, where it was not primarily the nature of *topos* which was at stake, sometimes speak as if he was committed to such a 'bastard' concept, the more so since his own 'classical' concept was not in all circumstances really adequate. Some occurrences of an unorthodox concept of *topos* in the *Phys.* and other related writings will be surveyed, and tentatively explained, in section 4.5. By way of a conclusion, I shall in section 4.6 try to fit the data of the other sections into a more or less coherent perspective.

4.1 *Topos and the category of quantity*

4.1.1 The *Categories* contains some brief passages concerning *topos* which appear in the course of chapter 5, as part of the discussion of the category *poson* (usually translated 'quantity').[2] They are what we shall primarily be dealing with in the present section. I shall first, in 4.1.2, give an outline of the way in which *topos* is presented in these passages. It will be shown why and in which respects this presentation does not seem to fit the 'classical' Aristotelian concept of *topos* as it is presented in *Phys.* Δ.

Subsequently, in 4.1.3, it will be established that the way *topos* is presented in the *Cat.* presents problems of its own when approached from the point of view of the overall ontology of the *Cat.* It will be shown that these problems are of such a nature as to suggest that Aristotle was not committed to any particular coherent theory about *topos* when he wrote the *Cat.* The findings of the present section will be summarized in 4.1.4.

4.1.2 In the course of his treatment of the category of quantity (*poson*) Aristotle draws a sharp distinction between discrete and continuous quantities. Among the latter are a line, a surface, a body, and, besides, time and place (*topos*). The reasons for including place in this list are specified at 5 a 8–14:

[2] On the names of the categories and their translation see Ackrill (1963) 77–78.

> (1) Again, place (*topos*) belongs to the quantities which are continuous. For the parts of a body which join together at a common boundary occupy a certain place. Therefore, also the parts of place which are occupied by the several parts of the body join together at the same boundary at which the parts of the body do. Therefore also place is seen to be continuous. For its parts join together at one common boundary.[3]

A second important passage containing a reference to *topos* appears when Aristotle is investigating whether the category *poson* admits of contraries. Whereas his conclusion is that in general this is not the case, he argues, at 6 a 12, that

> (2) ... most of all contrariety in quantity seems to appear in the case of place (*topos*). For 'up' is commonly accepted as contrary to 'down', in that the region (*chôra*) towards the centre is said to be 'down', because it is at the greatest distance from the periphery of the cosmos.[4]

As a preliminary remark it may be noted that, though quotation (2) is in this respect at least ambiguous (a point to which I shall revert later, in 4.1.3), quotation (1) seems to be not primarily concerned with the 'locative' aspect of place. As might be expected in view of the context (the category *poson*), place seems to be merely regarded under its quantitative aspect.

We may perhaps assume that *location* was in principle thought to be covered by the category *pou* ('where'). Unfortunately the text of the *Cat.* as it stands does not offer an extensive treatment of the category *pou* comparable to the discussion of the category of quantity. Aristotle's intentions may be plausibly inferred, though, from the examples used in connection with the category *pou* at the outset of the treatise (2 a 1–2: 'where: e.g. in the Lyceum, in the market-place').[5] I shall revert to this below, in 4.1.3.

---

[3] *Cat.* 5 a 8–14: Πάλιν ὁ τόπος τῶν συνεχῶν ἐστί· τόπον γάρ τινα τὰ τοῦ σώματος μόρια κατέχει, ἃ πρός τινα κοινὸν ὅρον συνάπτει. οὐκοῦν καὶ τὰ τοῦ τόπου μόρια, ἃ κατέχει ἕκαστον τῶν τοῦ σώματος μορίων, πρὸς τὸν αὐτὸν ὅρον συνάπτει πρὸς ὃν καὶ τὰ τοῦ σώματος μόρια. ὥστε συνεχὲς ἂν εἴη καὶ ὁ τόπος· πρὸς γὰρ ἕνα κοινὸν ὅρον αὐτοῦ τὰ μόρια συνάπτει.

[4] *Cat.* 6 a 12–15: Μάλιστα δὲ ἡ ἐναντιότης τοῦ ποσοῦ περὶ τὸν τόπον δοκεῖ ὑπάρχειν. τὸ γὰρ ἄνω τῷ κάτω ἐναντίον τιθέασι, τὴν πρὸς τὸ μέσον χώραν κάτω λέγοντες διὰ τὸ πλείστην τῷ μέσῳ διάστασιν πρὸς τὰ πέρατα τοῦ κόσμου εἶναι.

[5] It is interesting to note that the category here (and at *Top.* 103 b 23) labelled ποῦ is in the list at *Metaph.* Θ 1050 b 15 ff. called both ποῦ and τόπος whereas it is only called τόπος in the lists of *Metaph.* K 1068 a 8–10 and *EN* A 1096 a 27. This might be an indication that later—presumably after writing

Be this as it may, from the above quotations we may infer the following characteristics of *topos*:

(a) it is continuous; which means that

(b) it has parts (which correspond to the continuous parts of the body which is 'in place');

(c) it is three-dimensional (as is particularly shown by quotation (2));

(d) it may also be called *chôra* (quotation (2));

(e) it seems to be different from the extension of the occupying body and it can be regarded separately (in quotation (2) there is no mention of any occupying body).

That the concept of *topos* which these quotations present us with is radically different from that which is put forward as the conclusion of the dialectical inquiry in the fourth book of the *Phys.* becomes immediately apparent when we compare the famous final definition provided at *Phys.* Δ. According to this definition *topos* is 'the first unmoved boundary of the containing body'.[6] Earlier in this study (chapters 1 and 2) I suggested that this 'classical' Aristotelian definition should be regarded as the elaboration of one among several senses the term place (or, for that matter, the Greek word *topos*) could bear in common parlance. It seems that if a similar concept of place is to be looked for in the *Cat.* at all, it is most probably to be located in the category *pou*.

On the other hand, the concept of *topos* which appears in the course of the *Categories'* chapter on quantity figures in *Phys.* Δ as one of the concepts which are positively rejected. There, at 211 b 7 ff., Aristotle argues that there are four candidates for the identification of *topos*: (1) form, (2) matter, (3) the *extension between the limits of the body*, or (4) the limits themselves (which, as the context makes clear, should be understood as the limits of the encompassing body).

Now the third candidate, which amounts to a concept of place of the kind applied in the *Cat.*, is eliminated because it involves two absurd consequences. First, it forces us to assume that there is an infinite number of (partly overlapping) places, obviously

Phys. Δ—Aristotle had become aware of the inconsistencies involved by including τόπος among the category πόσον. The fact that these treatises are all later then *Phys.* Δ seems to be widely agreed upon.

6 *Phys.* Δ 212 a 20: τὸ τοῦ περιέχοντος πέρας ἀκίνητον πρῶτον.

because there is no natural limit to each place. In other words, each part of a place (being a three-dimensional extension) is itself also a place etc. *ad infinitum.*[7] Second, this view is vulnerable to something much resembling Zeno's paradox of place (= fr. 29 A 24 DK): place, this time presumably regarded as the inherent extension of the located body itself, moves along with this body and must for that reason itself be in a place, so that there will be more than one coextensive place.[8]

Quite in line with all this, the possibility that *topos* is just a three-dimensional extension is once more rejected in Aristotle's conclusion at *Phys.* Δ 212 a 2–6:

> If *topos* is no one of these three—neither form, nor matter *nor an extension distinct from the extension of the body which is being replaced*—it must needs be the fourth of the alternatives, namely the limiting surface of the surrounding body.

The conclusion seems warranted that the views on *topos* as they appear in the *Cat.* and in *Phys.* Δ respectively would at first blush seem difficult, if not impossible, to reconcile. The question whether and in what sense they are in the end still to be squared with each other will be dealt with in the subsequent sections of this

---

[7] *Phys.* Δ 211 b 19 ff. (Ross' reading of the text at b 19–20, based on Laas' reconstruction from Themistius' paraphrase, has to be accepted. Hussey's objections (1983) 115 are neither philologically nor philosophically compelling). Like the second argument (211 b 24) this argument has a Zenonean ring. It resembles those of Zeno's paradoxes which centre on the problem of infinite divisibility (e.g. the first two paradoxes of motion). One may compare the 'infinity of places' here with the paradoxical ἄπειρα διεξελθεῖν in Aristotle's treatment of Zeno's first paradox of motion (*Phys.* Θ 263 a 4 ff.). There it is said that whoever transgresses a (finite) physical distance κατὰ συμβεβηκὸς ἄπειρα διελήλυθεν, ἁπλῶς δ' οὔ· συμβέβηκε γὰρ τῇ γραμμῇ ἄπειρα ἡμίσεα εἶναι, *ἡ δ' οὐσία ἐστὶν ἑτέρα καὶ τὸ εἶναι.* As the latter words make clear, it was Aristotle's own physical conceptual apparatus—with its notions of form, matter, actuality and potentiality—which provided him with the means to dodge the consequences of Zeno's paradox of motion for his own physical system. Obviously, however, he thought that the concept of a self-subsistent threedimensional space, having of itself no form to limit it and to make it one *qua* οὐσία, could not be defended against a similar 'Zenonean' argument concerning a resulting infinity of places.

[8] *Phys.* Δ 211 b 24. The argument is not very clear. Hussey (1983) 115, rightly objects that it does not really seem to be pertinent as an argument against self-subsistent space. Perhaps, however, it should be viewed as a somewhat awkward version of Zeno's paradox of place, which after all—as I shall try to show later on in this chapter—was thought to be the clinching argument against the conception of place as a self-subsistent three-dimensional extension (διάστημα).

chapter. First, however, we shall turn to some additional problems inherent in the way in which *topos* is presented in the *Cat.*

4.1.3 The fact that the presentation of *topos* in the *Cat.* seems to be radically at odds with the classical concept of *topos* of the *Phys.* is certainly not the only difficulty the *Cat.* account confronts us with. The way *topos* (place or space) is presented as an item of the category *poson* seems also to be difficult to square with the derived ontological status, so to speak, of the non-substantial categories. Here we may recall the well-known sentence at *Cat.* 2 b 5 ff., in which Aristotle argues that

> If the primary substances did not exist it would be impossible for any of the other things to exist. For all the other things are either said of the primary substances as subjects or they are in them as subjects.[9]

Moreover, at *Cat.* 1 a 24 Aristotle clarifies his famous distinction between 'said of' and 'being in' in such a way as to make clear that the accidental categories (which are said to be 'in a subject') are ontologically dependent on the substantial category:

> 'By 'in a subject' I mean what is in something not as a part *and cannot exist separately from what it is in*'.[10]

If we take these passages seriously it has to be concluded that *if* place is *itself* a quantity, it has to be—*qua* inherent property—ontologically dependent on (a particular) substance. This presents us with an interesting problem since, as we saw, Aristotle is talking as if he wants us to believe that *topos* is something over and above the body occupying it, though exhibiting the same kind of continuity. In other words, *topos* is presented as if it were *independent* of the occupying body.

This reflects a general feature of the way in which Aristotle deals with the category of quantity. For instead of classifying quantitative properties or corresponding quantitative predicates, he lists and groups the *owners* of quantitative properties.[11] Both *topos*

---

[9] *Cat.* 2 b 5–6: Μὴ οὐσῶν οὖν τῶν πρώτων οὐσιῶν ἀδύνατον τῶν ἄλλων τι εἶναι· πάντα γὰρ τὰ ἄλλα ἤτοι καθ' ὑποκειμένων τούτων λέγεται ἢ ἐν ὑποκειμέναις αὐταῖς ἐστίν.

[10] *Cat.* 1 a 24 ff.: ἐν ὑποκειμένῳ δὲ λέγω ὃ ἔν τινι μὴ ὡς μέρος ὑπάρχον ἀδύνατον χωρὶς εἶναι τοῦ ἐν ᾧ ἐστίν, etc.

[11] Cf. Ackrill (1963) 91.

and the other examples of *posa,* lines, surfaces etc., are presented as if they were substances, as bearers of quantity rather than quantitative aspects of a body (substance). As we shall presently see, this curious presentation of the category of quantity already troubled some of the ancient commentators.

The problem, then, seems to be this: the mere fact that *topos* is subsumed under a non-substantial category would point to its having no independent status, whereas the way it is actually presented in the course of the chapter on quantity seems to suggest that it is a more or less self-subsistent entity. Now there are basically three possible ways to bring these apparently divergent claims concerning the ontological status of *topos* into a coherent perspective:

(1) *topos* should be taken as a *mere* quantitative property and the list of 'bearers of quantities' has to be taken as just an oddly expressed presentation of a list of quantitative properties;

(2) *topos* should be taken as a *recurrent* property in order to allow talk about things *changing place*;

(3) *topos* is not really a property but a 'thing' in its own right.

It seems worthwhile to examine the merits of each of these three solutions. As to the first one, this view was endorsed by Simplicius, as is shown by his reply to the criticisms of the 2nd century Platonists Lucius and Nicostratus. They reproached Aristotle for having subsumed *sôma,* i.e. a substance, under a non-substantial category.[12] In this connection Simplicius pointed out that, when discussing *posa,* Aristotle was really speaking about substances *qua* quantified.[13] As such these quantities, being in fact nothing but attributive names signifying certain quantitative aspects of substances, might—as a suitable heuristical device—very well be presented *as if* they were *themselves* substances. Within the Aristotelian corpus such a procedure can of course be paralleled from the studies concerning (the ontological status of) mathematical (or geometrical) objects which are to be found in *Metaph.* B and M.[14]

---

[12] Cf. Simp. *In Cat.* 125, 16 ff.; for general information concerning Lucius and Nicostratus see Moraux (1973–84) II, 528–561.

[13] Cf. Simp. *In Cat.* 125, 13: το δὲ σῶμα, καθόσον μὲν τριχῇ διαστατὸν καὶ μετρεῖσθαι πέφυκεν, ποσὸν ὑπάρχει· καθόσον μέντοι ὑποκείμενόν ἐστιν καὶ ταὐτὸν καὶ ἓν ἀριθμῷ μένον τῶν ἐναντίων ἐστι δεκτικὸν, κατὰ τοῦτο οὐσία ἐστι.

[14] On Aristotle's procedure of 'separating' mathematical or geometrical objects from physical bodies see Heath (1949) 66 and 212 with the examples quoted; see also Mueller (1970) esp. 158 ff. (repr. 98 ff.).

This would at least seem to provide a clue as to how we should understand Aristotle's presentation of numbers, lines and surfaces in the chapter of the *Cat.* we are here concerned with. Accordingly, one might, not implausibly, take the references to *sôma* at 4 b 24 and, particularly, at 5 a 6 in this way and so assume that Aristotle may have been thinking of *geometrical* solids—regarded in typically Aristotelian fashion as quantitative abstractions from physical bodies.[15] Even if, however, this interpretation would allow us to explain the substance-like presentation of bodies, surfaces and lines, it would hardly be of any help in the case of *topos*.

Indeed, at 5 a 8–14 (quotation (1) in our subsection 4.1.2), where Aristotle uses *sôma* in connection with *topos, sôma* seems to mean 'physical body' and it seems to be implied that there is, apart from the continuous extension of such a physical body, also a separate extension occupied by that body.[16] However, even if we, implausibly, suppose that *sôma* is here only intended to mean 'geometrical extension', it should be realized that the passage hypostatizes, so to speak, *both* the extension of a physical body itself (*sôma*) *and* the extension it occupies (*topos*). This would be an odd procedure if it were not backed by an ontology allowing for a self-subsistent three-dimensional space. In that case the supposed independent status of *topos* would seem to be at odds with its non-substantial ontological status as a quantitative property. On this line of thought, therefore, the occurrence of *topos* among the items of the category quantity would seem to be not easy to account for.

This brings us to the second solution proposed above. In a recent article it has been suggested by H. Mendell that a way to save the ontological coherence of the *topos* passages in the *Cat.* is to assume

---

[15] The text at *Cat.* 5 a 4–6 reads: ... ὡσαύτως δὲ καὶ ἐπὶ τοῦ σώματος ἔχοις ἂν λαβεῖν κοινὸν ὅρον, γραμμὴν ἢ ἐπιφάνειαν, πρὸς ἣν τὰ τοῦ σώματος μόρια συνάπτει; σῶμα, in other words, could at this place refer to geometrical solids rather than to physical bodies, the term itself of course being in this respect ambiguous. As appears from a quotation from Porphyry *ap.* Simplicius *In Cat.* 124, 33, this seems to have been the interpretation of the second century CE (orthodox) Peripatetic Herminus (on whom see Moraux (1973–84) II, 361–399, esp. 370) who took σῶμα to refer to τὸ μαθηματικὸν ὃ κατὰ ψιλὰς τὰς πάντῃ διαστάσεις λαμβάνεται.

[16] Cf. Ackrill (1963) 93: 'The proof given here that place is continuous treats it ... as filled by (or perhaps only fillable by) a body. This raises the question whether place has a right to count as an independent primary quantity in addition to body'.

that place is a *recurrent* property of substances, in other words, a property which they can exchange:

> According to the *Categories*, all predication must ultimately be tied to a predication of substance (*Cat.* 3 a 1–5): it would follow that the place belongs to the solid, its immediate occupant, and the solid to the substance, or some such thing. .... But only if each place is a recurrent property could another body occupy the same place as the initial occupant.[17]

In this connection Mendell refers to the position of G. E. L. Owen who in a famous—though not universally applauded—paper [18] defended the thesis that there are recurrent 'individual' properties within Aristotle's categorial system.[19]

In *Cat.* 1 a 23–4 the sentence 'by in a subject I mean what is in something not as a part and cannot exist separately from what it is in' should, according to Owen, not be interpreted in the 'traditional' way. This means that the words 'from what it is in' should not be taken to mean 'from the particular substance it is in' but rather 'from something which contains it'. Otherwise, Owen contends, a particular bit of linguistic knowledge cannot lodge in more than one head,[20] and

> two things cannot be said to occupy the same particular place at different times. Aristotle's account of place in the *Phys.* rests squarely on the assumption that A can move into the identical place vacated by B.[21]

However, on closer scrutiny Owen's views meet with a number of objections which seem to me to be quite fatal.

First, there is no indication whatsoever that Aristotle thought along these lines: a class of recurrent properties is nowhere outlined as such in his work. Secondly, Owen's interpretation is not even needed for the purpose for which it was conceived. For on closer scrutiny the problem it was supposed to solve turns out to be

---

[17] Mendell (1987) 222. Mendell himself, by the way, in the end rejects this suggestion because, he argues, in this view 'either the place is changing its occupants, or one substance is giving a property to another. In either case the place persists through change of its subject. Hence the place itself must be a substance'.

[18] Owen (1965).

[19] It should be remarked, though, that when writing his article Owen rather seems to have had individual instances in the category ποῦ in mind.

[20] Owen (1965) 99.

[21] Owen (1965) 102.

non-existent. Owen's interpretation was primarily put forward in opposition to what he called 'a fashionable dogma' concerning individuals in non-substantial categories—a dogma embodied in the work of, *inter alios*, Ackrill. According to this 'dogma' particular (in the sense of: numerically individual) attributes are not in more than one individual. Unfortunately, however, Owen appears to have mistaken this 'dogma', which is in fact arguably correct, for a quite different one, which is obviously not correct, viz. that e.g. a particular (in the sense of 'specific') shade of knowledge or a particular bit of linguistic knowledge cannot be present in more than one individual. In the latter cases we are dealing, not with particulars in the strict sense of the word, but with *infimae species*. One specific shade of pink may be present in (A) as well as in (B), but at the same time (A)'s particular pink and (B)'s particular pink are—though specifically identical—numerically different. It is only in the latter (numerical) sense of the words 'the same' (identical)—and in the corresponding strict sense of 'particular'—that different subjects indeed cannot share the same (identical) particular property.

As soon as this distinction between specific and numerical identity is properly observed, the reason for assuming that e.g. a particular place, or a particular bit of linguistic knowledge, is a recurrent individual property disappears: two persons (A) and (B) may be at the very same place (an *infima species* of the category *pou*) at different times, e.g. they may both, at different moments, be in the very same spot in the gymnasium, but that does not mean that (A)'s being in the gymnasium is not numerically distinct from (B)'s being in the gymnasium. Thus the whole theoretical basis for the assumption of recurrent individual properties collapses: the phenomena Owen wants to explain (a particular bit of grammatical knowledge lodging in the heads of both Peter and John, or Peter and John both being in the very same place) may also very well be explained by the traditional interpretation of 'inherence' so that there is no need to take recourse to this odd theory which finds no support in the texts.[22]

---

[22] See also the criticisms of Owen's view by Allen (1969); see also the pertinent criticisms of Moravcik's position—which is partly similar to Owen's—by De Rijk (1980) 20. Frede (1987a) esp. 57, on the other hand seems to join Owen in his opposition to the alleged 'dogma', though Frede's argument against the dogma seems to go no further than the

Finally, and more importantly in connection with our present investigation, Owen's interpretation still leaves us with a conception of place that is from the point of view of physics quite unworkable. Indeed, on the basis of an ontology which, instead of regarding place as an absolute extension or as a relation, treats place as a property—even if it is a *recurrent* property—any attempt to describe or explain locomotion would be bound to end up in a hopeless muddle. Consider e.g. how, on this line of thought, one would have to explain what happens if the place of a body (A) is subsequently occupied by two bodies which are each half the size of (A). In this connection Owen's reference to *Phys.* Δ is beside the point. For there Aristotle does all he can to make clear that *topos*, if it is to figure in contexts where locomotion is at issue, should be understood as independent of the located body.[23] The idea of place as a recurrent property is, moreover, very far removed from any ordinary conception of place and locomotion.

It would seem then that these attempts to fit *topos* as a property—whether simple or recurrent—into the substance/property scheme of the *Cat.* are unsuccessful. In this connection it may be interesting to note that the problems outlined here were apparently already sensed by some of the ancient commentators. As I take it, the fact that the ontology of the *Categories* seems to force us into the direction of the odd concept of *topos* as a recurrent property of substances, constitutes the background for Porphyrius' somewhat uncomfortable suggestion that *topos* does not 'belong to' the body in the same way as the other *posa*, like a line or a surface, which are 'unified with the body'.[24] For *topos* is said to 'follow [the existence of?] a body from outside'.[25]

How then should we conceive of the ontological status of *topos* and its fellow-*posa*? We are left with the third solution, which takes *topos* as a 'thing' in its own right. But what kind of thing? As

---

statement that it actually *was* a dogma among late antique and scholastic commentators.

[23] Cf. 209 b 29–30 where the choice of ἀγγεῖον as a model for τόπος is motivated by the statement that an ἀγγεῖον is οὐδὲν τοῦ πράγματος.

[24] Porph. *In Cat.* 104, 2: place, unlike other quantities, 'οὐχ ἥνωται τῷ σώματι'.

[25] Porph. *In Cat.* 103, 21: ἐπακολούθησε τῷ σώματι, and p. 104, 2: ὁ δὲ τόπος οὐχ ἥνωται τῷ σώματι, ἀλλ' ἔξωθεν ἐπακολούθησεν. See also Simp. *In Cat.* 125, 18–19. Dexippus' commentary on the *Categories* makes no mention at all of the occurrence of τόπος among the πόσα.

we saw, Owen suggested that according to Aristotle surfaces and the like 'appear as the standard possessors of dimensional properties, but are not themselves properties nor apparently anything else'.[26] And Ackrill argues that 'Aristotle does not discuss the status of lines etc. in his own categorial scheme. They are obviously not substances, though they have properties. Their relation to primary substances is not at all elucidated in the *Categories*'.[27] So both Owen and Ackrill exclude the possibility that the 'bearers of quantity' like lines, surfaces and *topos* are substances. That they are correct in doing so is likely if we take into account that all examples of substances adduced in the *Cat.* (but also elsewhere)[28] concern what we would call 'natural objects'. In addition, the anti-Platonic tendencies apparent elsewhere in the *Cat.* do not seem to lend much probability to the thesis that Aristotle at this stage still thought of mathematical objects as substances. Since *topos* is presented on the same footing as these, we should probably conclude that it was not Aristotle's intention to suggest that this was a substance either. Rather, the ontological status of *topos*, as well as of the other *posa*, was apparently left obscure.[29]

---

26 Owen (1965) 102; in the same passage Owen implies that for Aristotle this is a familiar problem: 'this is why Aristotle is exercised by the old question whether surfaces and other boundaries of a body should count as substances'. In this connection he refers to *Metaph.* 1002 a 4 ff. and 1071 b 17–21. However, it should be noted that at those places the problem of whether geometrical or mathematical objects are substances (which was, as is well-known, the common view of the early Academy) is brought up by way of an ἀπορία as part of a dialectical survey, and Aristotle's conclusion in the end is that they are not.

27 Ackrill (1963) 92.

28 Cf. e.g. M. Frede, 'Substance in Aristotle's Metaphysics', in Frede (1987b) 76: 'It is characteristic of ZHΘ that Aristotle tends to, or in fact does, restrict substances to natural objects ... it is not entirely clear whether this is supposed to restrict substances to animate things, but these certainly are paradigms of natural objects'.

29 De Rijk (1980) 1–62, has made a strong case for taking the categories as *names*, an interpretation which has the merits of being supported by the text and showing that the old controversy about whether the categories are primarily a classification of predicates or a classification of reality is rather pointless. However, when it comes to discussing the category πόσον, De Rijk ((1980) 48–49) too easily disposes of Ackrill's remark that Aristotle does not list or attempt to classify quantitative properties or corresponding quantitative predicates. According to De Rijk this is only natural since 'Aristotle prefaces his semantic treatment of *poson* ... with an enumeration of those things that can be named in their own right' [probably we should read 'those things

We may provisionally conclude, then, that the way *topos* is introduced as an item in the category quantity does not suggest the presence of a well thought-out and articulate concept of space or place on Aristotle's part. He is rather outlining the ways in which common parlance may be naming things as *posa*.[30] Probably the idea is that *topos* is among these things, since people tend to regard *topos* as an extension which may e.g. be measured. The ontological implications of this rather inarticulate common sense view are apparently not what primarily interests Aristotle in this connection.

4.1.4 There are some more features of the treatment of *topos* in the *Cat.* which seem to point to Aristotle's using a rather inarticulate concept of *topos*. In the first place the curious separation of *topos* as an item of the category quantity on the one hand and the independent category *pou* ('where') on the other commands some further attention.

If it is true that Aristotle took the ways in which people commonly name things as his starting point in writing the *Categories*, it is perhaps not too surprising to find *topos* and *pou* thus separated. For people *speak of* a *topos* as a quantity (*poson*) and they speak of substances as being 'somewhere' (*pou*). If, on the other hand, one rather focuses on the ontological side of the matter, questions will arise concerning the precise ontological status of *topos* as a *poson* and concerning the relationship between this *topos* and the category *pou*.

An attempt to answer these questions was made by the 'editor' of the *Corpus Aristotelicum*, Andronicus of Rhodes as well as by the anonymous author of ps.Archytas 'On the Universal Theory' (*Peri tou katholou logou*), a curious Neopythagorean forgery, originally written in the Doric dialect, which is probably to be dated to the second century CE.[31] Both Andronicus and *anonymus* turned *topos* and *chronos* (time) into *separate* categories and assigned *pou* a

---

that can be named *poson* in their own right', KA]. In other words, τόπος, on this line of thought, is just one of the 'things' which ordinarily may be called πόσα. True as this might be, it still leaves us with the problem of the ontological status of these 'things'.

30 See De Rijk (1980) 48 ff.

31 Cf. Szlezák (1972). See also Moraux (1973–84) II, 608–623 and Gottschalk (1987) 1102 and 1131 ff.

subordinate place within that category, whereas *pote* ('when') was given a subordinate place within the category *chronos*.[32]

This modification will, at least partly, have been called forward by the difficulties inherent in treating *topos* as a mere quantitative property, the substance-like presentation of *topos* in the course of the chapter dealing with quantity, and the fact that in enumerations of the categories to be found in other works Aristotle, possibly in the aftermath of his own discussions of *topos* in *Phys.* Δ, substitutes a category *topos* for the category *pou*.[33]

Secondly, another oddity which it is worthwhile to point out in this connection is the fact that *Cat.* 6 a 12 ff. (quotation (1) in 4.1.1) suggests that in the case of 'up' and 'down' there is contrariety of quantity.[34] Now the question whether within a particular category one may speak of contraries belongs to the standard issues raised by Aristotle in the *Cat.* In the present case, however, there is something strange. It has been rightly remarked by Ackrill that 'since in any case 'up' and 'down' would not give the quantity of anything (but rather its 'where' or 'whither') the view that they are contraries does not seem to justify the suggestion that there is after all contrariety of quantity'.[35] In fact, whereas in the present context (the category *poson*) Aristotle apparently treats of place *qua* extension, and as such *qua poson*, it has to be admitted that 'up' and 'down' are not contraries *qua* quantities, but only *qua* localizations. Strictly speaking, therefore, one would expect them to appear within the account of the category *pou*, not in connection with *poson*—a fact which, by the way, was already pointed out by the Peripatetic commentator Herminus.[36]

---

[32] Cf. Simp. *In Cat.* 357, 28 ff. See also Moraux (1973–84) I, 111–112, and Gottschalk (1987) 1106.

[33] See above, n. 5. Note that this change of wording shows that De Rijk's stress ((1980) 30 and 44) on the semantic element in Aristotle's approach to the categories and his concomitant statement that '*naming* things and things as *named* such-and-such are in the focus, not things by themselves' is not in all cases supported by the text. In those cases where Aristotle substitutes τόπος for ποῦ, τόπος is obviously not the 'thing named such-and-such', for that thing is a substance which is spoken of as being 'in such and such a place'. Neither is τόπος the name itself, for the thing in place is not called 'place'. In fact, one would perhaps rather have expected the denomination ἐν τόπῳ or κατέχον τόπον as a substitute for the ποῦ of the *Cat.*

[34] *Cat.* 6 a 12: μάλιστα δὲ ἡ ἐναντιότης τοῦ ποσοῦ περὶ τὸν τόπον δοκεῖ ὑπάρχειν.

[35] Ackrill (1963) 97.

[36] Quoted by Porph. *In Cat.* 107, 25 ff.

The separation, in other words, of *topos* as an item of the category quantity on the one hand and a particular localization, as an item of the category *pou*, on the other was not altogether waterproof. This again interestingly suggests that Aristotle's mind was not yet clear about the difference between place or space as *extension* and place as a localization.

4.1.5 We may summarize the findings of the present section as follows. The presentation of *topos* in the *Categories* appears at first sight to be fully at odds with the classical definition of *topos* put forward in *Phys.* Δ. At the same time we saw that the *topos* passages in the *Cat.* presented problems of their own. The ontological status of *topos* as an item in the category quantity remains obscure, and the relation between this *topos* and the category *pou* is left in the dark.

All this seems to point to Aristotle's not having been consciously and explicitly committed to any articulate 'concept of space or place' when he wrote the *Categories.* He rather seems to have been led by the conventions of common parlance in which, as we saw in the previous chapters the 'extensional' and the 'locative' functions of *topos* are not always clearly separated. Thus the attempts, to be discussed in the next two sections, to read the *topos* passages in the *Cat.* as a serious ontological or physical account of place, and to square them with the account put forward in *Phys.* Δ, seem to be qualified in advance as attempts to take the *Categories* too seriously *in physicis.* Nevertheless a fuller discussion and evaluation of these attempts will be in order if only because the two approaches towards the *Corpus Aristotelicum* to which they testify—the 'synthesizing' and the 'developmental' approach—are closely connected with two different, but still important, schools of thought in Aristotelian scholarship.

4.2 *The synthesizing approach*

4.2.1 We may now revert to the differences between the way *topos* is presented in the *Cat.* on the one hand and in *Phys.* Δ on the other. In the previous section I drew attention to the fact that these differences were so obvious and so far-reaching as hardly to allow of a reconciliation. Yet attempts to reconcile apparently inconsistent passages are as old as Aristotelian exegesis itself and many of the ancient and mediaeval commentators displayed a quite astonishing virtuosity in this respect. So we may well try and see how in the present case a synthesis has been reached and whether such a synthesis on closer examination proves acceptable. Attempts to harmonize the relevant passages from the *Cat.* and *Phys.* Δ have gone both ways. On the one hand, in late antiquity Ammonius and his two pupils Simplicius and Philoponus opted for reading the theory of the *Phys.* into the *Cat.* Their interpretations will be discussed in 4.2.2.

On the other hand, as recently as in 1950 H. King published an elaborate defence of what he called an 'Aristotelian theory of space' in which he argued that, contrary to what the *communis opinio* would declare, *Phys.* Δ positively advances the conception of *topos* as three dimensional extension which is also to be found in the *Cat.*,[37] a view which had been defended earlier, though with different arguments, by A. E. Taylor.[38] This view will be briefly discussed in 4.2.3.

4.2.2 The sixth century commentators Ammonius, Simplicius and Philoponus were all of the opinion that the passages from the *Cat.* quoted above can be read as allowing of an interpretation along the lines set out in *Phys.* Δ. In other words, they tried to show that the way *topos* was used in these passages was quite compatible with the definition—worked out in *Phys.* Δ—of *topos* as the first boundary, i.e. the outer surface, of the containing body. Thus they took Aristotle's remark (our quotation (1) in 4.1.1) that the parts of a body each occupy their own *topos* which is part of the overall *topos* of the body, to refer to the parts of the outer surface of the contained body being exactly matched by the parts of the outer surface of the containing body.[39] On closer inspection,

---

[37] King (1950) 76 ff.

[38] Taylor (1928) 664–677, esp. 674.

[39] Cf. Ammon. *In Cat.* 58, 16–26, where the *Cat.* account is explicitly linked to the views of the Φυσικὴ Ἀκρόασις. See also Phlp *In Cat.* 87, 7 ff. After having quoted the definition of τόπος which was given in *Phys.* Δ, Philoponus explains the statement in the *Cat.* that place is continuous by arguing that 'just as the contained (τὸ περιεχόμενον), so is the place', and he goes on to show (*op. cit.* 87, 11) that τόπος, in so far as it is a surface, is continuous *qua* being a surface, whereas, in so far as it is a place, it is continuous *qua* being a περιέχον. Thus he makes clear that he is reading the concept of place of the *Phys.* into the *Cat.* Finally, see Simp. *In Cat.* 125, 17 ff. Simplicius's position here is interestingly ambiguous. For on the one hand he

however, this view meets with some insurmountable difficulties.

First, though the curious interpretation of these Neoplatonist commentators cannot strictly speaking be proved wrong on the basis of the wording of *Cat.* 5 a 8–14. (quotation (1) in 4.1.1), it certainly does not fit the way *topos* is presented at *Cat.* 6 a 12 ff. (quotation (2) in 4.1.1). For there the fact that *topos* is a quantity which seems to admit of contraries is corroborated by a reference to what we might call *regions* of space,[40] so that it would seem that that passage cannot possibly be reinterpreted in terms of place-as-a-surface. Second, even if the evidence of *Cat.* 6 a 12 ff. were disregarded, the interpretation of Ammonius *c.s.* would not succeed in reconciling the views on *topos* of the *Cat.* and *Phys.* Δ respectively in all important respects. As an example we may take the way in which the *Cat.* on the one hand and *Phys.* Δ on the other speak of the *parts* of place.

As we saw, in the *Cat.* Aristotle apparently feels no qualms about saying that there are parts of place. Yet, at *Phys.* Δ 211 b 14 ff., we find, as one of the absurd consequences of regarding *topos* as a three-dimensional extension, the explicit statement that it will thus follow that also the parts of a body each have their own place. On this view, it is concluded, one is bound to end up with an infinity of (partly overlapping) places. Now it might be argued, following Ammonius c.s., that this is merely a polemical *reductio ad absurdum* of the view of Aristotle's opponents, applying only to the extension view of *topos,* whereas if we interpret the 'parts of a body' as being in fact the 'parts of the surface of the body' no such

maintains that Aristotle's views in the *Cat.* and the *Phys.* do not differ substantially and makes clear that he thinks that also in the *Cat.* τόπος should be taken to be the surface of the containing body (at 125, 21–22 he calls it περιέχον and πέρας). On the other hand he argues that Aristotle does not derive the continuity of τόπος from its being a surface—which would on this line of thought of course have been the most obvious thing to do—because he had at the time of writing the *Cat.* not yet fully investigated the nature of τόπος (125, 28: διότι οὔπω ἔγνωστο τί ἐστιν τόπος; note, by the way, that this shows that the application of a 'genetic' approach to the *Corpus Aristotelicum,* even if only on a fairly modest scale, was not fully and completely alien to the thought of all ancient commentators). Moreover, at 133, 35 ff., he argues, by way of a reply to unnamed critics (he might still have Lucius and Nicostratus, who were mentioned at 127, 30, in mind), that in the *Cat.* Aristotle does not intend to go into the *nature* of time and place because that kind of inquiry belongs to physics.

40 Note e.g. that Aristotle speaks of τὴν πρὸς τὸ μέσον *χώραν*.

absurd consequences follow. However, this will not do. For Aristotle goes further than just pointing to some difficulties which would arise exclusively if we take *topos* to be a three-dimensional extension. Rather, he emphatically denies that the parts of a body (substance) can be said to be in a place of their own *in any way whatsoever.*[41] This is why he takes pains to argue that only on his own definition of *topos* the *reductio ad absurdum* (an infinity of places) does not follow. On the contrary, on this definition of *topos* —i.e. when we regard it as the first boundary of the container—it follows that when a body is moving as a whole, its parts are not changing place.[42] So the whole notion of the parts of a body having a place of their own, which is central to the presentation of *topos* in the *Cat.*, is alien to the discussion of *Phys.* Δ.

It appears, then, that in the *Cat.* Aristotle cannot reasonably be held to have used *topos* according to the rules laid down in the *Phys.* Accordingly, the synthesizing approach advocated by Ammonius and his pupils seems to fail.

4.2.3 Still fewer reasons can be found for assuming that the conception of *topos* as a three-dimensional self-subsistent substratum, as it is encountered in the *Cat.*, is also positively defended in *Phys.* Δ. For there Aristotle is very straightforward and clear as to which

---

41 Cf. *Phys.* 211 b 25: Οὐκ ἔστι δὲ ἄλλος τόπος ὁ τοῦ μορίου, ἐν ᾧ κινεῖται, ὅταν ὅλον τὸ ἀγγεῖον μεθίστηται, ἀλλ' ὁ αὐτός.

42 Cf. Furley (1982) 20. As Furley points out, Simp. *In Cat.* 125, 24–32, was alert to this particular difficulty. Note, however, that Furley is wrong in assuming that Simplicius tried to iron out this inconsistency and that he had the *Categories* in mind when he said that '*here* [ἐνταῦθα; Furley's translation] Aristotle only meant to say that the continuous parts of a body are in place, indeed, but not in their own right'. For the Greek text reads: ... οὔπω ἔγνωστο τί ἐστιν τόπος· *φυσικῆς* γὰρ ἦν *θεωρίας* ἡ σκέψις. ἐπιστῆσαι δὲ ἄξιον, ὅτι *ἐνταῦθα* τὰ τοῦ σώματος μόρια τὰ συνεχῆ οὐ καθ' ἑαυτὰ ἐν τόπῳ εἶναί φησι· καὶ γὰρ κατέχει μὲν καὶ αὐτὰ τόπον, οὐ μὴν καθ' αὑτά, ἐπειδὴ οὐδέ ἐστιν ἐν τόπῳ καθ' ἑαυτὰ οὐδὲ περιέχεται καθ' ἑαυτά. The last part of this quotation clearly is a reference to the context of *Phys.* 211 b 25 and the word ἐνταῦθα which Furley took as referring to the *Cat.* and which he accordingly translated as 'here', should be translated as 'there' since it refers to the φυσικὴ θεωρία mentioned in the preceding part of the sentence. Simplicius, in other words, seems to acknowledge that the *Cat.* account differs from that of *Phys.* Δ, and he points out that *there* Aristotle had a different conception concerning the localization of the parts of a continuous body, viz. that they cannot be said to be in place in their own right. On Simplicius ambiguous attitude towards the differences between the *Cat.* and *Phys.* Δ in this respect see above, n. 39.

views on the nature of *topos* should be accepted and which not. As we saw, the conception of *topos* as a three-dimensional extension (διάστημα) is obviously among the latter.

Nevertheless, in the course of his elaborate defence of the Aristotelian theory of place H. R. King has it that

> ... you cannot speak of the bound of a container without implying some contained, not a particular contained body, but a διάστημα ὡς κενόν, receptive of *all* body. Of course, Place is not a mere διάστημα, an extension over and above the extended bodies, as Aristotle says; but τὸ μεταξὺ ὡς κενὸν cannot be divorced from the notion of Place as the bound of a container .... and it is this aspect of Place as a receptive διάστημα ... that Aristotle is thinking of when he calls Place 'continuous' in the *Cat.* [43]

King tries to substantiate this claim by pointing to Aristotle's remark in *Phys.* Δ 212 a 13–14 that 'Place appears to be *not only* [King's italics] the boundaries of the vessel, but also the extension (διάστημα) in between *regarded as empty*' [again, King's italics]. A brief glance at the context in which these lines occur may show, however, that they do not at all prove what King wants them to prove.

The passage from which King was quoting (*Phys.* Δ 212 a 8 ff.) reads:

> Place seems to be something profound and difficult to grasp, both because the notions of matter and form present themselves together with it (παρεμφαίνεσθαι), and because of the fact that change of position within a moving body occurs within a surrounding body which is at rest; for [from this] it appears to be possible (ἐνδέχεσθαι γὰρ φαίνεται) that there is an extension in between which is something other than the magnitudes which move. Air, too, contributes to this suggestion, by appearing to be incorporeal; place seems (φαίνεται) to be not only the limits of the vessel, but also that which is in between, which is considered as being void.

This passage more or less outlines the upshot of the preceding discussion. At 211 b 6–10 Aristotle has presented the following *tetralemma.* If we want to establish that '*topos* is (x)', there are four possible substitutions for (x), viz. (1) form, (2) matter, (3) the extension (διάστημα) between the extremities (scil.: of the surrounding body), or (4) the extremities themselves (i.e. taken as the limit of the surrounding body). Note, by the way, that here already

---

43 King (1950) 88.

Aristotle seems to imply that only *one* of these candidates can in the end turn out to be the right one.[44]

Next, from 211 b 10 onward, Aristotle considers the four candidates one by one. In the course of this investigation the third possibility is said to derive its plausibility from the fact that we are inclined to imagine that in the case of a vessel there is something, an extension, which remains when we take away the actual content of the vessel (e.g. water). His reply is straightforward: 'But this is not so'.[45] Follow the *reductiones ad absurdum* which show that this view, for all its initial plausibility, cannot stand up to closer dialectical investigation. A similar procedure is applied to the other candidates (form and matter).[46]

As I said, the passage quoted by King summarizes this procedure. The four candidates are listed once more and we are reminded that the difficulty of the whole subject is precisely constituted by the fact that the rejected candidates are not wholly extravagant alternatives, but rather belong to the way we are used to speaking about *topos*. They are connotations involved by our ordinary concept of *topos* which, in the context of a careful physical investigation into the *real nature* of place, are ultimately unacceptable. Here we see Aristotle confronted with ordinary language as a *Fehlerquelle*.[47] It is obvious, then, that since King did not notice the dialectical character of the passage on which he based his interpretation, his attempt to reconcile the views on *topos* of the *Cat.* and *Phys.* Δ should be regarded as unsuccessful.

Also A. E. Taylor, in his 'Aristotle's Doctrine of Space', a study which is still often referred to, mistakenly assumes that in *Phys.* Δ Aristotle endorsed a conception of space as a three-dimensional extension.[48] However, his reasons are not the same as King's.

---

44 211 b 6–7: ὧν ἀνάγκη τὸν τόπον ἕν τι εἶναι.

45 211 b 18: τὸ δ' οὐκ ἔστιν.

46 For a fuller discussion of these issues see below, subsection 4.4.4.

47 Cf. Wieland (1962) 280 and see also below, n. 125.

48 Taylor's treatment is heavily dependent on the distinction he draws between (1) space as a network of relations of situation between certain ultimate elements and (2) the metrical view of space: space as that of which a given body occupies given amounts. According to Taylor (1928) 668, the former view represents the *esprit juste* since, he argues with a reference to Russell's *Principia Mathematica*, you can build a whole system of geometry without introducing metrical ideas, though you cannot construct a metrical geometry without non-metrical fundamental notions. The second view, on the other hand, Taylor regards as ἡμῖν πρότερον owing to the close

Taylor really was aware of the dialectical character of Aristotle's treatment of the view according to which place is 'the extension between the limits of the surrounding body'. But he wrongly took this formula to refer to 'an interval between container and contained'. Consequently, it was this odd conception of *topos*, and not the view according to which *topos* is a three dimensional self-subsistent substratum, which on Taylor's line of thought was ruled out by Aristotle in the passages in *Phys.* Δ just referred to. Clearly, however, Taylor's interpretation is vitiated by the fact that Aristotle's wording as well as the examples he adduces unequivocally support the 'traditional' interpretation of the words 'that which is between the limits'.[49] This is why also this ingenious interpretation collapses.

4.2.4 We may thus conclude the present section by noting that all attempts to square the *topos* passages in the *Cat.* with those in *Phys.* Δ are seen to flounder. The somewhat uncomfortable attempts of the ancient commentators were obviously elicited by the desire to save the unity of Aristotle's philosophy as a whole, but even they proved unable to iron out all the differences.[50] The more recent attempts by Taylor and King were based on a misreading of a particular passage in the *Physics*. The 'synthetic' approach thus having been proved unfruitful, we may now turn to see whether a developmental approach would appear to be more successful.

---

connection, especially in early times, of geometry with the practical business of measurement in land-surveying. It is against this background that he also takes Aristotle's own definition—which of course rather points to his endorsing the first of these two conceptions (on which see chapter 1)—as providing a means for determining the magnitude of the volume occupied by a given body. Next, he states that Aristotle, in providing such a means, thinks he is describing the character of volume. However, Aristotle's discussion of τόπος in the *Phys.* has nothing to do with *volume*, and everything with *location*.

[49] Cf. 212 a 13–4: φαίνεται γὰρ οὐ μόνον τὰ πέρατα τοῦ ἀγγείου εἶναι ὁ τόπος, ἀλλὰ καὶ τὸ μεταξὺ (i.e. that which lies between the πέρατα τοῦ ἀγγείου, *not* that which lies between the extreme boundary of the container and that of the contained) ὡς κενόν. Also the example immediately preceding this quotation can be adduced: the argument that air—seeming to be incorporeal—contributes to the illusory view that τόπος is τὸ μεταξὺ ὡς κενόν would make no sense if the latter formula were to be taken in the way advocated by Taylor.

[50] See my remarks above, n. 39, on the ambiguity of Simplicius' position.

4.3 *The developmental approach*

4.3.1 Before we go on to investigate whether a development of Aristotle's thoughts on *topos* can be traced, and whether it is perhaps this development which is the cause of the significant differences between the treatments of *topos* in the *Cat.* and in *Phys.* Δ, we should make clear in our minds what in this connection we mean by 'development'.

There are basically three different ways is which we may hold a thinker's views to have 'developed'. For in using this expression we may either imply that a thinker at first *consciously* held a definite view on a certain subject and that he subsequently *consciously* changed it for a *substantially* different view. Alternatively, we may want to make clear that a thinker substantially held the same views at different stages of his 'career', these views having been relatively inarticulate at first and subsequently refined into a more sophisticated version. Finally, it is also possible that a thinker's thought on a specific subject was at first inchoate and inarticulate (e.g. because the subject did not interest him, or because he was unaware of the inherent problems), whereas he later developed a conscious and articulate theory.

A development of Aristotle's concept of *topos* in the first (strong) sense, viz. a development from a consciously Platonizing to the original Aristotelian view has recently been ascribed to Aristotle by H. Mendell. The second kind of 'development' was attributed to Aristotle by Simplicius who, as we saw, suggested that though the views on *topos* which were propounded in the *Cat.* were basically the same as those propounded in the *Physics,* they were at some points deficient because Aristotle had at this stage not yet investigated all details of the physical problems concerning space and place.[51] A development in the third sense is what I myself am willing to ascribe to Aristotle, as will appear below in 4.3.4.

Since no discussion of possible evolutions in Aristotle's work can manage without some information about the chronology of Aristotle's philosophical activities, 4.3.2 contains a brief outline of the *status quaestionis* concerning the problem of the internal development and the relative chronology of the works at issue. Next, the view according to which Aristotle's concept of *topos*

[51] See above, n. 42.

shows a development from Platonism towards the proper Aristotelian position will be expounded and criticized in 4.3.3. In 4.3.4, finally, I shall conclude the section by outlining my own views on the matter.

4.3.2 It is well known that until the beginning of the present century Aristotle's work was regarded as a more or less closed and monolithic system. This unified picture was probably smashed once and for all by Werner Jaeger's epoch-making studies [52] which purported to show that there had been a good deal more of a development in Aristotle's thought than had hitherto been assumed. As a result, nobody since could afford to ignore Jaeger's approach and the way back to the 'scholastic' view seemed definitely blocked. Most importantly, Jaeger's studies set the key for a new kind of Aristotelian research which was characterized by a fresh awareness of the fact that Aristotle's esoteric works bear the stamp of many years of his school practice and that, consequently, they contain *Schichten* of various dates. Thus many studies published during the present century have focused on Aristotle's development as a philosopher and on the relative chronology of (the parts of) his esoteric works.

In the meantime, however, serious doubts have been raised about the methodological validity of extreme versions of this 'genetic' approach and—ironically enough—most of the details of Jaeger's own arguments and conclusions are by now no longer universally accepted.[53] Especially his main thesis, Aristotle's development from Platonism (early exoteric writings) to empiricism (later biological works), as well as his ensuing relative chronology of the extant writings, have been proved untenable for various reasons. However, the important thing for our present purposes is that in the post-Jaegerian era it would in principle be possible to account for *prima facie* inconsistencies within the Aristotelian corpus by assuming that Aristotle's thoughts concerning the issue at stake underwent some changes in the course of the many years of his philosophical activity.

For the chronology of Aristotle's works I shall here only refer, as is now customary, to Düring's synthesis of much of the post-

---

52 Jaeger (1912) and (1923).
53 See now the survey of Flashar (1983).

Jaegerian studies.[54] According to this chronology the works which are relevant to the present inquiry are all to be dated in the first Athenian period (Aristotle's stay in the Academy: 367–347 BCE), the *Cat.* having been written at an early date, between 360 and 355, the *Phys.*, *Cael.*, and *GC* all somewhere between 355 and 347.

The latter group—to judge from the opening chapter of the *Meteorologica* [55]—seems to have been looked upon by Aristotle as a more or less coherent and continuous group of writings of which the *Meteor.* was to be the crowning piece. Together they were probably intended to form a suitable curriculum in physics, starting with the first causes of nature and natural motion (*Phys.*), next treating the movements of the stars (*Cael.*), the elements and their mutual transformations (*Cael.* and *GC*), growth and decay (*GC*), and the meteorological phenomena occurring in the sublunar sphere (*Meteor.*). This sequence was probably also observed by Andronicus of Rhodes' recension of the *Corpus Aristotelicum*, if it may be assumed, as is usually done, that Ptolemaeus' *pinax*, a post-Andronican list of titles, preserved in two 13th century Arab versions, is to be traced back to Andronicus' list of titles.[56]

It appears, then, that in principle there is room for a development of Aristotle's thought on the subject of *topos* between his writing the *Categories* on the one hand and the physical treatises on the other and perhaps—though this is at first blush not very likely—even between the earlier and later parts of the physical

---

54 Düring (1966) 48–52; and *idem*, 'Aristoteles', *RE*, Suppl. 11 (1968) 159–336.

55 *Meteor.* 338 a 20–26: Περὶ μὲν οὖν τῶν πρώτων αἰτίων τῆς φύσεως καὶ περὶ πάσης κινήσεως φυσικῆς, ἔτι δὲ περὶ τῶν κατὰ τὴν ἄνω φορὰν διακεκοσμημένων ἄστρων καὶ περὶ τῶν στοιχείων τῶν σωματικῶν, πόσα τε καὶ ποῖα, καὶ τῆς εἰς ἄλληλα μεταβολῆς, καὶ περὶ γενέσεως καὶ φθορᾶς τῆς κοινῆς εἴρηται πρότερον. Λοιπὸν δ' ἐστὶ μέρος τῆς μεθόδου ταύτης ἔτι θεωρητέον, ὃ πάντες οἱ πρότεροι μετεωρολογίαν ἐκάλον.

56 On which see Moraux (1951) 297 ff. See also Düring (1957) 241–246, and Moraux (1973–84) I, 85–94. That the 'bibliographical' remarks at the beginning of the *Meteor.* cannot have been interpolated by Andronicus or a post-Andronican hand is virtually certain, for the separate reference to the first causes of nature (περὶ τῶν πρώτων αἰτίων τῆς φύσεως) on the one hand and natural motion (περὶ πάσης κινήσεως φυσικῆς) on the other reflects the pre-Andronican division of what from Andronicus onward was going to be called the Φυσικὴ Ἀκρόασις, into Φυσικὰ on the one hand (*Phys.* I–V) and the three books Περὶ Κινήσεως (*Phys.* VI–VIII) on the other. On this division see Simp. *In Ph.* 923, 8 ff. and Moraux (1951) 104.

treatises themselves. Whether such a development actually took place is the question to which I shall now turn.

4.3.3 In section 4.2 we saw that place (*topos*) is in the *Categories* presented as a self-subsistent extension. In addition I have tried to make clear that this view of place presents problems of its own. More particularly, Aristotle does not bother to fit it in with the overall ontology of the treatise. It is primarily for this reason that the view according to which the *Categories* contain serious reflection on the *real nature* of place on Aristotle's part has little *prima facie* plausibility.

Thus it does not seem very likely that, as has been suggested by H. Mendell, the *Cat.* shows how at that early stage Aristotle consciously took over some of the features from the receptacle of Plato's *Timaeus* whereas later, in *Phys.* Δ, he explicitly criticized his own earlier position.[57] But perhaps we should not be too rash in drawing this conclusion and rather turn to the details of Mendell's argument.

That there is a kind of ancestry-and-descent connection between Plato's *Timaeus* and Aristotle's *Categories* Mendell takes as established by a study of J. Driscoll. In this paper the latter aims to show that the *Timaeus*' receptacle lent much of its distinctive features to the primary substances of the *Categories*. [58] In this connection he notes the following similarities between the Platonic receptacle and Aristotle's primary substance:

(1) both are the ultimate subjects of inherence;
(2) both lack contraries;
(3) both receive contraries, whereas
(4) both do so through a change in themselves, while nevertheless
(5) both preserve their own identity.

Driscoll then concludes that the similarity of the two treatises is 'too great to be merely accidental',[59] adding as 'purely verbal

[57] Unless, of course, it could be shown that the inconsistencies detected between the *Categories*' treatment of τόπος and the overall ontology of the treatise were themselves taken over from the *Timaeus*. However, these inconsistencies were caused by the typically Aristotelian substance ontology and have no counterpart in the *Tim.*

[58] Driscoll (1979) 253–269.

[59] Driscoll (1979) 258.

evidence' the fact that both in the *Tim.* and in the *Cat.* the term *dechesthai* and connected forms like *hupodochê* or *dektikon* are prominently used in connection with the receptacle or the primary substances: 'It is difficult to imagine how Aristotle's choice of words could fail to remind an Academic audience of the Receptacle, and more difficult still to imagine that Aristotle was unaware of the fact'. To be sure, Driscoll acknowledges that there are also important differences between the *Cat.* and the *Tim.*, the most important among them being:

(1) the *Cat.* allows for species and genera, the *Tim.* does not;

(2) the *Cat.* uses the word *onta* in a way inimical to the fine points of the *Timaeus* ontology;

(3) in the *Cat.* there are many subjects of inherence (primary substances), in the *Tim.* there is only one;

(4) the Platonic ontology contains transcendent Forms, the *Cat.* of course does not.[60]

At this point Driscoll makes a curious move which, to my mind, very much weakens his specific point concerning the particular influence of the *Timaeus*. For he argues that of the four points just mentioned only the last one represents a typically Aristotelian change of position, the former three outlining issues as to which the *Cat.* differ from the *Tim.* while being in agreement with the *Sophist.* Subsequently he takes sides in the debate between Owen and Cherniss concerning the relative date of the *Tim.*,[61] arguing with Owen that the *Tim.* must be earlier than the *Soph.* Accordingly, he concludes that the *Cat.* should be regarded as 'an attempt by the young Aristotle to bring the inherence ontology of the *Timaeus* up to date by making allowance for the multiple subjects

60 As a related point we may note that the *Timaeus*' receptacle is absolutely characterless, whereas the Aristotelian primary substances are not. Accordingly, in later works Aristotle tends to identify the receptacle with matter (ὕλη; see my notes in the previous chapter, section 3.3), *not* with substances, and it seems to me that one has to have very strong reasons for supposing that at the time of writing the *Cat.* Aristotle endorsed a different interpretation of the receptacle passages of the *Tim.* (The mere fact, at any rate, that the *Cat.* does not positively contain the notion of matter seems inconclusive). As an additional difference between the receptacle and the Aristotelian primary substances we may note that the receptacle is represented as a *constituent* of phenomenal sensible bodies (on which see again my remarks in the previous chapter), whereas the primary substances are *themselves* phenomenal sensible bodies.

61 Cf. Owen (1953) and Cherniss (1957).

of pronominal reference acknowledged by Plato in the *Sophist*'.[62]

This, however, leaves little room for Driscoll's original claim that the *Tim.* is at the basis of Aristotle's view on primary substances as it appears in the *Cat.* Rather, if the *Cat. qua* ontology resembles the (later) *Soph.* rather than the (earlier) *Tim.*,[63] whereas at the same time it rejects the most fundamental aspect of Plato's (earlier and later) ontology, viz. the theory of Ideas, the most logical and economical—though perhaps also a rather trivial—explanation will be that Aristotle was confronting problems with which also Plato had had to cope. In that case the resemblances Driscoll makes so much of are to be explained by reference to the simple fact that Plato and Aristotle were both working within the same tradition—not only in the narrow sense of a 'school tradition' but also in the wider sense of 'problem tradition'.[64] For the rest, Aristotle obviously opted for an ontological framework which — though in some respects resembling Plato's later ontology [65]— constituted a fundamental break with Platonism.

With this in mind it is not at all reasonable to assume that Aristotle took Plato's cosmological dialogue as his starting point in the *Cat.*, only to attempt to 'bring the inherence ontology of the *Tim.* up to date'. In other words, to defend—in spite of the

---

[62] Driscoll (1979) 265.

[63] Even this thesis, by the way, needs qualification. For though it may be true that the *Soph.* contains a theory of 'subjects of pronominal discourse' (though De Rijk (1986) *passim,* has made a strong case for considering the *Soph.* as not concerned with predicative sentences but rather with the assignment of names) it obviously does not contain the view that there are at the ontological level stable subjects of inherence comparable to Aristotle's primary substances. The ontological counterparts of 'subjects of pronominal discourse' (or of 'things named') in the *Soph.* are rather bundles of immanent properties, or, if you prefer, bundles of immanent forms. See on this e.g. the pertinent remarks of Rosen (1983) 300. For a convenient survey of the connection between Plato's semantics and metaphysics in this respect as well as for further references see De Rijk (1986) 327 ff.

[64] Also the in itself interesting correspondence (pointed out by Driscoll (1979) 265) between Plato's sentence Θεαίτητος κάθηται—adduced at *Soph.* 263a as a sample of statements which may be true or false—and Aristotle's example at *Cat.* 4 a 24, εἰ ἀληθὴς εἴη ὁ λόγος *τὸ καθῆσθαί τινα,* ἀναστάντος αὐτοῦ ὁ αὐτὸς οὗτος λόγος ψευδὴς ἔσται, may best be explained by assuming that within the Academy '(x) is sitting' had become a stock example in the philosophical discussions concerning truth and falsehood of statements.

[65] Driscoll (1979) 260, is certainly overstating the resemblances between the *Cat.* and the *Soph.* when he has it that 'If ... Owen is correct in holding that the *Tim.* antedates the *Soph.* ... then the *Cat.* accurately reflects Plato's thought concerning the physical world as it evolved after the *Tim.*'

differences between the two treatises—the existence of any direct influence from the *Tim.* on the *Cat.* is plainly question-begging if it is not supported by unambiguous textual evidence. I am afraid that the additional 'purely verbal evidence' (the use of *dechesthai* in both treatises) adduced by Driscoll is in this respect far from conclusive, since—as a quick glance at a dictionary will show—the verb *dechesthai* is a very common, and by no means technical, term.

If, then, Driscoll's conclusions which constitute the *protasis* of Mendell's claim that 'if the *Timaeus* is a background to the *Categories'* theory of substance, then it is also a background to the *Categories'* theory of place',[66] are thus shown to be not sufficiently supported by the evidence, it appears that also the force of Mendell's claim as a whole is considerably weakened. Nevertheless we shall go on and see whether Mendell can offer any evidence independent of Driscoll's general thesis on Platonic influence on the *Categories.*

Mendell construes three possible ways in which the differences between the presentation of *topos* in the *Cat.* and in the *Tim.* can be explained: 'Aristotle could have been unaware of the conflict between his position in the *Physics* and his argument of the *Categories.* Alternatively he may have changed his theory of place, or he may be speaking loosely at *Phys.* Δ'.[67] Mendell himself opts for the second possibility and adds that 'it is a plausible hypothesis that the early Aristotelian notion of place has the *Timaeus* as its Platonic ancestor'.[68]

Now on closer scrutiny this position is no less circular than that which was defended by Driscoll. To begin with, the concept of *topos* which appears in the *Cat.* is, as we noticed, the concept of place or space as an independent three-dimensional extension (the (c) kind of space outlined in chapter (1)). Though, as we saw in the previous chapter, there are traces of such a concept of space in Plato's presentation of the receptacle, it was also noted that Plato was in this respect not fully consistent and that Aristotle—at least

---

66 Mendell (1987) 208.

67 Mendell (1987) 210. The third alternative was, as Mendell rightly points out, endorsed by King, whose interpretation was discussed in the previous section. Note, by the way, that the most obvious alternative, viz. that Aristotle was speaking loosely in the *Cat.*, is not considered!

68 Mendell (1987) 216.

at the time of writing his physical works—knew this. Even on Mendell's line of thought it has to be assumed that also at the time of writing the *Cat.* Aristotle implicitly criticized Plato's conception of the receptacle for being ambiguous because, as Mendell argues, in making use of the *Timaeus,* Aristotle was separating the 'material' and the 'spatial' aspects of the receptacle.[69]

But if Aristotle had his reasons for separating, so to speak, the two main features of the *Tim.*'s receptacle and assigning them to primary substances on the one hand and *topos* on the other—thus dividing up the receptacle between different categories[70]—this would only point once more to his endorsing an ontology which is significantly different from that expounded in the *Tim.* In that case it is odd to assume that a concept such as 'the locative features of the receptacle' was consciously taken over *independently of its original physical and ontological setting,* unless there are clear and unambiguous indications of direct borrowings from the *Tim.* in the relevant passages of the *Cat.* In this respect the resemblances between the *Cat.* and the *Tim.* are limited to the mere fact that both use the expression *katechein topon* ('to occupy a place').[71] This, however, can be much more conveniently explained by assuming that both Plato and Aristotle were referring to one of the common ways of using the word *topos.*

In short, Mendell's hypothesis meets with the following objections. As we noted earlier in this chapter, the problems with which the treatment of *topos* in the *Cat.* confronts us do not suggest that Aristotle gave the problem of place and space much thought in writing this treatise. Thus it would not seem that he consciously took over any specific theory, e.g. Plato's, unless it can be shown that also the problems inherent in his treatment were taken over from Plato. This is not so. They are specifically connected with the Aristotelian substance ontology. Moreover, the problems connected with Plato's treatment of the receptacle in the *Tim.*, viz. the fact that it is made to fulfil both the function of matter

---

69 Mendell (1987) 218: '... it is not implausible that place should become, as it were, a watered-down receptacle. It is what is left over after we have introduced prime substance'.

70 Cf. Mendell (1987) 226: 'the locative features of *Tim.'s* place (sic!) become a property of the substantive feature of *Tim.'* s place, now prime substance'.

71 Cf. Mendell (1987) 216.

and that of space, have no counterpart in the *Cat.* In other words, the problems involved in the *topos* accounts of the *Tim.* on the one hand, and of the *Cat.* on the other, are of a quite different nature, which does not suggest any direct ancestry-and-descent relation.

So we had better conclude—in agreement with what was said in chapter 1 concerning the way in which Greek philosophies of space relied on ordinary language—that both Plato and Aristotle had recourse to everyday conceptions of space and place. In the case of Plato's *Timaeus* the metaphysical requirements ('the metaphysics of the image') seem to have furthered the ambiguities sketched in the previous chapter. In the case of the *Categories* ambiguities of a different kind arose, probably because the approach of name-giving in the end blinded Aristotle to the ontological and physical problems involved in the everyday concepts he was using.

4.3.4 In this section we inquired whether and in what sense we may speak of a development of Aristotle's concept of *topos* between the time of writing the *Cat.* and that of *Phys.* Δ. Though we did not go into the details of the problems concerning the relative chronology of the works of the *Corpus Aristotelicum,* it was concluded that, according to the relative chronology which is now commonly accepted, there is, in principle, room for such a development.

As we saw, however, a development in the strong sense, in this case a development from a consciously held Platonic position towards the clear-cut Aristotelian position of *Phys.* Δ cannot be defended on the basis of the texts. The—superficial—resemblances between the presentation of place or space in the *Tim.* and in the *Cat.*, are rather to be accounted for by reference to the fact that both treatises base themselves on quite common notions of *topos* (and *chôra*).

If we want to account for the differences between the presentation of *topos* in the *Cat.* and in *Phys.* Δ in terms of a *development* of Aristotle's thought, it appears that of the different kinds of development outlined in 4.3.1 it is the third one which best squares with the evidence. In this case, Aristotle's development may be described as a growing awareness of the problems linked up with the common sense notions of space and place. The *Cat.* account of *topos,* as we saw, does not seem to presuppose much conscious reflection on this issue on Aristotle's part. Rather, his inquiry seems to have been led by common ways of thinking and speaking (name-giving). In *Phys.* Δ, on the other hand, Aristotle shows himself very much aware of the fact that common usage of spatial terms does not involve a consistent commitment to one specific concept of space or place, but that it tends to oscillate between various such concepts. For that reason, he argues, it is easily confusing. In fact, as we shall see in the next section, his whole method in the *Phys.* may be described as a sifting out and refining of the initially confused data of common experience and common parlance.

Finally, I would like to draw attention to the fact that the non-technical character of the *Categories* in this and other respects was already pointed out by some of the ancient and mediaeval commentators. I have already referred to Simplicius' almost Jaegerish opinion that in the *Cat.* Aristotle had 'not yet' investigated the nature of *topos*,[72] and that such an investigation belonged to physics proper. As we saw, Simplicius was nonetheless of the opinion that the two treatises reflected basically one and the same conception of place on Aristotle's part. A position more like the one here defended was taken by Averroes, who in his *Middle Commentary on Porphyry's Isagoge and on Aristotle's Categoriae*, makes the following point about Aristotle's treatment of the category of relativity (*pros ti*):

> Aristotle also explains that it is easy to raise problems on this matter but difficult to solve them, the reason being that his examination of the subject was conducted from the point of view of commonly accepted notions.[73]

and concerning the problem whether the object of sensation precedes the existence of the sensation:

> Aristotle deferred the solution to this difficulty to another occasion since potential existence is not a commonly accepted notion, and he is discussing the present subject only from the viewpoint of what is commonly accepted.[74]

We may wind up the present section with a quotation from the late mediaeval commentator Joannes Buridanus in which the non-technical character of the *Cat.*'s discussion of *topos* is under-

---

[72] Simp. *In Cat.* 125, 28 and 133, 35; see also above n. 39 and 42.
[73] Davidson (1969) 5.
[74] Davidson (1969) 56.

lined. In his *Questiones* on the Aristotelian *Physics* Buridanus argues—apparently following Averroes, 'the Commentator'—that the differences between the account of *topos* in the *Phys.* and that in the *Cat.* is to be explained as follows:

> ... it seems that Aristotle in the book about predicaments spoke about place in accordance with that concept which is quite common. For there he says that the parts of place, which receive the respective parts of the body, come together at the same boundary at which also the parts of the body come together and for this reason the Commentator, in the fifth book of his commentary on the *Metaphysics* rightly argues that Aristotle often spoke in the predicaments in accordance with widespread usage (*secundum famositatem*), not in accordance with the real determination about those things of which proper investigation belonged to other parts of philosophy.[75]

4.4 *Aristotle's dialectical method and the topos discussion of* Phys. *Δ*

4.4.1 We shall now move to Aristotle's discussion of *topos* in the *Physics.* I shall try to show in the present section that the most fruitful way to approach this text is to focus in particular on Aristotle's dialectical method. In this respect our discussion will link up with a trend in contemporary Aristotelian research which has been set by studies of scholars like Le Blond and Weil,[76] but probably most of all by G. E. L. Owen's famous paper '*Tithenai ta Phainomena*'.[77] The main characteristic of this new trend may be described as a fresh awareness of the importance of dialectic as part and parcel of the scientific and philosophical method in most of the preserved Aristotelian treatises.[78] This new picture of Aristotle's method inevitably entailed abandoning some

---

75 J. Buridanus, *Questiones super octo Physicorum libros Aristotelis,* printed in J. Buridanus, *Kommentar zur Aristotelischen Physik* (Frankfurt a/Main 1964) (facsimile of the Paris edition of 1509) fol. lxxiii r, col. 2: '... et videtur quod secundum istam imaginationem que est satis vulgaris locutus est Aristoteles in libro predicamentorum de loco ubi dicit quod loci particule que obtinent singulas particulas corporis ad eundem terminum communem copulant ad quem et corporis particulae. Et ideo bene dicit Commentator quinto Metaphisice quod Aristoteles sepe in predicamentis locutus est secundum famositatem non secundum veram determinationem, scilicet de illis quorum propria perscrutatio pertinebat as alias partes philosophie'.

76 Le Blond (1939) 9 ff.; Weil (1975) 88–113; see also Evans (1977) 77–85.

77 Owen (1961).

78 Dissenting views, however, have been taken by Bolton (1991), who defends the traditional 'Baconian' picture of Aristotle's method in physics, and Irwin (1988), on whom see below, n. 82.

of the traditional views on the status of Aristotelian deductive syllogistic on the one hand, and on the role of empiricism and induction (what some have called the 'Baconian picture' of Aristotelian science) on the other.

In this section I shall briefly outline the role of dialectic in the *Corpus Aristotelicum* in general and in the discussion of *topos* in *Phys.* Δ in particular. As luck would have it, the latter discussion was given an important exemplary role in Owen's epoch-making study and kept that role in most of the subsequent scholarly discussions of the topic. For this reason, the present treatment of the issue cuts both ways.

For one thing, a correct assessment of Aristotle's dialectical method in general may further our understanding of *Phys.* Δ. It will in particular show how Aristotle conceived of the relationship between the concept of *topos* which he finally accepted and others (e.g. place as a three-dimensional independent extension) which he rejected. Furthermore, it will allow us to throw some light on the at first sight rather surprizing reappearance of the concept of place as a three-dimensional extension in some passages in the physical writings which are both chronologically and materially closer to *Phys.* Δ than the *Cat.*

For another thing, a closer investigation of Aristotle's dialectical method actually at work in the inquiry into the nature of *topos* may provide us with information which contributes to the clearing up of some of the controversial issues concerning the role of this method in general.

The procedure will be as follows. First, in 4.4.2, I shall briefly sketch what is by now the commonly accepted picture of the role of dialectic in Aristotle's philosophy of science—if that is what we may call it—in general. I shall deal in particular with Owen's view that dialectic, at least the way it works in the *Physics*, is primarily concerned with 'linguistic puzzles'. In 4.4.3 I shall present a critique of Owen's position. Although my criticisms do not really affect his main point, they will allow us to rephrase his interpretation of the role of dialectic in the *Corpus Aristotelicum* in a manner which seems to be more coherent and stays closer to Aristotle's own text. In the meantime it will be shown that the interpretation which Morsink[79] put forward as an alternative to

---

[79] Morsink (1982) 102–128.

Owen's has to be rejected. Getting these issues clear in our minds will prove to be a useful preliminary to the discussion of Aristotle's dialectical method in *Phys.* Δ which will be found in 4.4.4 and 4.4.5. The whole section will be concluded in 4.4.6.

4.4.2 Two presuppositions used to govern the picture which scholars have traditionally been drawing of Aristotelian science: (1) the assumption that science is basically a deductive system of *apodeixeis*, and (2) the assumption that the first principles, the *axiomata* of these *apodeixeis* are arrived at by a process of empirical induction with the help of *nous.*

In itself it is difficult enough to try to square, without further qualification, the 'Baconian' with the deductive aspect of Aristotle's philosophy of science. However, even more problems seem to arise in so far as, concerning the actual *application* of apodeictic syllogistic, Aristotle's practice seems to be hardly in line with his preaching. Even a casual acquaintance with the physical writings will show that deductive apodeictic syllogisms play no role there at all. Furthermore, the stress laid on the predominant role of 'Baconian' induction in the process of establishing the first principles (ἀρχαί) of science, primarily on the basis of *APost.* II, 19, was almost necessarily accompanied by a lack of attention to the remarks made by Aristotle in *Top.* I, 2 about the role of dialectic in that very same process. In the present century a number of studies have appeared which had the merit of more or less redressing the balance by qualifying the importance of deductive syllogistic for the Aristotelian scientific and philosophical practice and by trying to square with each other the passages concerning the foundation of the principles (ἀρχαί) as they are found in the *Analytics* and the *Topics* respectively.

Weil[80] and Barnes[81] have pointed out that we should differentiate between what might be called a *logic of acquisition* on the one hand, and a *logic of presentation* on the other. Apodeictic syllogisms are primarily meant to play a role in the context of the

---

[80] Weil (1975) 98 ff., esp. 110.

[81] Barnes, 'Aristotle's Theory of Demonstration', *Phronesis* 14 (1969) 123–152; reprinted in a revised form in Barnes/ Schofield/ Sorabji (1975–79) I, 65–87; a more or less similar view has been defended by Kullmann (1974). For some qualifications of Kullmann's (and Barnes') views see the review of Kullmann's book by J. Mansfeld, *Bibliotheca Orientalis* 38, 220–225.

latter. Thus the *Posterior Analytics* is not to be read as an essay in scientific methodology, but rather as an attempt to characterize the notion of an axiomatized deductive science as a model of how teachers should present and impart knowledge. On the other hand, scholars like Le Blond and Owen have stressed the importance of the role which dialectic was made to play in Aristotle's theory of the discovery of the first principles of science.[82] They pointed out, moreover, that the so-called doxographical surveys by which Aristotle prefaced many of his philosophical discussions should also be viewed against this background. It is especially Owen's study which I shall here take as a starting point because, for one thing, it has by now become more or less classic and because, for another, the *topos* discussion of *Phys.* Δ, which primarily interests us in the present connection, provided one of Owen's main examples.

Owen based his interpretation of Aristotle's philosophy of science on an observation on the ambiguity of the term *phainomenon*, both in Aristotle's work and elsewhere, e.g. in the philosophy of Protagoras. Thus, e.g., in a much quoted passage in *APr.* 46 a 17–22 Aristotle says that it falls to experience (*empeiria*) to provide the principles of any science. As an example he adduces astronomy, where it was only when the *phainomena* were sufficiently grasped that the relevant proofs could be discovered.[83] Here, as well as in a number of other passages,[84] it is implied that the *phainomena* in question are empirical observations. On the other hand, particularly in the ethical writings Aristotle seems to use the term *phainomenon* in a different sense: not so much the 'observed facts'

---

[82] For some qualifications of the 'new orthodoxy' as defended by Owen, see now Nussbaum (1986a) 240–263; see also below, n. 114. A radically different assessment of Aristotle's dialectical method in the *Physics* has been given by Irwin (1988), who argues that the 'weak dialectic' to be found in the *Physics* was compatible only with an un-Aristotelian coherence theory of truth, that Aristotle realized this, and that it is for this reason that in the *Metaphysics* (in particular in the fourth book) he swopped this 'weak dialectic' for a 'strong dialectic' which took the form of first philosophy. However, this distinction between two kinds of dialectic finds no support whatsoever in Aristotle's texts and Irwin's approach appears to me to be fatally invalidated by a post-Kantian analytical bias which expresses itself in a number of anachronistic conceptions being read into Aristotle's text. For a pertinent critique of Irwin's book see now Wardy (1991).

[83] *APr.* 46 a 20: ληφθέντων γὰρ ἱκανῶς τῶν φαινομένων οὕτως εὑρέθησαν αἱ ἀστρολογικαὶ ἀποδείξεις.

[84] *PA* 639 b 5–10 and 640 a 13–15; *Cael.* 306 a 5–17.

but the *endoxa*, the common conceptions on a particular subject, are what is at stake here. In order to show the methodological relevance of this kind of *phainomena* Owen quotes a passage by which Aristotle prefaces his discussion of incontinence in *EN* 1145 b 2–6:

> We must, as in all other cases, set the *phainomena* before us and, after first discussing the difficulties, go on to prove, if possible, the truth of all the reputable opinions (*endoxa*) about these affections, or, failing this, of the greater number and the most authoritative; for if we both resolve the difficulties (*aporiai*) and leave the reputable opinions undisturbed, we shall have proved the case sufficiently.[85]

According to Owen it is also in this particular sense of the word *phainomena* that Aristotle argued that dialectical arguments can be said to start from the *phainomena*.[86] Next, he argues that this ambiguity in the use of the term *phainomenon* is accompanied by a corresponding distinction in the use of various connected expressions. Thus, induction (*epagôgê*) does not necessarily start from the data of perception. It may also begin from *endoxa* and in this form too can it be used to find the principles of the sciences.[87] Similarly the *aporiai* which constitute, so to speak, the hard core of any scientific or philosophical discussion, may be what Owen calls 'questions of empirical fact' or they may be what he calls 'logical or philosophical puzzles generated, as such puzzles have been at all times, by exploiting some of the things commonly said'.[88] In so far as the *Physics*, being basically concerned with outlining the principles (ἀρχαί) of Aristotelian natural philosophy, sets out from a survey of the *phainomena*, it is, according to Owen, plainly this second, non-empirical, sense of the word that is appropriate.

Thus, he concludes, 'the *Physics* ranks itself not with physics, in our sense of the word, but with philosophy. Its data are for the most part the materials not of natural history, but of dialectic, and its problems are accordingly not problems of empirical fact but

---

85 *EN* 1145 b 2–6: Δεῖ δ' ὥσπερ ἐπὶ τῶν ἄλλων, τιθέντας τὰ φαινόμενα καὶ πρῶτον διαπορήσαντας οὕτω δεικνύναι μάλιστα μὲν πάντα τὰ ἔνδοξα περὶ ταῦτα τὰ πάθη, εἰ δὲ μή, τὰ πλεῖστα καὶ κυριώτατα (revised Oxford translation).

86 *APr.* 24 b 10–12; *Top.* 159 b 17–23.

87 *Top.* 101 a 36–b 4.

88 Owen (1961) 87.

conceptual puzzles'.[89] Quite in line with these conclusions, Owen next, in the second part of his paper, sets out to show that it was Plato's *Parmenides* which as a 'set of logical paradoxes'[90] supplied Aristotle with most of the central problems discussed in the *Physics*. The conclusion is that the impulse throughout the *Physics* is logical and Owen adds that 'it makes for better understanding to recall that in Aristotle's classification of the sciences the discussion of time and movement in the *Parmenides* are also physics'.[91]

4.4.3 Some further reflections on Owen's position—and on some of the objections that have been raised against it—appear to be in order. Without playing down the significance of Owen's study, I want to argue that some of his contentions are not supported or even contradicted by Aristotle's text, and that, hence, the general picture of Aristotelian dialectic which emerges from Owen's study is in need of revision at some important points. Though these discussions may at first blush seem to take us away from the proper subject of our study, they will nevertheless turn out to be relevant to it in that they allow us, in general, to put the arguments of *Phys*. Δ in perspective and, in particular, to clarify the nature of the concepts of place which Aristotle there discusses and rejects. This, in its turn, will allow us to substantiate some of our claims concerning the nature of the philosophical discussion of place and space in Greek thought in general.

My contention is that Owen's account needs to be qualified at a number of points. First, there is the issue of how dialectic stands to deductive syllogistic reasoning. The conclusions that Aristotle's 'logic of presentation' is not the same as his 'logic of acquisition' and that the role of 'logic of acquisition' was played by dialectic seem inescapable indeed. Aristotle's actual practice in the physical writings definitely forces us into this direction. Still it may be observed that it is not as easy as seems to be suggested by Owen and others to connect Aristotle's 'scientific practice' with the mandate given to dialectic in the *Topics*. For sure, among the applications of dialectic which are enumerated at *Top*. I, 2 (101 a 25 ff.) the possibility of reaching the principles (*archai*) of the sciences

[89] Owen (1961) 88.
[90] Owen (1961) 88.
[91] Owen (1961) 103.

by way of dialectically scrutinizing the *endoxa* at issue is explicitly mentioned.[92] It should be realized, however, that only a few pages further on Aristotle argues that there are two kinds of dialectical argument (*logismos*), viz. induction (*epagôgê*) and syllogistic reasoning (*syllogismos*). Now it is at this point that problems arise, for the actual 'dialectical' arguments in e.g. the *Phys.* are usually such as not to fall within either of these two official categories of dialectical argument.[93] This suggests that Aristotle's 'logic of acquisition' was not only looser than his 'logic of presentation' in that it took its starting point from the merely probable (*endoxon*), but also because its *formal* requirements were less strict. Nevertheless, we may well follow Owen and take it as established that Aristotle's usual method in the school writings was a kind of rather informal dialectical discussion. In many cases the *endoxa*, i.e. the views held by the many or by the happy few, by ordinary people or by distinguished philosophers, constituted the raw material of his investigations.

---

[92] In this connection it may be added that the exposition concerning dialectical problems and dialectical theses which is to be found at *Top.* 104 b 1–105 a 10 very well fits the Aristotelian scientific practice in writings like the *Phys.*

[93] The difficulty may be illustrated by the following two quotations. Ross (1923) 57, argues: 'The statement that the first principles of science are approached by way of dialectic is nowhere brought into relation with the other statement that they are approached by induction [viz. in *APost.* II, 19, K. A.]; but we must remember that induction is one of the two modes of dialectic'. To this Morsink (1982) 107, objects that 'clearly the mandate spelled out in chapter two of the *Topics* refers to the reasoning mode of dialectic, i.e. to that mode which starts from *endoxa.* The mandate is not given to the inductive mode which Aristotle does not mention till ten chapters later' [a statement which is somewhat misleading since the chapters at issue are indeed very small: *epagôgê* is introduced as early as *Top.* 105 a 10 ff.]. Now the only way to square Ross' suggestion with Aristotle's actual scientific and philosophical practice would be to assume that the term *epagôgê* refers to a variety of more or less informal methods. It is true, there are passages where *epagôgê* seems to mean nothing more than 'a sampling of cases' (cf. the use of ἐπακτικῶς at *Phys.* Δ 210 b 8) and it has been pointed out more than once that *epagôgê* may be used in different ways (cf. e.g. K. von Fritz, *Die ἐπαγωγή bei Aristoteles*, SBAW 1964. 5 (München 1964) 22–25). But I have not been able to find any passage where Aristotle himself refers to his own practice of scrutinizing the *endoxa* as a kind of *epagôgê.* Alternatively, Morsink's suggestion seems to ignore the fact that at *Top.* 105 a 10 ff. Aristotle does not speak of a such a vague thing as 'the reasoning mode of dialectic ... which starts from the *endoxa*', but explicitly refers to *syllogistic* dialectical reasoning. This description, however, clearly seems to be too specific to fit the dialectical passages in the school writings.

This having been said, we shall have to face an objection which is of a more fundamental character since it involves the general picture drawn by Owen of the nature of Aristotle's starting points in his dialectical inquiry, in particular his stress on their logical or *a priori* character. As we saw, he concludes that 'the impulse throughout the work [i.e. the *Phys.*] is logical and the restriction of its subject-matter to movable bodies and their characteristics does not entail a radical difference of method from other logical inquiries'. Time and again Owen makes clear that he views the difference between this perspective and the more empirical perspective which is, as he argues, usually 'expressly associated with natural science (*phusikê*)',[94] as basically a difference between the logical and the empirical or between the analytic and the synthetic.

However, this interpretation seems to run into difficulties at two important points. In the first place it was already argued by Morsink[95] that it is not clear which particular version of the analytic/synthetic distinction Owen has in mind when he speaks of 'logical puzzles' as opposed to 'questions of empirical fact'. In the second place, on any conception of the analytic/synthetic distinction, Owen seems to be mistaken in implying that Aristotle himself regarded all arguments which take their start in *endoxa* as *logical* arguments.

To begin with the former point, it actually proves difficult to determine the nature of the concept of analyticity which seems to underlie Owen's distinction between the two kinds of *phainomena*. He appears to oscillate between two different conceptions of analyticity: the 'traditional' one and a more 'pragmatic' one. On the one hand Owen seems to go a long way in the direction of the 'pragmatic' version of the analytic/synthetic distinction when he argues that 'the analyses of the *Physics* proper are preliminary to other *more empirical* inquiries and consequently *must be justified* in the last resort by their success in *making sense of the observations* to which they are applied' [my italics].[96] As was already suggested by Morsink,[97] the italicized phrases seem to point to a view according to which the difference between the analytic and the

[94] Owen (1961) 84.
[95] Morsink (1982) 110–111.
[96] Owen (1961) 91.
[97] Morsink (1982) 110.

synthetic is only gradual. On such a view there is, so to speak, also an empirical component in the *a priori* schemes governing science. They may be changed as soon as they no longer fit our experience. Such a conception of what might be called the relativity of the boundary between the merely 'conceptual' and the 'empirical' components of scientific theories was put forward by pragmatist philosophers like C. I. Lewis.[98] More recently the whole discussion seems to have been revived by Quine who, in his 'Two Dogmas', argued:

> The totality of our so-called knowledge or beliefs, from the most casual matters of geography and history to the profoundest laws of atomic physics or even of pure mathematics and logic, is a man-made fabric which impinges on experience only along the edges. Or, to change the figure, total science is like a field of force whose boundary conditions are experience. A conflict with experience at the periphery occasions readjustments in the interior of the field.[99]

On this line of thought no statement is immune to revision: 'revision even of the law of the excluded middle has been proposed as a means of simplifying quantum mechanics'.[100] As a matter of fact, Quine even opted for taking leave of the analytic/synthetic distinction altogether and, in general, a typical feature of the pragmatist approach to the problem of analyticity has been to stress that drawing the borderline between the analytic and the synthetic is merely a matter of perspective.

It is precisely for this reason that on closer view Owen's version of the analytic/synthetic distinction does not seem to be the pragmatic one. For he seems to assume that a rigid boundary exists between the 'conceptual puzzles' and the 'more empirical problems'. In this respect he rather seems to have the more traditional version of the analytic/synthetic distinction in mind according to which synthetic statements hold contingently on experience whereas analytic statements hold, in virtue of their meaning, come what may (i.e. in any conceivable physical system whatever). It was the 'traditional' conception of the analytic/synthetic

98 C. I. Lewis, *Mind and the World Order: Outline of a Theory of Knowledge* (New York 1929); see the quotations from this work in Morsink (1982) 126, n. 49.

99 Quine (1953) 42.

100 Quine (1953) 43.

distinction which was upheld by the 'modern empiricists' who were so forcefully attacked in Quine's now classic 'Two Dogmas of Empiricism'.[101] If we may cite Quine's words, these empiricists were committed to

> ... the feeling that the truth of a statement is somehow analyzable into a linguistic component and a factual component. The factual component must, if we are empiricists, boil down to a range of confirmatory experiences.[102]

The use of the words 'boil down' in this connection is significant. For they refer to what Quine called the dogma of 'reductionism', i.e. the dogma that all 'synthetic' discourse could in the end be *reduced* to a kind of sense-datum language, a language describing direct experience.

On closer scrutiny, however, this does not seem to be the kind of analytic/synthetic distinction Owen has in mind either. For he seems to regard *all* non-logical *phainomena* as straight *phainomena* of perception.[103] As a corollary, all problems which do not directly bear on sense-perception are *logical* problems. Now Owen acknowledges that the *phainomena* which constitute the starting point of Aristotle's discussion of incontinence in the *EN* or of his discussion of *topos* in *Phys.* Δ are not simply empirical observations. Thus he concludes that these discussions are actually logical discussions.

I think the conclusion is inescapable that Owen's account, for all its merits in other respects, is flawed by a rather confused application of the labels 'analytic' and 'synthetic'. In addition, by restricting the class of 'physical' arguments to those that take *direct* recourse to perception, he might be said to be committed to a somewhat naïve account of the role of the 'strictly empirical' as a component in the formation of physical theories. But even if we are prepared, for the sake of the argument, to accept his distinction between physical arguments—as strictly empirical—on the one

---

101 Quine (1953) 20–46.

102 Quine (1953) 41.

103 Cf. Owen (1961) 90: '...it is not a peculiarity of the phainomena in the second [i.e. the 'logical'] sense that they may fail to stand up to examination; for so may the phainomena of perception ...'; cf. also *op. cit.* 84–85, where *empirical observations* (the one kind of *phainomena*) are contrasted with conceptual analyses (the other kind of *phainomena*) and the contrast drawn—*op. cit.* 87—between 'data of *observation*' and logical or philosophical puzzles' [my italics].

hand, and logical arguments—as all arguments which are not strictly empirical—on the other, we are not allowed to assume that Aristotle made the very same distinction. Now this, apparently, is precisely what Owen does. On the one hand he suggests that Aristotle's own criteria for differentiating between physical (*phusikos*) and logical (*logikos*) arguments were such that all arguments starting from *endoxa* turned out to be *logical* arguments (in Owen's sense of the word, i.e. arguments taking no recourse to experience at all). On the other hand Owen himself has to admit that Aristotle explicitly refers to a number of arguments adduced in the *Phys.* as *physical* arguments, whereas they are nevertheless based on *endoxa*. From this he does not draw the rather obvious conclusion that Aristotle was apparently committed to a *different* conception of the distinction between the physical and the logical, but he infers, instead, that Aristotle did not apply the distinction in a very coherent manner. The problem seems to be, according to Owen, that what Aristotle would call 'physical' arguments partly rely 'on quite general definitions' and on 'quite general premisses'[104] and do not all depend, as they apparently should, on what Owen calls 'special empirical claims'.[105]

At this point, however, we are seriously in danger of losing contact with Aristotle's own thought. For it is one thing to use 'our'—essentially post-Kantian—labels 'analytic' and 'synthetic' (or, for that matter, 'a priori' as opposed to 'physical') in describing Aristotle's philosophy, but it is quite another to assume, as a matter of course, that Aristotle's own terms (*logikos* and *phusikos*) had the same meanings as these 'modern' terms. We may here recall our warnings, in chapter 1, against what I called 'essentialist' presuppositions. As in chapter 1, we may here also advocate a closer contextualist investigation—in this case an investigation of Aristotle's own use of the terms—as a methodologically preferable alternative approach.

---

104 Owen (1961) 103.

105 As an example of an argument which does depend on special empirical claims Owen adduces 'the unfortunate hypothesis of natural places' ((1961) 103). At the same time he adds (*ibid.* n. 37) that also this hypothesis is introduced by Aristotle as an *endoxon*. Since on Owen's line of thought *endoxa* are exclusively logical arguments, the fact that Aristotle treats the 'physical' hypothesis of natural places as an *endoxon* is supposed to show that Aristotle's distinction between logical and physical arguments was not altogether waterproof.

A survey of the way in which Aristotle contrasts physical and logical, or general (*katholou*), problems and arguments shows indeed that to his mind the distinction did not boil down to the contrast between 'special empirical' arguments on the one and more general or theoretical arguments on the other hand, but rather to a contrast between arguments (either directly empirical or of a more theoretical character) which are, so to speak, embedded in a theory about the physical world, and, on the other hand, those which are of a purely abstract character, taking no recourse to the world as it actually appears to us or even flatly contradicting common appearances.[106] Among the latter kind he ranked the arguments of the philosophers of the Eleatic tradition.

As an example we may take the 'proofs' put forward at *Phys.* Γ 204 b 4 ff. in order to show that there can be no infinite physical body. First, at 204 b 4–10, the 'logical' argument (λογικῶς σκοπουμένοις) is given. The idea of an infinite body is said to be at odds with the very definition of body as 'that which is limited by a surface'. Next, at 204 b 10 ff., follow the physical arguments (φυσικῶς θεωροῦσιν). These arguments are derived from our knowledge of the physical world as a whole. Thus, e.g., Aristotle

[106] Note that the term λογικός is used by Aristotle in several ways which are not as closely connected as seems to be suggested by Ross (1936) 540. In the first place it may be used as a synonym for διαλεκτικός. Cf. *Top.* 162 b 27: εἰ μὲν γὰρ ἐκ ψευδῶν ἐνδόξων δέ, λογικός [scil.: ὁ λόγος]. Second, it may, when it is opposed to φυσικός, mean 'logical' in the sense outlined here in the text (i.e. 'merely theoretical', 'having no recourse to experience') as in *Phys.* Γ 204 b 4 ff. (on which see my remarks in the text) and *GA* 747 b 28 (on which see the next note). Finally it may just mean 'more general' i.e. 'not applying to some specific cases', as e.g. at *APost.* 88 a 19 and 30, where the statement that syllogisms cannot all have the same principles is argued first by an appeal to quite general considerations (88 a 19: λογικῶς θεωροῦσιν) and subsequently by an appeal to a number of individual cases (88 a 30: ἐκ τῶν κειμένων). Simp. *In Ph.* 440, 21 seems to have recognized the same three kinds of 'logical' arguments in his commentary on *Phys.* Γ 202 a 21, where Aristotle speaks of an ἀπορίαν λογικήν: 'λογικὴν [scil.... ἀπορίαν] καλεῖ, ἢ ὡς ἐξ ἐνδόξων προιοῦσαν ἢ ὡς ἐν λόγῳ μόνῳ τὸ πιθανὸν ἔχουσαν καὶ οὐκ ἀπὸ τῶν πραγμάτων βεβαιουμένην (οὕτως γὰρ λόγοι λέγονται οἱ τοῦ Ζήνωνος οἱ πιθανῶς τὴν κίνησιν ἀναιροῦντες), ἢ λογικὴν λέγει τὴν κοινοτέραν καὶ οὐ προσεχῆ οὐδὲ ἰδίαν τοῦ προκειμένου οὐδὲ ἐξ οἰκείων ἀρχῶν'. The first of these possible uses of the term is of no concern to us here. As to the latter senses, the basic meaning is probably in both cases 'theoretical'. When opposed to 'physical' this may mean 'logical' (in the sense of 'not connected with the πράγματα'), when opposed to 'specific' it may mean 'general' (see also *Top.* 105 b 19–26 where the λογικαὶ προτάσεις are opposed to both φυσικαί and ἠθικαί [scil.: προτάσεις]).

argues that an infinite body cannot be a compound, for we know that the elements (στοιχεῖα) have to be more than one in number and that they balance each other, i.e. that the elementary changes always take place according to the same ratio. If any of them were infinite the ordered nature of the cosmic cycle, which is a central feature of the physical world as it is experienced by us, and hence of any physical theory, would be destroyed. Neither can it be simple. For in that case it would either have to be one of the known elements, or something in addition to them. In the former case our theorem would conflict with another basic theorem of our physical system, viz. that things in the phenomenal world always change from one opposite to another and that it is never observed that, as Heraclitus thought might happen, all things become one (205 a 1 ff.). In the other case our theorem would be no less at odds with physical experience, since, in the sublunar sphere, we do not know of an element in addition to fire, air, water and earth (204 b 23 ff.).[107]

---

107 Cf. also *Cael.* 280 a 30 where the discussion of the questions whether something generated can remain imperishable and whether something ungenerated can perish is concluded with the statement that as yet the problem has been dealt with in physical terms, i.e. with arguments concerning the actual nature of the world (φυσικῶς μὲν περὶ τοῦ οὐρανοῦ μόνον εἴρηται). Aristotle adds that more general arguments are to follow (καθόλου δὲ περὶ ἅπαντος σκεψαμένοις ἔσται καὶ περὶ τούτου δῆλον). What then follows is a quite general discussion of the concepts at issue (280 b 1: διαιρετέον πῶς ἀγένητα καὶ γενητά φαμεν καὶ φθαρτὰ καὶ ἄφθαρτα). See also 283 b 17 and Le Blond (1939) 204. Owen (1961) 102, n. 34, in addition refers to *Phys.* Θ 264 a 7–9 as a passage where Aristotle differentiates between physical and logical arguments (see also Ross *ad loc.* for a similar view) and argues that this passage shows that the distinction between physical and logical arguments is not clear-cut, the logical arguments being 'hardly ... marked by their generality ... nor the physical by their reliance on the special theorems of physics'. However, the text at issue contains no hint that Aristotle is thinking of some of the arguments as *physical*—as contrasted to *logical*—arguments. The arguments all seem to be of a logical nature, the only difference being that some are more specifically pertinent to the problem at issue than others. That is why they are called more οἰκεῖοι. At 261 b 27 the problem has first been stated in the form of the thesis that there is a uniform eternal motion, viz. rotation (*i.c.* the rotation of the heavenly spheres). The arguments adduced to establish this—and to establish at the same time that no rectilinear movement can be eternal—are at 264 a 7 referred to as οἰκείοις λόγοις (i.e., presumably, arguments which are specific for the subject). Aristotle adds that he will next provide 'logical arguments', the first of which turns out to be an argument applying to all kinds of change (264 a 14 ὁμοίως δὲ καὶ ἐπὶ τῶν ἄλλων) of one opposite into the other, whereas the second is explicitly introduced as a consideration concerning change in general (264 a 21: καθόλου μᾶλλον περὶ πάσης κινήσεως). The next

In general, Aristotle seems to have accorded a higher status to 'physical' arguments and to have regarded them as really appropriate (*oikeioi*) for the physicist.[108] Still, also logical problems do play a role in physics. In this connection we may adduce an interesting passage on the Eleatics from the first book of the *Physics*. After having argued, at 184 b 26 ff., that the questions raised by the Eleatics do not really belong to physics, and after having added, at 185 a 13, that we must take our starting point from the datum that the things of nature move—either all of them, or at least some of them—he motivates his brief examination of the Eleatic position in the following words:

> And yet since, although their theories do not concern nature, they nevertheless sometimes raise problems which are relevant to the study of nature (φυσικάς ἀπορίας), it will perhaps be apposite to examine them briefly. For this inquiry has got a philosophical interest.[109]

The upshot of this passage is clear. Though strictly speaking the theorems defended by the philosophers of the Eleatic tradition do not belong to the realm of physics, which after all deals with the phenomenal world *qua* moving, they are nevertheless to be taken seriously because they raise problems with which also the physicist will have to cope. Since among the *aporiai* raised by the Eleatic school the problem of the intelligibility of plurality, motion and

---

argument, 264 b 2 ff., focusing on change from not-white to white, is again introduced as more specifically appropriate to what was said (μᾶλλον οἰκεῖος τῶν εἰρημένων; not to be translated as 'more fitting *than* what was said', as Wagner, *Physikvorlesung*, 259, seems to suggest: the argument is connected with the problems raised in connection with this kind of qualitative change at 263 b 15 ff.; for this use of οἰκεῖος with a genitive, with examples from Aristotle, see LSJ, *s.v.* iv, 2a). There is, in other words, no reason to assume, with Owen, that the term οἰκεῖος is here to be taken as a synonym of the term φυσικός nor, accordingly, that it is the physical/logical distinction that is at issue here. The reason why people have nevertheless been led to think so, is probably the fact that at *GC* 316 a 13 Aristotle speaks of Democritus having used οἰκείοις καῖ φυσικοῖς λόγοις. There, indeed, the suggestion seems to be that Democritus' arguments are more appropriate (οἰκείοι) because they are physical arguments (cf. also *GA* 747 b 28 where the logical argument is said to be ὅσῳ καθόλου μᾶλλον, πορρωτέρω τῶν οἰκείων ἐστὶν ἀρχῶν). However, it does not follow that in Aristotle 'more appropriate' *always* means 'more physical'. In *Phys.* Θ 264 a 7–9 the suggestion seems to be rather that the arguments are more appropriate because they are more *specific.*

[108] Cf. *GC* 316 a 5–14.

[109] *Phys.* A 185 a 17–20: οὐ μὴν ἀλλ' ἐπειδὴ περὶ φύσεως μὲν οὔ, φυσικὰς δὲ ἀπορίας συμβαίνει λέγειν αὐτοῖς, ἴσως ἔχει καλῶς ἐπὶ μικρὸν διαλεχθῆναι περὶ αὐτῶν· ἔχει γὰρ φιλοσοφίαν ἡ σκέψις.

infinity ranked high, and since the physicist will have to deal with these concepts, he cannot ignore the Eleatic *aporiai* and paradoxes. Thus, as we shall see, one of the main reasons for Aristotle to prefer one physical *endoxon* concerning the nature of *topos* to others is the fact that this alone enables one to escape the vicious consequences of Zeno's so-called paradox of place.

Thus far I have formulated two main objections to Owen's interpretation of Aristotle's dialectical method in terms of the analytic/synthetic distinction. First of all I argued that this distinction is itself not uncontroversial and that several versions of it are current, so that we are in danger of trying to explain *obscurum per obscurius*, as is indeed witnessed by the confusion involved in Owen's account. Secondly, I have tried to show that there does not appear to be a one-to-one correspondence between the way in which Aristotle himself uses the terms 'logical' and 'physical' and any clear-cut modern version (let alone Owen's rather confused version) of the analytic/synthetic distinction. We may now add, as a third and more general objection, that Aristotle does not in any way connect his own distinction between the 'logical and the 'physical' with his dialectical method or with his conception of what counts as a *phainomenon*. In other words, the *phainomena* or *endoxa* that constitute the starting points of the dialectical discussion in the *Physics* are neither exclusively logical, nor exclusively and straightforwardly empirical. In fact Aristotle appears to have availed himself of a rather wide range of data.[110] They comprise:

---

[110] My opting for the rather neutral term 'data' here is deliberate. For we are not merely dealing with kinds of *phainomena*, but also with kinds of *aporiai* and kinds of arguments (as Owen recognized, on which see the text above, 157). In general, it is not as easy as it may seem to determine what exactly Aristotle would himself be prepared to call *phainomena* or *endoxa*. Thus, in view of the rather vague meaning of the word φαίνεται it might well be that he would treat what he called 'logical' propositions as φαινόμενα. Yet, the class of *endoxa* seems to have a more limited extension. At any rate, we may infer from *Top.* 104 a 10–12 that *wrongheaded*, or at least *paradoxical* views (e.g. those logical views which clearly conflict with common conceptions) were *not* to be called *endoxa*. All the same, they could very well figure in the dialectical discussion, viz. as *theseis* (on such *theseis* see *Top.* 104 b ff.; see also Evans (1977) 79). So Zeno's paradoxes—which evidently do play an important role in the discussion—were perhaps strictly speaking not to be called *phainomena*, but *theseis*. It should be kept in mind, however, that Aristotle's dialectical procedure in the *Phys.* etc. is a kind of *dialectica utens* rather than *docens*, so that perhaps we need not make too heavy weather of these terminological matters.

(1) the 'immediate' data of experience, or
(2) the *endoxa* arrived at either by the masses or by some philosophers, *on the basis* of immediate experience, or
(3) purely logical ('logical' in Aristotle's sense of the word) theorems or *aporiai*, like those which were put forward by the Eleatics.

We may see clearly now that, contrary to what was suggested by Owen, *endoxa* which constitute the *phainomena* of the second kind are in the end rooted in experience. Therefore the arguments starting from these *endoxa* cannot without further qualification be called 'logical' or *a priori*. In this connection we may refer to Le Blond who spoke of the difference between our first and second class of *phainomena* as the difference between *l'expérience personelle* and *l'expérience indirecte*.[111] On the other hand, it now also becomes apparent that we would be equally mistaken in following Morsink all the way in his criticisms of Owen and to assume that in what Owen called the logical or conceptual puzzles of the *Phys.* Aristotle 'does have *direct recourse* to experience'.[112] The φαίνεται or δοκεῖ ('it appears') by which such general or theoretical concepts like place and time present themselves is usually a 'we think', not a 'we perceive'.[113] This will appear more clearly in the next subsections when we shall see Aristotle's dialectical method in *Phys.* Δ actually at work.

The fact that Aristotle himself probably never consciously and explicitly made the above threefold distinction of *phainomena* or data, should keep us from concluding that, in actually using the three kinds *promiscue*, he was guilty of *ambiguity*.[114] In fact, in

---

[111] Le Blond (1939) 222–268. Cf. also Bourgey (1955) 79: 'L' observation scientifique se trouve donc confondue soit avec la simple donnée immediate, soit avec l' interprétation spontanée qu' en propose la pensée vulgaire ...'.

[112] Morsink (1982) 112 [my italics]; cf. also *op. cit.* 116: 'I conclude that in his discussion of place Aristotle does have direct recourse to experience ... it is clear that it is not a discussion in which conceptual problems and puzzles are the focus of attention'. Contrast my remarks in the next subsections.

[113] Note that, though this shows how the naïve Baconian picture of Aristotelian science is mistaken, we are not allowed to argue that in the dialectical practice advocated in the *Topics* Aristotle meets us as contemporary Popperian anti-inductivist. (This is how Morsink (1982) 112, depicted the consequences of Owen's view).

[114] This particular point is well brought out by Wardy (1991) esp. 95, and Nussbaum (1986a) 244–245. At the same time, however, I find it difficult to go along with Nussbaum's general opinion concerning Aristotle's method, which is rather uncritically lyrical. According to Nussbaum 'Aristotle would

practice the borderlines between the three kinds of data would have been difficult to draw. If we realize that, as Le Blond has pointed out,[115] for Aristotle experience (*empeiria*) was not just what in German might be called *Erfahrung*, but—in the absence of instruments for exact measurements etc.—often amounted rather to what might be called *Erlebnis* or *experientia vaga*, it will be clear that an appeal to direct experience (i.e. to data of the first class) was in Aristotle's mind perhaps not sharply distinguished from an appeal to established ways of speaking (data of the second class). Moreover, though I have tried to show that where Aristotle contrasts *physical* and *logical* theorems or arguments he probably has the contrast between data of the second and of the third class in mind, it may be that his thoughts were not always altogether clear on this subject.[116] At any rate we may conclude that for Aristotle physics, as part of philosophy, was not naturally and exclusively confined to any one of these three kinds of data and that, accordingly, his conception of physics, and of the method proper to it, was quite different from ours.[117]

By way of a summary we may draw the following conclusions.

---

be reminding us of the humanness of good science' and he offers us a 'rich account of philosophical procedure and philosophical limits' (*op. cit.* 245). Nussbaum's attitude is no doubt partly to be explained by the fact that—quite in line with the general tenor of her book—she is mainly interested in drawing up a contrast between Aristotle and the more rigidly 'rationalistic' strand of Greek thought exemplified by the Eleatics and Plato.

115 Le Blond (1939) 249.

116 Cf. *Top.* 105 b 25–29, where, after having divided problems and propositions into logical, ethical, and physical, Aristotle adds that it is difficult to give a strict definition of each of these classes: ποῖαι δ' ἕκασται τῶν προειρημένων, ὁρισμῷ μὲν οὐκ εὐπετὲς ἀποδοῦναι περὶ αὐτῶν, τῇ δὲ διὰ τῆς ἐπαγωγῆς συνηθείᾳ πειρατέον γνωρίζειν ἑκάστην αὐτῶν, κατὰ τὰ προειρημένα παραδείγματα ἐπισκοποῦντα.

117 Owen (1961) 88 goes so far as to conclude that 'by such arguments the *Physics* ranks itself not with physics in our sense of the word, but with philosophy'. This appears to me to be only partly correct, since I believe that from a twentieth century point of view neither Aristotle's subject matter in the physics nor his (partly empirical) method would naturally belong to philosophy. Also Charlton (1970) xi, speaking about about *Physics* A and B argues that in the *Physics* 'the arguments are logical, the method is dialectical, and the discussions are philosophical' and he apparently assumes that this explains why 'the first two books of Aristotle's *Physics* do not deal with problems in what we today call physics'. The interesting point, however, is that for Aristotle 'logical' or 'dialectical' arguments *were* relevant to physics so that we should not conclude that, by using such arguments, he was not really doing physics, but rather that he was doing physics differently.

First of all, Aristotle had a rather broad conception of the kind of *phainomena* or data on which his philosophical investigations were based. Secondly, in describing Aristotle's method the anachronistic and equivocal terms 'analytic' and 'synthetic' are of little help. This being so, *we* may of course distinguish between three kinds of data or 'apparent facts' from which the Aristotelian investigations depart, viz. purely empirical data, more theoretical data which are nevertheless in the end connected with the data of direct experience, and purely theoretical considerations, which Aristotle would himself call 'logical'. And indeed, this threefold distinction may allow us to show that both Owen's stress on the strictly logical and conceptual character of the *topos* discussion in *Phys.* Δ and Morsink's opposite stress on the strictly empirical character of that same discussion are one-sided and therefore misleading. Yet, we should keep in mind that the distinction between these three classes would not seem to correspond to any specific terminological distinction made by Aristotle himself, although Aristotle would probably have made a distinction between thr first and second classes on the one hand (as *physical* data) and the third class on the other (as *logical* data).

4.4.4 Now that a general picture of Aristotle's 'scientific method' has been drawn, we may turn to the specific way in which this method is applied in *Phys.* Δ in the discussion of *topos*. The present subsection will give a survey of the composition and structure of 208 a 27–213 a 11, i.e. of the first five chapters of the book, the sections dealing with *topos*. The next subsection will discuss the nature of some of the arguments used in these passages, as well as the nature and provenance of the several concepts which are dialectically discussed.[118]

As a whole the composition of the first five chapters of *Phys.* Δ is not very smooth. The surviving text gives the impression of having been patched up from a number of smaller notes each dealing with specific aspects of the problem of place. Apparently no attempts have been made to connect these individual sections

118 Note that, contary to what is suggested by Moreau (1965) 31, who strangely divides the inquiry into 'considérations mathématiques' (presumably the first part) and 'considérations dialectiques' (the rest), the procedure is dialectical from the very beginning.

in an explicit manner. Nevertheless, the text as a whole reflects a fairly coherent theory and it contains a sufficient amount of methodological pointers to enable us to reconstruct, without too many problems, the way in which Aristotle arrived at his conclusions.

The current division of Aristotle's text into chapters is of course not genuine. Yet it more or less corresponds with the main elements of Aristotle's discussion, and for the sake of convenience I shall here take it as my starting point. Each of the first two chapters (208 a 27–210 a 13) indeed constitutes a more or less coherent whole. They start the inquiry by adducing some arguments for and against the existence of *topos* (chapter 1), and by introducing two first candidates for its identification without arriving at any definite conclusion (chapter 2). The third chapter (210 a 14–210 b 31) is largely devoted to an inquiry into the several uses of the term 'in' in common Greek parlance. The last part of this chapter (210 b 27–31) is a separate, more or less self-contained, passage which substantially repeats a point also made earlier (209 b 17–210 a 13).

The fourth chapter leads up to Aristotle's final definition of *topos* by gradually eliminating all other possible candidates for this role. All the same its structure is somewhat odd in so far as the survey of established data about place with which it opens (210 b 32–211 a 11) already seems to presuppose the conclusion of the discussion which is to follow (211 b 5–212 a 30). The fifth chapter (212 a 31–213 a 11), again, is a more or less self-contained unit, discussing some additional problems concerning place.

The contents of the first five chapters may then be summarized as follows:

A. Chapter 1 (208 a 27– 209 a 30)

A.1. Motivation of the study of topos by the natural philosopher (208 a 27–32) and statement of the *aporiai* (208 a 32–b1).

A.2. *Prima facie* arguments for the existence of place (208 b1–27):

- a. its existence seems to be clear (δοκεῖ δῆλον εἶναι) from the phenomenon of replacement (208 b1–8);
- b. the phenomenon of natural motion (208 b 8–25);
- c. those who assert the existence of void seem to mean topos (208 b 25–27).

A.3. This is why the idea of a self-subsistent topos is as old as Hesiod (208 b 27–209 a 2).

A.4. If we thus assume that it exists we must next try to discover what it is; this involves the following *aporiai*:

a. what is its *genos*? it has 3 dimensions but it is not a body (209 a 4–7);
b. if *topos* is an underlying extension, points and surfaces, being parts of a body, will also have to be in a place, but they cannot (209 a 7–13);
c. *topos* is neither an element, nor a compound of elements, nor a body, nor an incorporeal (209 a 13–18);
d. *topos* is not one of the four causes (209 a 18–22);
e. Zeno's paradox of place (209 a 23–26);
f. what happens to a body's place when the body grows? (209 a 26–28);
g. conclusion: the difficulties are such that we are not only forced to doubt *what* place is, but even *whether* it is (209 a 28–30).

B. Chapter 2 (209 a 31–210 a 13)
B.1. If place is regarded as a first surrounder it will seem to be form (209 a 31–b 6).
B.2. If, on the other hand, place is regarded as the extension (διάστημα) of the body, it will seem to be matter, as in Plato's theory (209 b 6–17).
B.3. Topos cannot be form or matter (209 b 17–22), since
a. it must be separable. Therefore it is rather like a vessel (ἀγγεῖον). 209 b 22–32);
b. special arguments against Plato: why are the Ideas not in a place? (209 b 32–210 a 2);[119]
c. natural motion is difficult to explain if place is matter or form (210 a 2–9);
d. place would seem to perish (presumably because it would be a kind of property of the body) in the process of elementary change[120] (210 a 9–13).

C. Chapter 3 (210 a 14–b 31)
C.1. The senses of 'in' distinguished (210 a 14–24).
C.2. Nothing can be 'in itself' in the primary sense (210 a 25–b22).
C.3. Solution of Zeno's paradox (210 b 22–27) by applying the distinctions made in 210 a 14–24.
C.4. Appendix: it is clear that place cannot be matter or form (210 a 27–31).

D. Chapter 4 (210 b 32–212 a 30)
D.1. Positing of the apparent data (ὅσα δοκεῖ ἀληθῶς καθ'αὑτὸ ὑπάρχειν αὐτῷ) about place (210 b 32–211 a 7):
a. it is a first surrounder and not a part of the located body;
b. it is of the same size as the located body;
c. it is separable;
d. it has 'up' and 'down'.

[119] On this argument see the previous chapter, section 3.3.
[120] On the nature of this difficult argument see the illuminating comments by Hussey (1983) 107.

D.2. Statement of the methodological principles (211 a 7–11):
a. ultimately we must arrive at a definition (τὸ τί ἐστιν) of place;
b. the *aporiai* must be solved;
c. the apparent data (τὰ δοκοῦντα ὑπάρχειν) must be accounted for;
d. a reason must be given for the inherent difficulties of the subject.

D.3. Relation between *topos* and motion (κίνησις): things moving on their own and things moving accidentally (κατὰ συμβεβηκός). Parts and whole. Kinds of continuity (211 a 12–b 5).

D.4. What place is, has now become clear. Fresh start: the four candidates for being place (211 b 5–10):
a. place is not form (211 b 10–14);
b. place is not the interval between the limits of the surrounding body (211 b 14–29);
c. place is not matter (211 b 29–12 a 2);
d. therefore, place must be the limit of the surrounding body (212 a 2–7);
e. Explanation of the inherent difficulty of the subject: the four candidates are once more enumerated (212 a 7–14).

D.5. Further qualification: place must be immobile (212 a 14–21).

D.6. This solution fits a number of important data about place (212 a 21–30).

E. Chapter 5 (212 a 31–213 a 11)

E.1. Some further problems concerning being in place (212 a 31–b22).

E.2. The difficulties raised can all be solved by the new account of *topos* (212 b 22–29):
a. place does not grow when a body grows;
b. there is indeed no place of a point (or a surface etc.);
c. there are not two bodies in the same place;
d. there is no self-subsistent three-dimensional extension (σώματος διάστημα);
e. Zeno's paradox solved: *topos* is 'in something else' but not in the strictly local sense of 'in'.

E.3. The solution offers a reasonable explanation of the phenomenon of natural motion (212 b 29–213 a 11).

Even if the overall structure of the text as it stands seems somewhat messy, the underlying dialectical procedure can, I believe, be coherently reconstructed. This is what we shall attempt now.

4.4.5 As luck would have it, in *Phys.* Δ Aristotle is more explicit about his own dialectical method than elsewhere, with the exception of *EN* 1145 b 6–7. For this reason these two texts figured prominently in Owen's account of Aristotle's dialectical method in '*Tithenai ta Phainomena*'. At *Phys.* Δ 211 a 7–11 (= item D.2. in the scheme of our subsection 4.4.4) Aristotle provides us with what might be called the methodological key passage of the whole *topos* account:

> We must try to make our inquiry in such a way that the 'what-it-is' is provided, the *aporiai* are solved, the apparent facts about place are accounted for, and, finally, so that the reason for the difficulty and for the problems around it are clear.[121]

This passage may be taken as a starting point for our reconstruction of the way in which Aristotle eventually arrived at his own concept of place. It will be seen that the *topos* discussion meets all four demands made here, for

(1) it provides a definition;

(2) it accounts for the apparent facts;

(3) it solves the *aporiai*; and

(4) it makes clear the reason for the difficulties inherent in the subject.

As to the first point, the 'what-it-is' is, of course, in the end provided in the form of what we called the 'classical' Aristotelian definition of *topos*.[122] In order to get a clearer picture of how Aristotle envisages the relationship between such a final definition and the data from which the inquiry takes its start, it seems worthwhile to adduce another important methodological passage from the same treatise, viz. *Phys.* A, 184 a 26–b 14.

After having made the familiar point that the more immediate and confused data of experience are more accessible (γνωριμώτερα) to us than the physical principles themselves, Aristotle argues that the inquiry into the principles will have to take its start from these 'apparent facts'. Next, the relationship between the principles on the one hand and the vague conceptions of common sense is elucidated as follows:

> The relation of names to definition will throw some light on this point; for the name gives an unanalysed indication of the thing

[121] *Phys.* Δ 211 a 7–11: δεῖ δὲ πειρᾶσθαι τὴν σκέψιν οὕτω ποεῖσθαι, ὅπως τὸ τί ἐστιν ἀποδοθήσεται, ὥστε τά τε ἀπορούμενα λύεσθαι, καὶ τὰ δοκοῦντα ὑπάρχειν τῷ τόπῳ ὑπάρχοντα ἔσται, καὶ ἔτι τὸ τῆς δυσκολίας αἴτιον καὶ τῶν περὶ αὐτὸν ἀπορημάτων ἔσται φανερόν.

[122] On the connection between ὅρος (definition) and τὸ τί ἦν εἶναι, or τὸ τί ἐστι cf. Guthrie (1967–81) VI, 147. On the role of the definition in Aristotle's methodology in general see Le Blond (1939) 269–291.

> ('circle', for instance), but the definition analyses out some characteristic property or properties. A variant of the same thing may be noted in children, who begin by calling every man 'father' and every woman 'mother', till they learn to sever out the special relation to which the terms properly apply.[123]

As is shown by the beautiful example of the small children, Aristotle is here not—or at least not exclusively—dealing with the 'classical' function of the definition, viz. to provide the *genus* and the *differentia specifica.* Rather, he is presenting the definition as a means for distinguishing the various meanings of a term which is used in an ambiguous and undifferentiated manner.[124] This interpretation perfectly fits the context of the remark, which, as we noticed, deals with the method of physics, viz. the analysis of the sometimes diffuse data of experience (in its broadest sense) into their elements. In a number of cases this amounts to analyzing and scrutinizing rather inarticulate *endoxa.*[125] If we now return to the *topos* discussion in *Phys.* Δ, it will appear that this procedure of extracting a final and refined definition from the rather indefinite common usage—i.e. the several 'things' people in general may call '*topos*'—is clearly witnessed (items B.1–B.3 and D.4. in our scheme of subsection 4.4.4). So much then for the first methodological demand, the providing of a definition.

This takes us to the next two demands. How are the *aporiai* solved and the 'apparent facts' accounted for? These questions are interconnected, since—contrary to what might seem to be suggested by the above quotation—not *all* apparent facts survive the dialectical inquiry.[126] Some of them are dropped because they

---

123 *Phys.* A 184 a 26–b14.

124 Cf. Ross' remark about this passage (1936) 458: 'Though it is essentially the business of definition to state the logical elements of a complex term, incidentally in doing this it will distinguish the various meanings of a term if this happens to be ambiguous. Only on this interpretation, apparently, will the remark about definition serve to illustrate, even remotely, what it is put forward as illustrating, viz. the transition from the recognition of the generic nature of an object to the recognition of its specific nature'.

125 This picture of the relationship between 'name' and definition may also help to put Aristotle's frequent appeals to linguistic usage into the right perspective. Linguistic conventions may be used as starting-points in a dialectical inquiry, but they may in the end turn out to have been rather a source of error. Cf. Wieland (1962) 280 about language as *Fehlerquelle*; Le Blond (1939) 251–267, esp. 266–267, only seems to envisage those cases where linguistic conventions are regarded as informative.

126 This point is explicitly made at *EN* 1145 b 1–7, a text quoted above, 157,

involve insoluble *aporiai*. Those that eventually do survive are listed at 210 b 32–211 a 7 (item D.1. in the survey of 4.4.4).[127] Accordingly, not all *aporiai* are solved either. Those that are not may serve as an indication that there is something wrong with the data which gave rise to them.

The 'apparent fact' about the *existence* of place, viz. *that it exists* (item A.2.), is initially doubted because of the *prima facie* difficulties inherent in attempts to determine the *nature* of place (item A.4.g). When these *aporiai* are finally solved, it can be stated with confidence that place exists (213 a 12). As to the 'apparent facts' concerning the *nature* of place, we witness a gradual sifting out of those which involve *aporiai* which cannot be solved. Thus chapter 2 opens with the two main ways in which place may be envisaged, viz. either as a surrounding something (i.e. what in chapter 1 was called the (b) concept of place) or as an extension of the body (i.e. the (a) concept or the (c) concept of place of chapter 1). At 209 a 31–209 b 17 (items B.1 and B.2) Aristotle argues that these 'apparent facts' might at first blush be translated into the identification of *topos* with form and matter respectively.

These identifications, however, involve a number of fatal *aporiai*, which are, it should be noted, of a *physical* nature: 209 b 17–210 a 13 (item B.3). They show, in other words, that such an identification would make no sense in the context of the physical system as a whole. Thus the system requires that matter and form are physically speaking inseparable, whereas place, if it is to play any significant role in the explanation of locomotion, must be separable (item B.3.a). Furthermore, the physical phenomenon of natural motion and natural places could not be accounted for if place were matter or form (item A.4.c).

When matter and form have thus been rejected as possible candidates for the identification of place, there are two candidates left which meet the criterium of separability—given, that is, the fact that we basically conceive of place as either a surrounding something or a three-dimensional extension. In that case place will have to be either a separate surrounding something or a

---

where it is said that, in investigating incontinence, the philosopher will have to prove, *if possible*, the truth of *all* the reputable opinions or, *failing this*, of the greater number and the most authoritative.

[127] On the curious fact that at the moment when they are enumerated the inquiry itself is not yet finished, see my remarks above, 171.

separate extension. The arguments explicitly raised against the latter conception (at 211 b 14–29, item D.4.b) are fairly difficult to follow,[128] but at any rate they are of what Aristotle would regard as a *logical* nature. They do not purport to show that the concept of *topos* as an independent three-dimensional extension does not fit the system as a whole, but they attempt to show that it involves some intrinsic conceptual difficulties. However, if we may assume, as seems plausible, that also the *aporiai* listed as items A.4.a, c, and d concern the concept of place as a three-dimensional extension, it appears that, apart from being vulnerable to purely logical arguments, this concept of place was also thought to be difficult to fit into the physical system as a whole.

This whole procedure of playing off some apparent facts against others finally leaves Aristotle with the list of 'established facts' which is provided at 210 b 32–211 a 7 (item D.1). These, in other words, are the 'apparent facts' which are not affected by the unsolved *aporiai*.[129] Thus, to take the most famous example, it is obvious that Aristotle thinks that Zeno's paradox of place shows the untenability of the concept of place as a self-subsistent three-dimensional extension. At the same time he shows that the infinite regress involved in the Zenonean argument does not apply to his own concept of place. This he can do on the basis of the survey of the senses of 'in' which was provided in 210 a 14–b 27 (items C.1–C.3). Thus the ultimate limit of the surrounding body is itself in something else, though not in the strictly local sense of 'in', but in the sense of 'in as a part in the whole'. This sense of 'in' does not involve an infinite regress. In the same fashion the problem of the localization of points and surfaces is solved (item A.4.b): they also are not in a place on Aristotle's own final account, but they are in something as a part in the whole.

Of all other *aporiai* listed at 209 a 3–30 (item A.4) and 209 b 17–210 a 13 (item B.3.) most—i.e. those repeated at 212 b 22–29, (item E.2)—appear to be solved *only if* we opt for the concept of place as the limit of the surrounding body. The only *aporiai* which are potentially damaging to Aristotle's own final concept of *topos* and which are merely raised but not solved in the text as it stands are the problems raised at 209 a 13–22 (items A.4.c and d), viz. the

[128] See above, n. 7 and 8, and text thereto.
[129] On this 'revised' list of apparent facts see also below, ch. 5, 204.

problem of the precise ontological status of place, and the problem whether and in what sense it figures as a cause.[130] I shall revert to these issues in the next chapter.

All in all the dialectical structure of Aristotle's line of argument concerning *topos* is fairly clear. A number of apparent facts about place are taken as a starting point and they are subsequently confronted with other apparent facts which may take the form of what, as we saw, Aristotle would call physical arguments ('the system does not allow for such a concept of place because...') or of what in Aristotle's sense of the word would count as logical arguments ('the concept in itself involves some logical absurdities'). From among the several concepts of place which have a *prima facie* plausibility only the concept of place as the first boundary of the surrounding body survives the whole procedure.

We are left with the final demand made of the dialectical inquiry: it should explain the difficulties arising in connection with the subject. This, in fact, is done at 212 a 7–14 (item D.4.e). This passage, which I already quoted above, in subsection 4.2.3, is of the utmost importance because it shows how Aristotle thought of the relation between his own final concept of place and the rejected candidates. I shall therefore render it here again in full:

> Place seems to be something profound and difficult to grasp, both because the notions of matter and form present themselves together with it (παρεμφαίνεσθαι), and because of the fact that change of position within a moving body occurs within a surrounding body which is at rest; for [from this] it appears to be possible (ἐνδέχεσθαι γὰρ φαίνεται) that there is an extension in between which is something other than the magnitudes which move. Air, too, contributes to this suggestion, by appearing to be incorporeal; place seems (φαίνεται) to be not only the limits of the vessel, but also that which is in between, which is considered as being void.[131]

The difficulty of the subject, in other words, is that in ordinary usage the term *topos* is used in different ways. Many times it

---

130 That these two *aporiai* remain unsolved was recognized by Simplicius *In Phys.* 600, 13–39.

131 *Phys.* Δ 212 a 7–14: Δοκεῖ δὲ μέγα τι εἶναι καὶ χαλεπὸν ληφθῆναι ὁ τόπος διά τε τὸ παρεμφαίνεσθαι τὴν ὕλην καὶ τὴν μορφήν, καὶ διὰ τὸ ἐν ἠρεμοῦντι τῷ περιέχοντι γίνεσθαι τὴν μετάστασιν τοῦ φερομένου· ἐνδέχεσθαι γὰρ φαίνεται εἶναι διάστημα μεταξὺ ἄλλο τι τῶν κινουμένων μεγεθῶν. συμβάλλεται δέ τι καὶ ὁ ἀὴρ δοκῶν ἀσώματος εἶναι· φαίνεται γὰρ οὐ μόνον τὰ πέρατα τοῦ ἀγγείου εἶναι ὁ τόπος, ἀλλὰ καὶ τὸ μεταξὺ ὡς κενόν.

involves connotations which on closer scrutiny would lead us to think that *topos* is actually matter or form. At the same time, in so far as place seems to be something that remains when the located body is removed, it would appear to be an independent three-dimensional extension. This suggestion concerning the real nature of *topos* is reinforced by the apparent incorporeality of air: if things move through air they seem to move through a pre-existing absolute extension.

We may well connect this passage with Aristotle's own words about 'name' (ὄνομα) and definition (λόγος) in *Phys.* A (in the passage quoted above, 174–175). Since, apparently, the word *topos* in ordinary usage involves a number of different connotations, establishing a *physically meaningful* definition on the basis of ordinary usage amounts to ruling out those connotations which do not meet the demands made of a concept of *topos* in a serious *physical theory,* either because they involve confusion with other concepts of that theory (matter and form), or because they present problems of their own (independent extension). The difficulty of the subject is constituted precisely by the fact that the views which are ultimately to be ruled out play such an important role in ordinary thinking and common parlance.

Incidentally, it should be noted that this passage definitely proves that Morsink's claim, as against Owen, that Aristotle, in so far as he took the *phainomena* as a starting point in his discussion of *topos,* had *direct recourse* to 'experience or empirical observations',[132] is unjustified. It would of course be absurd to maintain that the 'appearing together' παρεμφαίνεσθαι of matter and form meant that they 'are perceived at the same time'. For one thing, how could matter be perceived? For another, how could place be perceived as two different things at the same time? The idea is

---

[132] Morsink (1982) 115. Two factors may have particularly fostered Morsink's interpretation. Firstly, Morsink seems to have taken Owen's position as implying that for Aristotle established ways of speaking were 'authoritative'. This, as the present section aims to show, is not true. The conventions of common parlance have to be scrutinized, sifted, and, if necessary, refined. Secondly, Morsink apparently did not differentiate between a situation where a concept is directly derived from observational experience, and a situation where empirical data are merely 'relevant to the acceptence of' a certain concept. Place or space are not directly perceived. Nevertheless a concept of space or place will in the end be judged according to the extent to which it makes sense of our empirical observation of motion and replacement.

rather that the common notions of matter and form are diffusely intermingled with that of place and that in some contexts we may use the term *topos* to denote, e.g., what actually should be called matter. Similarly the φαίνεται ('it appears') in the sentence 'it appears to be possible...etc.' of course means '*we conceive* it to be possible...etc.'. Also the meaning of the φαίνεται in the final sentence is perfectly clear. Aristotle does not imply that place presents itself empirically in two different forms, but rather that we may *think of* place in two different ways.

Of course all this is on closer thought not very surprising. 'Place' is not an observational term. Nevertheless it is part and parcel of a theoretical framework which in the end will have to fit the data of direct experience. Therefore problems concerning place or space are not purely conceptual problems—if we use this term in Owen's sense, viz. as referring to problems which have no connection with observational experience at all. Neither, on the other hand, are they directly linked to experience, as was in fact Morsink's opinion. If we may use an image provided by Papineau, we may observe that a scientific theory might be conceived of as a pyramid made of struts:

> The observational terms are the points at which the framework is moored to the ground. The theoretical terms are the points at which the struts making up the framework join. The struts then represent the generalizations linking the theoretical terms to the observational terms and each other. Some theoretical terms will be linked directly to a number of observational terms. Other more abstract theoretical terms, higher up the structure as it were, will be linked indirectly to observational terms through being linked to several of the former theoretical terms.[133]

In Aristotle's physical system *topos* figures as a theoretical term. As such it is given its place in the theory as a whole by the specific way in which it is related to other theoretical terms, in particular matter and form. Furthermore it has to be differentiated from a concept of *topos* (viz. the concept of an independent three-dimensional extension) which could not be fitted into the system at all.

Of course for twentieth century philosophers the contrast between theoretical and observational terms entails difficulties of its own. As is well known, it has proved immensely difficult to

---

[133] D. Papineau, *Theory and Meaning* (Oxford 1979) 9.

account adequately for what are here metaphorically called the 'struts' linking observational and theoretical terms, the failure in this respect of Carnap's *Der logische Aufbau der Welt* still being a classical example. However, such problems obviously did not bother Aristotle. He just, so to speak, took the 'theoretical terms' of the rather vague and primitive physical theory of common parlance as his starting points and turned them into 'refined' theoretical terms which fitted into his physical system.

We may conclude, then, that in spite of the fact that *Phys.* Δ 1–5 apparently consists of different patches, and that its composition is not very elegant and systematic, it happens to have been written as a whole from a coherent dialectical perspective. Starting from the somewhat confused data of common thinking, and by subsequently sifting these data in the light of the *aporiai* which they produce, Aristotle ended up with a definition of what he regarded as the only defensible concept of place. As a paradigmatic case of Aristotle's dialectical method actually at work the *topos* discussion seems to deserve even more attention than it has been given by Owen.

4.4.6 We may summarize the findings of the present section as follows. We inquired into Aristotle's dialectical method in his school treatises in general, and into the way in which this method determines the form and contents of the *topos* discussion in *Phys.* Δ in particular. Taking the studies of scholars like Owen, Le Blond, Barnes, and Weil as a spring-board, the investigation allowed us to assess some claims that had been made concerning the nature of the data from which Aristotle's dialectical procedure takes its start. We saw that both Owen's stress on the logical and purely conceptual character of the discussion and Morsink's alternative stressing of its empirical nature were misleading.

We concluded that Aristotle took the common use of *topos* as his starting point, and that he recognized that the name (ὄνομα) '*topos*' was used ambiguously in common parlance, viz. to denote either a three-dimensional extension, whether separable or not, or a surrounding limit. This confused concept derived from ordinary language was subsequently refined so as to leave only one candidate: the first unmoved boundary of the surrounding body.

4.5 *Unorthodox conceptions of topos in the physical works*

4.5.1 Earlier in this chapter I drew attention to the fact that the *Categories* contains a concept of *topos* which is at odds with the classical Aristotelian concept arrived at in *Phys.* Δ. It was concluded that the context suggests that Aristotle did not wittingly and willingly introduce this concept as a physically correct one. Rather, he stuck to the conventions of common parlance without bothering about the ontological or physical implications. The *Cat.*, in other words, was not written from a physical perspective and should be read accordingly.

In the previous section we saw how in *Phys.* Δ Aristotle seemed to be well aware of the limitations of such a vulgar use of spatial terms. He recognized that it might entail ambiguities and equivocations. Therefore he tried his best, first, to separate the various connotations apparent in the common usage of the term *topos* and, next, to eliminate those connotations which could be shown to be inadequate in a more sophisticated physical context. Ideally the 'refined' physical theory which resulted from this procedure should have permitted the translation of all instances of the use of *topos* in a non-adequate sense into other refined physical concepts (e.g. matter) or into the refined *topos* concept. Aristotle did not come up to this ideal standard, as the present section aims to show.

I shall survey a number of passages in works which are, both as to contents and as to probable date of composition, close to *Phys.* Δ. In these passages Aristotle nevertheless seems to disregard the final definition of *topos* of *Phys.* Δ and, instead, still talks as if he is committed to the concept of place (or space) as a three-dimensional extension. I shall begin by looking at some texts which at first sight might seem to be not interesting at all, since they use *topos* in the fairly trivial 'geographical' sense of (two-dimensional) 'region, place, stretch'. However, since it can be shown that the difference between this use and the 'physical' use of *topos* in the sense of 'three-dimensional extension' is merely gradual—as was also implied in chapter 2—I think it will be apposite to start with some examples of this 'original' use (4.5.2). Next, I shall turn to some texts which use the concept of *topos*-as-three-dimensional-extension in a physical context dealing with locomotion, where one would perhaps rather have expected the concept of *topos* of Phys. Δ to appear.

4.5.2 The purely geographical use of *topos* by Aristotle is evidenced by the following observation, to be found at *Meteor.* 368 b 13:

> Earthquakes are of a local character and are often confined to a small region (*topos*), but winds are not.[134]

A quick glance at Bonitz' *Index Aristotelicus* shows that this 'geographical' use of *topos* is widespread in some treatises of the *Corpus Aristotelicum.*[135] That some of the uses of *topos* as a three-dimensional something are still rather close to the 'geographical' use is shown by *Meteor.* 353 b 6 ff., where Aristotle is referring to views about the nature and origin of the sea. After having mentioned the views of the ancient 'theologians', he turns to 'those who were more versed in secular philosophy (ἀνθρωπίνη σοφία)':

> They say that at first the whole region (*topos*) about the earth was wet, and that as it dried up the water that evaporated became the cause of winds and the turnings of sun and moon.[136]

Here the implication is not necessarily that *topos* is something which exists on its own, apart from the bodies occupying it: 'the wet region' simply stands for 'water'.[137] However, at *Meteor.* 355 b 2, i.e. in the same context, we find *topos* used as a mere (part of) space *occupied* by the sea.[138] The use of the verb 'to occupy' (*katechein*) in this connection may remind us of the use of *topos* (also in connection with *katechein*) in the *Cat.* At *Meteor.* 340 a 6 we find the expression 'if the space/region (*topos*) between earth and air were filled with two elements ... etc.'. Here the context clearly

---

[134] *Meteor.* 368 b 12–14: Κατὰ μέρος δὲ γίγνονται οἱ σεισμοὶ τῆς γῆς, καὶ πολλάκις ἐπὶ μικρὸν τόπον, οἱ δ' ἄνεμοι οὔ κατὰ μέρος.

[135] Cf. e.g. *Meteor.* 343 a 8, 360 b 18–19, 363 a 12 ff., 368 b 15; see also *Pol.* 1260 a 41, 1276 a 20, 1280 b 14.

[136] Εἶναι γὰρ τὸ πρῶτον ὑγρὸν ἅπαντα τὸν περὶ τὴν γῆν τόπον κτλ. (transl. H. D. P. Lee, Loeb edition).

[137] Note that the two-dimensional, 'geographical' counterpart of this use of *topos* is found in expressions like ἀπὸ τοῦ κατακεκαυμένου τόπου πνέοντα ἄνεμόν (*Meteor.* 363 a 12–3). The possibility of using τόπος in this way may have furthered the confusion in the *Tim.* between the material and the spatial aspect of the receptacle. If the word τόπος was thus ambiguous between 'a part of space' and 'a stretch or region *of something*', e.g. a region of fire, it becomes more understandable how Plato could call fire not (only) an occupant of the receptacle, but the 'ignified part (πεπυρωμένον μέρος) of the receptacle' (on which see above chapter 3, 95).

[138] *Meteor.* 355 b 2: κατέχουσαν τόπον τὴν θάλατταν.

shows that *topos* is used as a synonym for διάστημα (three-dimensional extension).[139] Also this use of *topos* seems to be quite common, particularly in the *Meteor.*[140] In these cases the use of *topos* as an independent three-dimensional extension seems to have been a more or less natural development of the ordinary geographical use of *topos* as 'region'.

4.5.3 In the passages discussed so far the presentation of *topos* as a three-dimensional extension might still be excused by pointing out that Aristotle is here giving a more or less static description of the universe and its different regions, or the regions occupied by the different elements. Location—apart from the more or less vague conception of natural places or regions for the different elements—and locomotion are not what is primarily at issue. However, there are also a number of passages where *topos* is used in this same unorthodox sense in a context which is in this respect more demanding.

Thus at *Meteor.* 339 a 25 ff. the celestial element is contrasted with the terrestrial bodies in the following words:

> ...the celestial element is eternal and the spatial path (*topos*) through which it moves is endless, though always complete, while the terrestrial bodies each have their distinct and limited regions (*topous*).[141]

Here the motion of the celestial sphere is at the focus of interest. Aristotle is concerned with outlining the differences between the celestial and the terrestrial elements. He argues that the opposition between eternal circular motion on the one hand and

---

139 Compare *Meteor.* 340 a 5 ff.: εἰ δύο στοιχείων πλήρης ὁ μεταξὺ γῆς καὶ οὐρανοῦ *τόπος* ἐστίν, and *ibid.* 340 a 1: εἰ γὰρ τά τε *διαστήματα* πλήρη πυρὸς κτλ.

140 Cf. *Meteor.* 346 b 15–16, 347 b 12; but also *GC* 330 b 30: Ὄντων δὲ τεττάρων τῶν ἁπλῶν σωμάτων, ἑκάτερον τοῖν δυοῖν ἑκατέρου τῶν τόπων ἐστίν. Cf. also *Metaph.* K 1067 a 11–13 where Aristotle envisages the counter-factual state of affairs of earth being infinite. In that case it would be difficult to explain whether a singular clod of earth would always move or always be at rest, since its natural place would be infinite (i.e. it would everywhere be in its natural place): ὁ γὰρ τόπος τοῦ συγγενοῦς αὐτῇ σώματος ἄπειρος. καθέξει οὖν τὸν ὅλον τόπον; (this passage is substantially the same as *Phys.* Γ 205 a 10 ff.).

141 *Meteor.* 339 a 25 ff.: ἡ μὲν ἀίδιος καὶ τέλος οὐκ ἔχουσα τῷ τόπῳ τῆς κινήσεως. ἀλλ' ἀεὶ ἐν τέλει· ταῦτα δὲ τὰ σώματα πάντα πεπερασμένους διέστηκε τόπους ἀλλήλων.

rectilinear motion on the other serves as one of the features which endow the celestial sphere with its unique position in the physical system. The way *topos* is used in this connection is singularly at odds with the remarks made in *Phys.* Δ about the outer sphere of the heavens being not in a place at all.[142] This may be due, however, to the fact that it is the *motion* of the heavenly sphere, not the problem of its *location* as such, which is focused on here. In the latter case Aristotle could perhaps have stuck to his classical concept of *topos.* In the present context, i.e. for the purpose of describing the trajectory of the motion of the heavenly sphere, that concept would not seem to have been of much help anyway.

As another example of this 'loose' use of *topos* we may adduce a passage from the *De Caelo.* That at the time of writing this work Aristotle had already developed the specific theory of place which is to be found in *Phys.* Δ can be gauged from his remark, *Cael.* 310 b 7 ff., that 'place is the limit of the surrounding body'.[143] Nevertheless, at *Cael.* 287 a 12–23 we find a curious argument purporting to show that the cosmos is spherical. This is proved by a reference to the fact, here taken for granted, that beyond the outermost circumference there is neither void (*kenon*) nor place (*topos*). For, it is argued, if the cosmos were bounded by straight lines, this would involve the existence of place, body and void outside its present boundaries. In that case the cosmos would, in revolving, no longer occupy the same space (*chôra*) all the time.

---

142 Cf. *Phys.* Δ 212 b 8–10: ὁ δ'οὐρανός, ὥσπερ εἴρηται, οὔ που ὅλος οὐδ' ἔν τινι τόπῳ ἐστίν...etc. This is a partial retractation of the apparent fact, noted at 211 a 13–14 that ...καὶ τὸν οὐρανὸν μάλιστ' οἰόμεθα ἐν τόπῳ, ὅτι ἀεὶ ἐν κινήσει. Aristotle's eventual solution seems to turn on a distinction between the two kinds of motion (rotation and translation, the outer heavenly sphere only rotating, so that it does not need an *external* place: 212 a 34) and on a distinction between the heavens as a whole and their parts: the circular motion of the whole forces us to conclude that the parts of the whole (which itself has no place) are in a way in a place: 212 b 10–11; ἐφ' ᾧ δὲ κινεῖται, ταύτῃ καὶ τόπος ἔστι τοῖς μορίοις· ἕτερον γὰρ ἑτέρου ἐχὸμενον τῶν μορίων ἐστίν. Though these passages have given rise to much controversy—on which see the next chapter, 193, n. 5—I think we may at any rate conclude that (1) the heavens as a whole are not in a place and that (2) the heavens rotate and for that reason the parts change place. This place of the parts, whatever it may be, is represented as a surrounding something, not as an underlying three-dimensional extension. So, in speaking of *topos* as a path for the celestial element to move through, the passage from the *Meteor.* we are here dealing with certainly offers a different perspective..

143 'Επεὶ δ' ὁ τόπος ἐστὶ τὸ τοῦ περιέχοντος πέρας etc.

Owing to the changing position of the corners there will at one time be no body where there was body before, and there will be body again where now there is none. In other words, this would

> involve the existence of place (*topos*) and void (*kenon*) outside the revolution, because the whole does not occupy the same space (*chôra*) throughout.[144]

The connection with the verb 'to occupy' (*katechein*) and with the terms *chôra* and *kenon*, as well as the general context, make clear that Aristotle is here thinking of *topos* as a three-dimensional extension.[145] Here again the focus is not so much on *location* but rather on a pre-existent trajectory for the motion (*phora*) of the outer sphere.

As a final example I would like to adduce a passage from the discussion of time, i.e. from the very book of the *Phys.* which contains the 'classical' *topos* discussion. At *Phys.* Δ 219 a 11 ff. Aristotle derives the continuity of time from the continuity of motion, which is in its turn derived from the continuity of magnitude or distance (μέγεθος).[146] In this connection he argues that:

> The primary sense of before-and-afterness is found in the case of place (*topos*). For there it has to do with position. Now since there is before-and-after in spatial extension (μέγεθος), there must also be before-and-after in motion, by analogy with them. But then there is also a before-and-after in the case of time, in virtue of the dependence of time on motion.[147]

---

[144] *Cael.* 287 a 21–23: ... συμβήσεται καὶ τόπον ἔξω καὶ κενὸν εἶναι τῆς φορᾶς, διὰ τὸ μὴ τὴν αὐτὴν χώραν κατέχειν τὸ ὅλον.

[145] Another instance of this use of *topos* can be found at *Cael.* 275 b 15 ff., where one of the arguments against an infinite body is that the infinite cannot move since this would require in addition another infinite place for it to move into: δεήσει γὰρ ἕτερον εἶναι τοσοῦτον τόπον ἄπειρον εἰς ὃν οἰσθήσεται. Here Aristotle might be excused by pointing out that he introduces the argument as 'more logical' (λογικώτερον, 275 b 12) which means that he is not primarily thinking about any specific concept of physical *topos*, but is rather stressing the 'logical' fact that something infinite, being everywhere, cannot have something else next to it to move into.

[146] As a translation 'distance', is perhaps preferable to 'magnitude', since what is at stake is obviously the extension which is to be traversed, whereas 'magnitude' might seem to refer to the moving sensible body itself. Cf. also *Phys.* Z 231 b 19–20 (a passage also dealing with the relation between μέγεθος, κίνησις and χρόνος), where the Loeb translation of Wicksteed and Cornford translates μέγεθος as 'spatial extension' and 'distance', whereas at Δ 219 a 11 it strangely prefers 'magnitude'.

[147] *Phys.* Δ 219 a 14–19: τὸ δὲ δὴ πρότερον καὶ ὕστερον ἐν τόπῳ πρῶτόν ἐστιν. ἐνταῦθα μὲν δὴ τῇ θέσει· ἐπεὶ δ' ἐν τῷ μεγέθει ἐστὶ τὸ πρότερον καὶ ὕστερον,

The conclusion seems inescapable that Aristotle here uses *topos* as a synonym for μέγεθος (extension).[148] For all his talk elsewhere of motion not presupposing an underlying independent extension, this passage shows that in his account of the connection between motion and time he could not really do without such a concept of place (or rather: space). One need only imagine (or rather imagine how hard it is to imagine) what the argument would have looked like had Aristotle been forced to apply his own theory of motion as *antimetastasis* and his classical concept of *topos*.

G. E. L. Owen has suggested that in this case Aristotle is consciously providing a *geometrical model*, and that this may account for some of the oddities in the presentation.[149] But this suggestion does not seem to be very helpful in explaining the unorthodox use of *topos* here. For if we were to regard this passage as providing a geometrical model, we might still expect that such a vital element in the model as extended *topos*, being in the end responsible for the continuity of motion as well as the continuity of time, corresponds to a feature of physical reality. It would not do, in other words, to say that though the model suggests the existence of an independent three-dimensional extension, there is no such extension in physical reality. For in that case the whole account of the continuity of time would collapse.

I would rather suggest that Aristotle here stuck to a common sense conception of *topos* which in the context of a discussion of time rather easily presented itself since the 'spatial counterpart' of measurable time and of measurable motion seems to be *measurable extension*, rather than the merely 'locational' concept of *topos* of *Phys.* Δ.

What the passages collected in the present subsection suggest—and perhaps this should not really surprize us—is that Aristotle's classical concept of *topos*, though introduced in order to allow the

---

ἀνάγκη καὶ ἐν κίνησει εἶναι τὸ πρότερον καὶ ὕστερον, ἀνάλογον τοῖς ἐκεῖ. ἀλλὰ μὴν καὶ ἐν χρόνῳ ἐστὶ τὸ πρότερον καὶ ὕστερον διὰ τὸ ἀκολουθεῖν ἀεὶ θατέρῳ θάτερον αὐτῶν.

[148] On the whole passage see the excellent comments of Owen (1976) esp. 22–25. By contrast Hussey's comments (1983) 143–149 are extremely cluttered and hardly clarifying; Wagner (1967) deals with the passage in a few brief and neutral remarks; Ross (1936) pays no attention to it at all.

[149] Owen (1986), 'Aristotelian Mechanics', in Nussbaum (1986b), esp. 317 (Nussbaum's pagination; I have not had recourse to the original version and Nussbaum in this case does not provide the original pagination).

physicist to explain physical motion,[150] was in the end merely useful to account for the *location* of *static* bodies. Whenever the *distance* or path which a moving body had to traverse was at issue, it was, even for Aristotle himself, rather a different concept of *topos* which came to mind. It may be that Aristotle, rather implausibly, thought that statements—like those here quoted—which used such a 'bastard' concept of *topos* could in the end be translated into physically more exact statements containing the only really admissible concept of *topos* or perhaps into statements which accounted for the distance to be traversed merely in terms of *matter* without implying the existence of any self-subsistent extension. However, all this is mere guesswork. The only stable fact for us to rely on is that Aristotle sometimes showed himself unfaithful to the conclusions of his own theoretical discussion of *topos* in *Phys.* Δ in at least appearing to be committed to a concept of an independent three-dimensional *topos.*

4.5.4 We may conclude the present section by making the following observations. We saw that quite a number of passages in the physical writings contain references to a concept of *topos* as a three-dimensional extension, a concept which was unambiguously rejected in the *Phys.* Though some of these cases merely concerned a rather harmless non-physical 'geographical' use of the term, we also pointed to some others which figured in more sophisticated contexts where physical motion was at issue and where, for that reason, one would rather have expected the classical concept to appear.

This shows, first, that it is one thing to develop a kind of philosophical technical language, but quite another to use it in a consistent manner. Second, it corroborates our conclusions of the previous section, viz. that the alternative candidates for the identification of place were not rejected as fully absurd conceptions. In fact they represented common ways of speaking about *topos* which were in specific contexts at first glance adequate and which were only on second thoughts to be rejected in the context of a physical theory.

---

150 Cf. the programmatic statement at the beginning of *Phys.* Δ, where the subsequent treatment of *topos* is motivated by the claim that the physicist deals with the problems connected with motion and that ἄνευ τόπου καὶ κενοῦ καὶ χρόνου κίνησιν ἀδύνατον εἶναι (200 b 21–22).

That Aristotle nevertheless at times also had recourse to an unorthodox concept of place or space in a physically sophisticated context dealing with motion may have been due to the fact that he was in the cases at issue not *primarily* dealing with *topos*, so that we may account for his 'cavalier' use of the term as a more or less unconscious slip of the pen. On the other hand, we may note, his 'classical' definition of *topos* would in these cases hardly have been of use.

Conclusions

The investigations of the present chapter took the different concepts of place (*topos*) as they appear in the *Corpus Aristotelicum* as their starting point. First, in sections 4.1–4.3, I discussed the relationship between the concept of *topos* which appears in the course of the discussion of the category *poson* in the *Cat.* and the famous definition of *topos* established in *Phys* Δ. Though scholars like Duhem and Jammer, and more recently, King and Mendell have taken these passages seriously as containing an unambiguous account of physical place[151]—and have consequently tried their hardest to establish in what way these passages were related to the account in *Phys.* Δ—I concluded that they present enough problems of their own to invalidate such claims. If we take the now more or less universally accepted relative chronology of the surviving school works as established—and I have not been able to find reasons for not doing so—and if we may thus assume that the *Cat.* was written some five or ten years earlier than *Phys.* Δ, we may conclude that in so far as we might speak of a *development* of Aristotle's philosophy of place between the *Cat.* and *Phys.* Δ, this development should not be described as the substitution of one articulate view by another, but rather as a growing awareness of the problems inherent in the common sense notions of place and space.

This seemed to be confirmed by the findings of section 4.4. There I investigated Aristotle's dialectical method in general and in *Phys.* Δ in particular. Against Owen on the one hand, and Morsink on the other, I argued that the data from which Aristotle's

---

[151] Cf. Duhem (1957) I, 197–198; Jammer (1954) 15; Mendell (1987) *passim*; King (1950) 88.

dialectical procedure in *Phys.* Δ took its start were for the most part what might be called the 'theoretical terms' of the 'physical system' of everyday thought. Concerning such a theoretical physical term as *topos* which is not *directly* linked to experience, Aristotle took apparent facts, i.e. views endorsed by the world at large or by some individual philosophers, as his starting point. We might call this, with Morsink[152], a process of 'conjectures and refutations', as long as it is kept in mind that in Aristotelian dialectic such 'conjectures' usually do not spring forth from the genius of the individual physicist, but are largely determined by the conventions of everyday thought and common parlance[153].

We saw that the whole further process boiled down to the scrutinizing and refining of these 'apparent features'. A number of them were rejected for involving insoluble *aporiai.* Those features that survived the dialectical investigation were incorporated in Aristotle's eventual 'physical' concept of place. All this involved the recognition that ordinary thought and common parlance did not use the term *topos* in a very coherent manner and that the actual task of the physicist was to eliminate those connotations of the term which, for all their *prima facie* plausibility, turned out to be of no use in the context of physical theory as a whole. Thus the relation between the account of *topos* of the *Cat.* and that of *Phys.* Δ could be explained. In the *Cat.* Aristotle was using *topos* in one of the at first sight plausible senses of common parlance which were reviewed and rejected in *Phys.* Δ.

On the other hand, as section 4.5 showed, this unorthodox concept of *topos* as a three-dimensional self-subsistent extension crops up in a number of passages in the more sophisticated physical writings as well, probably because as an inveterate *façon de parler* it was still hard to banish altogether and probably also because Aristotle's own orthodox concept did not prove to be useful in all circumstances.

As a whole the present chapter seems to corroborate our thesis of chapter 1, viz. that Greek philosophical theories of space and place were closely linked to—and indeed started off from—the ways in which spatial terms might be used in ordinary language.

---

152 Morsink (1982) 105.

153 See, moreover, my qualifying remarks concerning such a comparison with Popper, above, n. 113 and thext thereto.

As I concluded in chapter 3, it was a more or less unreflective use of some of the ambiguities of common parlance which was partly responsible for the obscurities in Plato's receptacle account. In the present chapter we noticed that in the course of his philosophical career Aristotle did develop an awareness of the ambiguities and equivocations of everyday thinking and speaking and that for him the conventions of ordinary language and the difficulties which they involved constituted the raw material for his dialectical inquiries into the nature of such theoretical entities as place and space.

CHAPTER FIVE

PROBLEMS IN ARISTOTLE'S THEORY OF PLACE AND EARLY PERIPATETIC REACTIONS

Introduction

In the previous two chapters the theories of place of Plato and Aristotle have been compared in a number of respects. Each time Aristotle's theory came out best. First of all it was clearly designed as a *physical* rather than a *metaphysical* theory. In addition it gave evidence of Aristotle's being more sensitive than Plato to the ambiguities and problems inherent in everyday speaking and thinking about place, and that, unlike Plato, he tried his best to disentangle the various underlying concepts of place. Nevertheless, his own discussion of the subject in *Phys.* Δ and the theory of place in which it resulted, are certainly susceptible of criticism. It has often been observed, for example, that the arguments against the most important rival conception (place as an underlying three-dimensional extension) are very unsatisfactory.[1] Moreover, a general drawback of Aristotle's theory, one already signalled in the previous chapter, is that it appears to have been designed primarily as a theory of the *location* of *static* bodies, while it can hardly play the role it should play in the explanation of the *motion* of bodies. Ironically, that is probably why in such a context, i.e. where the focus is on the *trajectory* of a moving object, Aristotle sometimes consciously or unconsciously resorted to the very concept of place as a three-dimensional extension which he had rejected for theoretical reasons.[2] This problem was possibly acknowledged by the third century BC dialectician Diodorus Cronus,[3] and it certainly figures among the arguments against

---

[1] For a brief discussion of the two arguments against place as a three-dimensional extension (to be found in *Phys.* Δ 211 b 19–23 and 211 b 23–25) see the previous chapter, 125–126. Aristotle's arguments were pertinently criticized by Philoponus at the beginning of his *Corollarium de Loco* at *In Ph.* 557, 12–563, 25.

[2] On which see the previous chapter, section 4.5, esp. 4.5.3.

[3] It has been suggested by Sorabji (1983) 18 and 347, that the arguments of Diodorus recorded at S. E. *M.* 10, 85–90 and 143 (= frs. 13 and 15 Giannantoni), according to which a body cannot move in the place where it is, were

Aristotle's conception of place, listed in the sixth century AD by Philoponus in his *Corollarium de Loco*.[4]

In addition Aristotle's own conception of place as the limit of the surrounding body involved some rather counter-intuitive consequences. Thus, on this account the outer sphere of the heavens turned out not to be in a place of its own (hence neither the heavens as a whole nor the whole cosmos could be said to be in a place in the relevant sense), because it did not have a surrounding body.[5] Similarly the parts of homogeneous substances could not be said to be in a place in their own right, because they were not surrounded by separate substances.[6] True, these problems were discussed by Aristotle himself, but particularly in the case of the emplacement of the heavens the details of his own solution remained unclear. We need not be surprized, accordingly, to find that problems such as these remained on the agenda of Aristotelian commentators.[7]

In the second half of the present chapter I shall discuss what evidence we have of the early Peripatetic discussions of place against the backdrop of the problems inherent in Aristotle's theory. But first, in sections 5.1 and 5.2, we shall study two problems that are particularly relevant to our understanding these early Peripatetic discussions, but that have not been sufficiently studied in the extant scholarly literature. It will appear on closer view that we are dealing with the two *aporiai* that Aristotle did not explicitly solve in *Phys* Δ.[8] The first one concerns the question

---

intended to show that, even in Aristotle's own view, motion would have to occur in jerks (although not necessarily in atomic jerks), and that also Aristotle would have to subscribe to the conclusion that a thing cannot be said to move, but only to 'have moved'. In that case Diodorus has signalled a gap between Aristotle's theory of place as a 'discrete' location on the one hand and his theory of motion as something continuous requiring a continuous stretch of 'space' or distance.

[4] Cf. Phlp *In Ph.* 567, 8–29.

[5] See the previous chapter, 185, and below 235–236. See also S. E. *M.* 10, 30–36; the solutions of the Greek commentators are discussed in Sorabji (1988a) 194–196; for some medieval solutions to this problem see Grant (1981b) 72–79.

[6] On which see below, 240–241. Some brief discussions of this problem are also to be found in Sorabji (1988a) 195 and Grant (1981b) 59.

[7] See below, 234–237 and 255–258.

[8] On these unsolved *aporiai* see above, chapter 4, 177–178. They were recognized as such by Simplicius, as may appear from his words at *In Ph.* 600, 13–17: Ἐπιστcησαι δὲ ἄξιον ὅτι δύο παρῆκεν ἀλύτους ἀπορίας ὁ Ἀριστοτέλης τῶν πρὸς τὸν τόπον ἀπορηθεισῶν τήν τε τρίτην καὶ τὴν τετάρτην· ὧν ἡ μὲν ἔλεγε 'τί

whether place may count as a cause, in particular whether natural places are in any way the causes of natural motion. This question will be addressed in section 5.1. Whether and to what extent the early Peripatetics dissented from Aristotle at this point is not immediately evident. The situation is of course the more complicated because Aristotle's position itself is not very clear at first sight. He deals with the subject at various places, but the passages at issue cannot be easily fitted into a coherent picture, and Aristotle himself nowhere explicitly puts all the pieces together. This has led to quite divergent interpretations in the later Aristotelian tradition, especially among scholastic commentators. Since we are here dealing with an interesting and important element of Aristotle's physical world picture—one, moreover, which is intimately connected with ideas that are central to his philosophy, such as causality and teleology—it seems worthwhile to try and see whether anything like Aristotle's 'original position' can be reconstructed, and to try to determine how the testimonies concerning the views of the early Peripatetics on natural place relate to this position.

The second *aporia* that was not explicitly answered in *Phys.* Δ, viz. the problem of the ontological status of Aristotelian place, will play an important role in section 5.2. Actually, the section deals with a problem which proved to be one of the notorious stumbling blocks for later Aristotelians, viz. the fact that Aristotelian place would seem to lack the required immobility.[9] But I shall argue that this problem is closely connected with the problem of the ontological status of Aristotelian place-as-a-surface.

The second half of the chapter, sections 5.3 and 5.4, will be devoted to early Peripatetic reactions to Aristotle's theory. In section 5.3 we shall study two important and interesting Theophrastean fragments dealing with place, while section 5.4 will concentrate on what remains of Eudemus of Rhodes' discussion of *Phys.* Δ. I shall try to show that both Theophrastus and Eudemus were, each in their own way, sensitive to the problems inherent in Aristotle's theory which are discussed in the first sections of this chapter. In the concluding section I shall try to draw some

---

γὰρ ἄν ποτε καὶ θείημεν εἶναι τὸν τόπον κτλ.' [...] τὴν δὲ 'ἔτι δὲ καὶ τίνος ἄν τις θείη τοῖς οὖσιν αἴτιον τὸν τόπον'.

[9] On medieval treatments of the problem of immobility see Grant (1981b).

threads together and to sketch a more general picture of early Peripatetic thought on place. Throughout this chapter references will be made to some of the later Greek and medieval commentators whenever such references appear to be apposite in the context of the subject of our enquiry. The reader should bear in mind, however, that this subject is the way in which the early Peripatetics reacted to some of the problems involved by Aristotle's theory of place, and that the present chapter does not pretend to provide anything like a comprehensive discussion of the study of place in the Aristotelian tradition.

5.1 *A problem in Aristotle: natural place and the explanation of natural motion*

5.1.1 That the elements move naturally to their own natural places is probably one of the most well-known features of Aristotle's world picture. Yet the details of Aristotle's explanation of natural motion, and in particular his ideas about the physical role of natural place in this process cannot very easily be extracted from his texts. His remarks on the subject are fairly brief and sometimes enigmatic, and there appears to be much that is left implicit. It should come as no surprise, accordingly, that on what seems to be a quite central issue, the question whether natural place functions as a *cause* of natural motion, Aristotle's theory has been variously interpreted. On closer view, there appear to be two main reasons for this exegetic *diaphonia*. Firstly, the various statements on natural place in the discussion of *topos* in *Phys.* Δ can at first glance not easily be fitted into a coherent picture. It seems particularly difficult to square Aristotle's statement that natural place has a certain power (*dunamis*) with his quite explicit remark that place is not one of the four well-known causes. Secondly, also the discussions of natural place and natural motion that can be found elsewhere in the *Physics* and in the *De Caelo* involve problems of their own. In fact, the details of the Aristotelian theory of the natural motion of the elements are controversial, so that their implications for the interpretation of *Phys.* Δ and for our reconstruction of Aristotle's original conception of natural place, are not easy to establish.

An extensive discussion of the various interpretations of Aristotle's theory of natural place which have been developed in the course of the Aristotelian tradition would of course be too much, but a brief survey of the main positions that have been taken, will at this point be appropriate. In their commentaries on *Phys.* Δ 208 b 8–25 (where we are told that place has a certain *dunamis*) and 209 a 18–22 (where we are told that place is not a cause) Themistius, Simplicius and Philoponus simply paraphrase Aristotle's argument without in any way implying that these passages are inconsistent or otherwise problematic.[10] In those passages, on the other hand, where for purposes of their own Simplicius and Philoponus go beyond a mere explication of Aristotle's text—Philoponus in order to criticize Aristotle's concept of place, Simplicius rather in order to supplement it—they gravitate towards crediting Aristotelian place with a causal status. In his corollary on place Philoponus is careful not to use the word 'cause' in describing the alleged role of natural place in Aristotle's theory of natural motion. Instead, he dutifully sticks to the word 'power' (*dunamis*). At the same time, however, he makes clear that he understands this power as a power *to move bodies*, and that natural place attracts those bodies as an object of desire.[11] In doing so he comes close to regarding place as a final cause of natural motion. Simplicius goes one step further. Although he appears to believe that, from Aristotle's point of view, place is not a cause of whatever kind, he adds that the subject deserves further enquiry, because the other commentators too readily pass it over.[12] Further on in his commentary, moreover, he adds on his own behalf that place can, in view of its role in the natural motion of the elements, very well be regarded as a final cause.[13]

---

[10] Them. *In Ph.* 103, 5 in addition appears to play down the importance of Aristotle's statement of the *dunamis* of place: τοῦτο τοίνυν ἐπιδείκνυσιν οὐ μόνον ὅτι ἔστι τι ὁ τόπος, ἀλλ' ὅτι καὶ ἔχει τινὰς διαφορὰς οἰκείας καὶ *οἷον* δυνάμεις.

[11] Phlp. *In Ph.* 579, 27–580, 3: καὶ εἰ δεῖ τὸν τόπον τινὰ *φυσικὴν δύναμιν* ἔχειν, καθ' ἣν τὰ κοῦφά τε καὶ βαρέα *τῶν ἰδίων ἐφιέμενα τόπων* ἕκαστον ὑπὸ τῆς φυσικῆς ὁρμῆς ἐπὶ τὸν οἰκεῖον φέρεται, τὸ δὲ τοιῦτον διάστημα κενὸν τῷ ἰδίῳ λόγῳ ὑπάρχον οὐδεμίαν δύναμιν ἔχειν οἷόν τε, τίνι λόγῳ τὰ μὲν ἐπὶ τόδε αὐτοῦ τὸ μέρος φυσικῶς ὁρμᾷ, τὰ δὲ ἐπὶ τὸ ἕτερον; That Philoponus thinks of the *dunamis* of Aristotelian natural place as a kind of causal power is even more apparent when he finally contrasts this position with his own, 581, 29–31: οὐχ ὁ τόπος οὖν ἔχει *τὴν δύναμιν τοῦ φέρεσθαι τὰ σώματα ἐπὶ τοὺς οἰκείους τόπους*, ἀλλὰ τὰ σώματα ἔφεσιν ἔχει τοῦ τὴν αὐτῶν φυλάττειν τάξιν. See also *In Ph.* 581, 18–21, quoted below, n. 21.

[12] Simp. *In Ph.* 533, 31–32: τοῦτο δὲ καὶ ἔτι ζητητέον προχείρως αὐτὸ παραδραμόντων τῶν ὑπομνηματιστῶν.

[13] Simp. *In Ph.* 600, 30–38.

This view of place as having a power to attract in the sense of being a final cause of elemental motion became widespread in the Middle Ages. It can be found in thinkers who are as divergent as Averroes, Thomas Aquinas and John Buridan.[14] Others, such as Bonaventura and Roger Bacon, went even further and came close to regarding natural place as an *efficient* cause working, albeit together with other causes, as a force attracting the elements the way a magnet attracts iron.[15]

Also among modern scholars opinions differ. Some of them, like Peter Machamer and David Furley, are quite willing to take Aristotle's own explicit statement that place is not a cause at face value.[16] Others, however, maintain that natural place is a cause after all, either a formal cause (Pierre Duhem),[17] or a final cause (Michael Wolff, Richard Sorabji).[18] Sorabji's interpretation is of particular importance for the purpose of the present chapter, for he argues that it is precisely this Aristotelian conception of the dynamic character of natural places that came under attack in the early Peripatos. According to Sorabji not only Strato—who, as is well known, rejected the whole Aristotelian theory of the dynamics of elemental motion—but also Theophrastus moved away from Aristotle, in rejecting the view that place is a final cause.

---

14 Cf. Averroes in *Aristotelis de physico auditu libri octo cum Averrois Cordubensis variis in eosdem commentariis*, Venezia 1562, 122 K: 'et apparet ex hoc, quod locus est *perfectio moti ad ipsum*'; and Thomas Aquinas, *De physico auditu sive physicorum Aristotelis Commentaria liber IV*, lectio i (ed. A. M. Pirotta, Napoli 1953): 'non autem ex hoc ostenditur quod locus habeat virtutem attractivam nisi *sicut finis* dicitur attrahere'. However, on the fact that both Averroes' and Thomas' interpretation were in the end more subtle than might appear from these quotations (and indeed more subtle than is usually being assumed) see below, 220. For Crescas' interpretation, largely based on Averroes, see Wolfson (1929) 141, and 337–38, n. 22 (on Averroes).

15 Cf. Bonaventura *Sent.* II, dist. 14, pars I, art. III, qu. 2: 'dicendum quod ad motum gravis non sufficit solummodo gravitas sive qualitas propria, immo concurrit virtus loci attrahentis et virtus loci expellentis' etc.; for references to relevant passages in Roger Bacon's *Quaestiones supra libros IV Physicorum*, the *Quaestiones supra libros VIII Physicorum*, and the *Communia Naturalium*, for references to other scholastics, and for an excellent discussion of this view of natural place and the difficulties it entails, see Maier (^{2}1952) 174–182. See also Grant (1981a) 297, n. 126 on Buridanus' reaction to this position, with further references.

16 Furley (1976) 92; Machamer (1978) 377–387.

17 Thus e.g. Duhem (1913) 208.

18 Thus e.g. Dijksterhuis (^{4}1980) 44 and 195; Wolff (1987) 96, n. 44, and 111–113; Sorabji (1988a) 186–187.

Before we proceed to study Aristotle's own statements on the subject, we may briefly reflect on some problems we would seem to run into, if we credit Aristotle with the view that natural place is a cause of natural motion. Apart from the fact that this interpretation can only be defended at the cost of either ignoring or explaining away Aristotle's own explicit statement, at *Phys.* Δ 209 a 18–22, that place is not to be identified with any of the four causes,[19] it may be doubted whether it is a charitable interpretation from a philosophical point of view. Admittedly, the account of elemental motion as found in *Phys.* Θ clearly implies that natural place has at least *some* role to play in the explanation of natural motion. But given the theory of place-as-a-surrounding-surface of *Phys.* Δ, it would hardly be a good idea to treat natural place as a final cause. How, for one thing, could a surface have any power at all? It might be argued that this way of putting the question is misleading, in so far as it suggests that natural place works as an efficient cause, whereas after all it should be seen as a final cause, in the sense of a goal which the elements strive after by their own nature.[20] Even so, however, the theory does not seem to make sense. For in what sense could a surface serve as a goal for a body?[21]

But there is more. Place as described in *Phys.* Δ is independent of what is (going to be) in it, whereas the natural tendency of the elements towards their own natural places, of which *Phys.* Θ and other texts speak, is clearly supposed to be part and parcel of the elements' own constitution. In Solmsen's words, 'the place of an element belongs to its *phusis* and cannot be as extrinsic to it as the "place" which Aristotle investigates in the *Physics*'. Solmsen's

---

19 Thus Sorabji (1988a) 187, n. 6; and Sorabji (1987a) 16, n. 93, argues that Aristotle's rejection of the idea that (natural) place figures as a *cause* 'is merely part of a puzzle or *aporia*'. But see below, 202.

20 Weisheipl (1985b) 106 points out that 'final cause is said to "move" the body to act only metaphorically. That is to say, a final cause does not move the agent in the same way as the efficient cause moves the effect, and so the final cause is said to "move" metaphorically, since it moves not by efficiency but by being desired'. He is thus reviving a distinction that was also made by some medieval thinkers, on which see Maier ([2]1952) 180, esp. n. 73. Of course the final cause is no less a cause for all that.

21 This objection is also raised by Philoponus in his *Corollarium de Loco,* at *In Ph.* 581, 18–21: τὸ δὲ λέγειν δύναμίν τινα ἔχειν αὐτὸν καθ' αὐτὸν τὸν τόπον γελοῖον πάνυ· οὐ γὰρ ἐπιφανείας ἐφιέμενα φέρεται ἐπὶ τοὺς οἰκείους τόπους ἕκαστα, ἀλλὰ τῆς τάξεως ἐφιέμενα, ἧς παρὰ τοῦ δημιουργοῦ τετυχήκασιν.

own conclusion is that 'once more departmentalization appears as a significant characteristic of Aristotle's scientific work'.[22] The implication is that the concept of place of *Phys.* Δ does not prove to be useful in those contexts where natural place is at the focus of attention. In *Phys.* Δ, however, Aristotle is only too eager to link the two subjects of place and natural place,[23] and he even seems to claim that the theory of place which he there puts forward tallies best with the *phenomenon* of natural motion.[24] Drawing the conclusion that in actual fact he failed to link these two subjects would thus amount to saying that Aristotle did not live up to his own standards. That would at any rate be a high prize for buying a theorem—the theorem of the causal power of natural place—which Aristotle himself nowhere explicitly mentions. Before accepting this conclusion without further ado, we should therefore first see whether perhaps there is something wrong with the premise that natural place is a cause.

That is what the present section aims to do. The analysis in the next two subsections will show that Aristotle never credited place with any proper causal power and that his account of the role of natural place in the explanation of natural motion was coherent. First, in 5.1.2, I shall discuss the account of *Phys.* Δ. It will appear that, once its dialectical structure is properly observed, the *prima facie* inconsistencies, which it may seem to contain, vanish. Next, in 5.1.3, it will be shown that basically the same position underlies the accounts of natural motion in *Phys.* Θ and *Cael.* Δ, but that at the same time these accounts allow us to complete the picture by putting in a number of further details. The conclusions will be formulated in 5.1.4.

5.1.2 There are five passages in the discussion of *topos* in *Phys.* Δ that any interpretation of Aristotle's theory of natural place will

---

22 Solmsen (1960) 128. Cf. also *ibid.* 443: 'the definition of place as given in *Phys.* IV does not prove useful anywhere else'.

23 This may also count as an objection against Sorabji's claim, (1988a) 186, that 'Aristotle offers at least four views on place', viz. (1) place as a quantity in the *Cat.*, (2) natural place (in the *Cael.*), (3) in biology, the conception of up, down etc. being defined in terms of biological function, (4) the 'classical' conception of place outlined in the *Phys.* For at least in Aristotle's 'classical' discussion of *topos*, i.e. in the *Physics*, (2) and (4) are regarded as connected.

24 Cf. *Phys.* Δ 212 b 29–213 a 1, on which see below, 205–206.

have to take into account. First, at 208 b 8–25, Aristotle mentions the apparent fact that place has a certain attractive power (*dunamis*) —after all each of the four elements moves to its own place when not impeded—as one of the *phainomena* that point to the existence of place as something in its own right. Secondly, at 209 a 18–22 he puts forward an *aporia* which claims that place is not one of the four known causes. Thirdly, at 210 a 2 he works out his view that place is at any rate not to be identified with either matter or form. In the fourth passage to be taken into account Aristotle presents a revised list of *phainomena* in which the alleged *dunamis* of place is no longer mentioned, but in which it is nevertheless implied that (natural) place is in some way connected with the natural motion of the elements. Finally, at 212 b 29 ff., he alleges that on his own theory a reasonable account can be given of the phenomenon of the natural motion of the elements, the implication (once again) clearly being that (natural) place has some role to play in the explanation of that motion. How are these four passages to be reconciled?

Here I believe it will be helpful to realize that these apparently partly contradictory statements should not all be given the same weight, since we are obviously dealing with statements reflecting various stages of the dialectical procedure. On the basis of the investigations of the previous chapter into Aristotle's dialectical method, we may try to throw some light on these passages and their interrelations. I shall discuss them one by one.

Aristotle's remark about the phenomenon of natural motion—viz. that this phenomenon shows that place has a certain *dunamis* —is part of a list of *prima facie* arguments for the existence of place as such.[25] In the dialectical procedure of *Phys.* Δ these arguments merely constitute the first stage (the collecting of the *phainomena* or 'apparent facts'). Although the rules of this procedure determine that the 'apparent facts' will in the end have to be accounted for in some way, this does not mean that all *phainomena* concern-

[25] The argument on the *dunamis* of place is connected, by the word ἔτι, to the preceding one, which in its turn is introduced (208 b 1) by ὅτι μὲν οὖν ἔστιν ὁ τόπος δοκεῖ δῆλον εἶναι ἐκ κτλ. The whole section is resumed at 208 b 27 by the statement ὅτι μὲν οὖν ἐστί τι ὁ τόπος παρὰ τὰ σώματα, καὶ πᾶν σῶμα αἰσθητὸν ἐν τόπῳ, διὰ τούτων ἄν τις ὑπολάβοι. In other words the common conception that there are natural places is introduced as one of the *phainomena* which are relevant in connection with the study of the nature of place as such.

ing place that are listed in this first section of *Phys.* Δ will stand up to Aristotle's further dialectical examination.[26] Thus, he does not himself ultimately subscribe to the idea that there is a void, nor to the idea that place is prior to all other things (a supposition which, as he claims, underlies Hesiod's famous line about *chaos*). In these cases apparently his final account will have to show how such ideas could arise in the first place, while at the same time proving them to be wrongheaded.

But apart from the fact that Aristotle's dialectical method does not commit him to the endorsement of *all* initial *phainomena*, it does not commit him to fully and uncritically accept those he does take over either. In this connection we should recall what he says at the beginning of the first book of the *Physics* about the philosopher's obligation to proceed from 'what is better known to us' to 'what is better known by nature'.[27] There he makes clear that the initial apparent facts lie to the eventual refined concepts as the often ambiguous and vague terms of ordinary language lie to proper definitions.

These observations may serve to show that the statement that place has a certain power (*dunamis*) should not be accepted just like that, as something Aristotle was himself firmly committed to. Of course it *could* in principle just mean that place is a cause, although Aristotle's opting for the cautious formula of place 'having a certain influence' (ἔχει τινὰ δύναμιν), instead of simply speaking of place being a cause, should make us pause. But the statement could just as well represent a mistaken view which Aristotle himself does not subscribe to after all (like Hesiod's view), or it could represent a casual and vague formula behind which there is some basic truth that Aristotle's physical theory would in the end express in more adequate and more refined terms. It will be clear that the point of Aristotle's statement cannot be correctly assessed outside the (dialectical) context. We shall therefore now take a closer look at this context, i.e. at the further dialectical procedure, to decide which of these interpretations is the most likely one.

---

26 That not all initial *phainomena* need always be saved is explicitly stated by Aristotle at *EN* 1145 b 1–8, a programmatic statement introducing his discussion of temperance and incontinence (the passage is quoted and briefly discussed in the previous chapter, 157.

27 The passage at issue has been discussed in the previous chapter, 174–175.

First, the *aporia* of 209 a 18–22. Aristotle here explicitly rejects the idea that (natural) place figures as a *cause*:

> Also we may ask: of what in things is place the cause? None of the four modes of causation can be ascribed to it. It is neither cause in the sense of the matter of existents (for nothing is composed of it), nor as the form and definition of things, nor as end, nor does it move existents.[28]

The passage is part of a list of *aporiai* that starts at 209 a 2 with the question what place actually is, i.e. to what *genos* it belongs. This list of *aporiai* is linked by a 'however' (ἀλλὰ μὴν) to the list of initial apparent facts, and it exhibits a similar structure (with each subsequent *aporia* being introduced by the word ἔτι),[29] so that we may safely assume that it was meant to counterbalance the latter. Thus the *aporia* we are dealing with offers us a second thought about the alleged *dunamis* of place. We are not allowed to dismiss cavalierly such a second thought, or to play down the denial that place is a cause because it 'is merely part of a puzzle or *aporia*'.[30] In fact, only would we have a reason to assume that in formulating this *aporia* Aristotle does not have to be taken seriously or that he is not speaking *in propria persona*, if it could be shown that the thesis at issue is elsewhere explicitly contradicted. For, as we have seen in the previous chapter, Aristotle's procedural rules determine that an *aporia* has to be solved. So as long as it is not solved, the initial position against which it was raised in the first place will have to be regarded as wrongheaded. In this case, however, Aristotle nowhere shows that the *aporia* is harmless, and he *nowhere* speaks of natural place as a cause.[31]

Returning to the question of the interpretation of the statement

---

[28] *Phys.* Δ 209 a 18–22: ἔτι δὲ καὶ τίνος ἄν τις θείη τοῖς οὖσιν αἴτιον εἶναι τὸν τόπον; οὐδεμία γὰρ αὐτῷ ὑπάρχει αἰτία τῶν τεττάρων· οὔτε γὰρ ὡς ὕλη τῶν ὄντων (οὐδὲν γὰρ ἐξ αὐτοῦ συνέστηκεν) οὔτε ὡς εἶδος καὶ λόγος τῶν πραγμάτων οὔθ' ὡς τέλος, οὔτε κινεῖ τὰ ὄντα.

[29] See also above, n. 25.

[30] This objection is to be found in Sorabji (1988a) 187, n. 6; and Sorabji (1987a) 16, n. 93.

[31] Sorabji, who takes Aristotle's natural place to figure as a final cause, admits this (1988a) 186: 'The most likely explanatory role, *although he never says this*, is as a final cause or goal (not consciously sought) of motion' (my italics). In fact the only passage which comes close to explicitly identifying place with cause (but then rather with formal cause) is a somewhat obscure phrase at *Cael.* Δ 310 a 34–35: τὸ δ' εἰς τὸν αὑτοῦ τόπον φέρεσθαι ἕκαστον τὸ εἰς τὸ αὑτοῦ εἶδός ἐστι φέρεσθαι. But see on this below n. 33.

concerning the *dunamis* of place, we may now safely conclude that it was at any rate not meant to convey the idea that place figures as a cause. The second possible interpretation mentioned above, viz. that the statement represents a view to be rejected entirely, is not very likely either. For Aristotle apparently did not take the trouble to explicitly attack it, as he definitely did in the case of other *endoxa* that were to be rejected, such as the existence of void. On the other hand, it is not explicitly repeated anywhere either. Now this very fact may suggest that the third possible interpretation mentioned above is the one to be preferred, and that the statement concerning the *dunamis* of place should be considered as a rough and untechnical way of putting something which Aristotle's own theory would in the end express differently.

We now come to the third passage to be discussed here: *Phys.* Δ 210 a 2–9. As part of his attempt to show that place cannot be matter, Aristotle adduces two arguments which in actual fact prohibit the identification of place with either matter or form:

> Further how could a body be carried to its own place, if place was the matter or the form? It is impossible that that which has no reference to motion or the distinction of up and down can be place. So place must be looked for among things which have these characteristics.
> If the place is in the thing (it must be if it is either form or matter) place will have a place; for both the form and the indeterminate undergo change and motion along with the thing, and are not always in the same place, but are where the thing is. Hence the place will have a place.

The first argument focuses on the fact that matter and form are not of themselves connected with motion, nor with the fixed directions associated with natural motion. Natural motion is said to become inexplicable when place is identified with either of these two causes. The second argument turns on the same point as the objection we have already put forward against identifying place with a final cause, viz. that matter and form are intrinsic to that of which they are matter and form, whereas place as defined in *Phys.* Δ clearly is extrinsic.[32] It should be stressed that Aristotle in no way qualifies these arguments and that, given the overall

---

[32] Thus this argument partly repeats the claim of 209 b 22–23: τὸ μὲν γὰρ εἶδος καὶ ἡ ὕλη οὐ χωρίζεται τοῦ πράγματος, τὸν δὲ τόπον ἐνδέχεται.

approach of *Phys.* Δ, which repeatedly stresses the extrinsic character of place, they certainly make sense. So two of the four Aristotelian causes are here clearly excluded from being candidates for the identification of place *tout court*, and of natural place *a fortiori*.[33]

We shall now turn to our fourth passage. At *Phys.* Δ 210 b 32–211 a 5, Aristotle provides a brief recapitulation of the properties that appear to belong to place.

> Let us take for granted the various characteristics which are supposed truly to belong to it. We assume first that place is what contains that of which it is the place [...] and in addition that all place admits of the distinction of up and down, and each of the bodies is naturally carried to its appropriate place and rests there, and this makes the place either up or down.

What Aristotle here offers, as he himself indicates, is a *revised* list of phainomena: a list not just of those properties that are commonly believed to belong to place, but of those phainomena that appear *truly* to belong to place (ὅσα δοκεῖ ἀληθῶς καθ' αὑτὸ ὑπάρχειν αὐτῷ). I take it to be significant that in this revised list the idea of place having a *dunamis* no longer occurs. But it is equally significant that the fact that the difference between 'up' (ἄνω) and 'down' (κάτω) is in some way to be connected with the natural motion of the elements *is* explicitly mentioned. So this *is* a phainomenon of which the theory will eventually have to take account. This becomes even more clear in the next (and fifth) passage to be discussed.

Towards the end of the discussion of *topos* in *Phys.* Δ, Aristotle claims that the concept of place as a surrounding surface allows one to solve (a number of) the *aporiai* that have been put forward earlier (212 b 22 ff.). He then rather abruptly leaves the topic of the solution of the *aporiai*, to apparently attempt to show how his own concept of place fits in with the phenomena concerning natural

---

[33] The explicit statement that place can *not* be form carries an important implication for our interpretation of a famous line in the *De Caelo*, where it is said that a thing moving towards its own (natural) place is tantamount to its moving to its own form (τὸ δ' εἰς τὸν ἑαυτοῦ τόπον φέρεσθαι ἕκαστον τὸ εἰς τὸ ἑαυτοῦ εἶδος φέρεσθαι). We shall have a closer look at this passage below, but for the moment we may conclude that—unless we assume, with little likelihood, that when writing the *De Caelo* passage Aristotle had completely abandoned the position of *Phys.* Δ—this passage *cannot* simply mean that natural place just *is* form.

motion and natural place.[34] This issue is addressed in the following words (*Phys.* Δ 212 b 29–213 a 1):

> Also, it can be explained that each kind of body should be carried (φέρεται ... εὐλόγως) to its own place. For a body which is next in the series and in contact (not by compulsion) is akin, and bodies which are united do not affect each other, while those which are in contact interact on each other. Nor is it inexplicable that each should remain (μένει ... οὐκ ἀλόγως) naturally in its proper place. For parts do, and that which is in a place has the same relation to its place as a separable part to its whole, as when one moves a part of water or air: so too air is related to water, for the one is like matter, the other form—water is the matter of air, air as it were the actuality of water; for water is potentially air, while air is potentially water, though in another way.

The account is difficult and compressed. In fact Aristotle himself admits that he is merely forced by the present subject briefly to touch upon these matters, and he promises to give a fuller account elsewhere.[35] Nevertheless the train of thought can be followed in rough traits. The idea seems to be that the elementary masses, when in their proper order ('next in the series and in contact, not by compulsion'), are so akin as to be, in a sense, united so that they do not affect each other. This (their not affecting each other) is what causes their natural rest. Conversely, their motion to their proper places is apparently brought about when they are not in their proper order, and hence in contact (instead of forming a unity) and interacting. Admittedly, this condensed account may be regarded as containing problems of its own,[36] but we are allowed to draw two conclusions that are significant in the present context. Firstly, there is no hint that natural places are themselves, *qua* places, supposed to act as causes of the phenomena of natural

---

[34] Cf. Simp. *In Ph.* 597, 5–9: Τοῦτο οὐκ ἀπορίας λύσιν ἔχει, ἀλλὰ τελευταῖον ἔκειτο τῶν περὶ τόπου ἀξιωμάτων τὸ φέρεσθαι φύσει καὶ μένειν ἐν τοῖς οἰκείοις τόποις ἕκαστον τῶν σωμάτων. *δείκνυσιν οὖν καὶ τοῦτο ἀκόλουθον ὂν τῷ τοῦ τόπου ὁρισμῷ* ἅμα καὶ ἔνστασιν ὑποφερομένην οἶμαι λύων τινά· διὸ καὶ τοῦτο ἥρμοσε ταῖς τῶν ἀποριῶν διαλύσεσι.

[35] *Phys.* Δ 213 a 5: ἀλλὰ διὰ τὸν καιρὸν ἀνάγκη μὲν εἰπεῖν, ἀσαφῶς δὲ νῦν ῥηθὲν τότ' ἔσται σαφέστερον. The reference is usually assumed to be to *GC* I, 3.

[36] For example, the idea that natural motion and natural rest are caused by the interaction or lack of interaction of the elements cannot be squared very easily with the accounts of natural motion to be found elsewhere in Aristotle, on which see below, section 5.1.3. It may have been passages like this one in Aristotle that induced Strato of Lampsacus to frame an altogether different theory of the dynamics of natural motion, on which see Strato fr. 80 Wehrli.

motion and rest that are being described. Secondly, Aristotle's use of the terms εὐλόγως and οὐκ ἀλόγως seems to indicate that his own theory allows a meaningful explanation of the phenomena described. I surmise what he has in mind is this: the fact that his own definition of place makes use of the notion of the limit *of the surrounding body* makes it possible to differentiate between natural place (the limit of the right body, i.e. of the one next in succession) and non-natural place (the limit of the wrong body). In other words, it is precisely because the natural motion or rest of the elements is in one way or another dependent on the bodies that are surrounding them, that Aristotle's concept of place—in contradistinction to e.g. the isotropic space-void of the early atomists—allows for meaningful talk about natural places. *Qua* place, i.e. *qua* surface, it may be inert (it does not act in any way, nor is it as a surface the cause of natural motion), but at the same time it is not isotropic. For it *does* make a difference *to what kind of body* the surrounding surface actually belongs.

This interpretation is corroborated by comparison of these passages on place and natural motion with Aristotle's critique of the early atomist theory of the void. In his discussion of void, which follows immediately upon his study of *topos* in *Phys.* Δ, Aristotle wants us to believe that the early atomists introduced the void as a cause of motion, and he tries to show that it cannot play that role.[37] He is partly guilty of polemical distortion, since for all we know the atomists never claimed that place or the void figures as a cause of motion. They rather seem to have thought of place as a necessary condition for motion to occur.[38] That, however, need not concern us now. For our present purpose it is more important to look at the arguments Aristotle adduces to show why the concept of an infinite void space does not work.

Aristotle argues that all motion takes place either naturally or

---

[37] *Phys.* Δ 214 b 12 ff.; 214 b 28 ff.; 215 a 6 ff.

[38] Aristotle indeed takes the principle that void is a necessary condition for motion as his starting-point but reformulates it as the statement that place is a cause. He then turns the tables on his opponents by claiming that the void actually precludes natural motion and hence any kind of motion whatsoever. I believe that it is important to realize that this argument is polemical, and perhaps even positively sophistical, and that it should therefore not be taken as implying that Aristotle himself thought that void *qua* place (or space) should act as a cause. For a different view, however, see Sorabji (1988a) 142–143, 186.

by force. Since forced motion, being unnatural motion, is logically posterior to natural motion, the impossibility of the latter entails the impossibility of the former, and hence of all motion. He then continues as follows:

> But how can there be *natural* movement if there is no difference throughout the void or the infinite? For in so far as it is infinite, there will be no up or down or middle, and in so far as it is void, up differs no whit from down; for as there is no difference in what is nothing, there is none in the void—for the void seems to be a non-existent and a privation. But natural locomotion seems to be differentiated, so that the things that exist by nature must be differentiated. Either, then, nothing has a natural locomotion, or else there is no void.[39]

So one of the main objections against the early atomist concept of the infinite void is, according to Aristotle, that the void can play no role in the explanation of natural motion, since there is no 'up', 'down' or centre in the infinite, and no difference between up and down in the void which is equivalent to 'nothing'.[40]

We may now try to put the pieces together and see what picture emerges from Aristotle's discussion of place and void in *Phys.* Δ. The initial *phainomenon* that place has a certain power (*dunamis*) was nowhere explicitly contradicted, but neither was the *aporia* which was raised against it, viz. that place is not one of the four causes. That at any rate (natural) place cannot be a material or a formal cause was confirmed by the third passage discussed, which claimed that natural motion would be inexplicable if (natural) place was to be identified with either of these two causes, because form and matter are inseparable from the substance to which they belong, whereas place should be separate. The remaining three texts here discussed all fit in with this picture. They contain no suggestion that place should be regarded as a cause, but at the same time they all do suggest that place should play *some* role in the explanation of natural motion. All in all, we may conclude that the text of *Phys.* Δ does not support the traditional identification of natural place as the final cause of natural motion.

---

39 *Phys.* Δ 215 a 6–13.

40 These Aristotelian arguments were discussed and refuted by Philoponus in his *Corollarium de Loco*, at *In Ph.* 579, 21–582, 19. On Aristotle's arguments against motion in a void see Furley (1976) 83–100, esp. 94.

This leaves us with two questions. First of all, what final cause of motion *does* Aristotle offer as an alternative—for given the obviously teleological character of natural motion it is hardly conceivable that there is no final cause at work. Secondly, what then *is* the role which Aristotle is willing to accord to natural place, if it is not that of being a final cause? I shall try to provide a full answer to both questions on the basis of a closer study of some passages in *Phys.* Θ and *Cael.* Δ, where Aristotle further develops what we might call his theory of the dynamics of natural motion. What is more, on the basis of the passages from *Phys.* Δ discussed thus far, we may already provide a preliminary answer to the second question. What these texts imply is that the role of the concept of Aristotelian natural place is just to allow a meaningful application of the notions of natural motion and natural rest. It does so by making it possible to differentiate between natural and unnatural places, in a way rival theories of place do not. After all, for an element to be in its natural or unnatural place is tantamount to its being surrounded by respectively the right or the wrong elementary mass. It is for this reason that a theory defining place in terms of the *surrounding body* proves the most useful.

5.1.3 We shall now leave the account of *Phys.* Δ to turn to what are in fact Aristotle's fullest accounts of the dynamics of natural motion. They are to be found in the third section of the fourth book of the *De Caelo,* in the context of a general discussion of the heavy and the light, and in the fourth section of the eighth book of the *Physics,* in the context of a general discussion of the causes of motion. First the discussion of the causes of natural motion in the fourth chapter of *Phys.* Θ. The chapter as a whole offers a complicated account and the interpretation of some of the details is controversial. It will not be possible—nor necessary, for that matter—to study the text in detail here or to go into the *minutiae* of the scholarly debate concerning its interpretation.[41] It will be enough for the purpose of our present investigation first to

[41] Important modern accounts are to be found in: Maier (1952) 143–151 (who moreover in the sequel to these pages gives a useful overview of the most important medieval interpretations); Dijksterhuis ([4]1980) 26–35; Solmsen (1960) 253–286 (but Solmsen hardly considers *Phys.* Θ); Bourgey (1955) 383–392; Weisheipl (1985a) 75–79 and (1985b) 101–108; Furley (1976) 91–93, Machamer (1978) 377–387, Sorabji (1988a) 219–226.

paraphrase the argument in its outlines and, next, to show what it tells us about Aristotle's views of the causal status of natural place.

The question dealt with by Aristotle is basically this one: what, given the famous proposition at the beginning of the seventh book of the *Physics*, viz. that 'everything that is moved is moved by something', is that by which the elements are moved when they exhibit their natural motion? In the case of things that are moved by force the mover is extrinsic and obvious. In the case of things moved naturally *by themselves*, like animals, the mover is present in the thing itself, but is less obvious than the mover of things that exhibit unnatural motion, because the distinction within the moving thing itself between mover and moved is more difficult to make.[42] The movers responsible for the natural motion of the elements are even less obvious, and Aristotle admits that this case presents the greatest difficulty.

The elements are not moved by an extrinsic cause in the way things in forced motion are. Neither, however, are they moved by themselves in the way animate things are, for that way of being moved is characteristic of living things. In fact Aristotle offers several reasons for differentiating between animate and inanimate natural motion. To mention only one of them, living things that move themselves can move wherever they please, but it is not up to the elements to move one way rather than another.[43] Aristotle therefore has to take a rather intricate stand in explaining the natural motion of the elements. On the one hand he does claim that they have an intrinsic cause, or rather principle or source (*archê*) of motion:

---

[42] I here paraphrase *Phys.* Θ 254 b 25–32.

[43] I here paraphrase *Phys.* Θ 254 a 32–255 a 19. Note that for Aristotle the crucial difference apparently concerns the capacity to move actively, a capacity which animate beings exhibit, and the capacity to move passively, i.e. to be unable to do anything else but move, when unimpeded. Accordingly, there is no particular need to assume that Aristotle was forced to add external causes in the case of elemental motion, for the sole reason that otherwise there would have been no need to invoke a first unmoved mover to explain the rotation of the celestial sphere. Clearly, to his mind the internal causes were sufficiently different in kind, in the cases of animate beings on the one hand and inanimate beings on the other, to warrant the claim that external movers are required only in the case of the latter. For a different view see Sorabji (1988a) 219–222.

> 'So it is clear that in all these cases the thing does not move itself, but it contains within itself the source of motion—not of moving something or of causing motion, but of suffering it'.[44]

On the other hand it is in virtue of the fact that this internal source of motion is only a capacity for 'suffering' motion that an additional external cause is required. Indeed some cause will have to be found which is responsible for the fact that natural motion does occur at all, for why shouldn't the elements simply stay in their own natural places? Now the reason why they do not, is the fact that they can change into each other, as the following passage from the *De generatione et corruptione* makes clear:

> These considerations serve at the same time to explain what is to some people a puzzle—viz. why the bodies, since each of them is travelling to its own place, have not become dissevered from one another in the infinite lapse of time. The reason is their reciprocal transformation. For, had each of them persisted in its own place, instead of being transformed by its neighbour, they would have got dissevered long ago. They are transformed, however, owing to the motion with its dual character; and because they are transformed, none of them is able to persist in any fixed place.[45]

A cosmos without elementary change would be composed of static, immobile layers of elements. It is the occurrence of processes such as water turning into air that is responsible for the fact that, from time to time, a certain quantity of an element turns out, so to speak, to be in the wrong place.

Aristotle identifies two kinds of external cause. In order to be able to sketch their respective roles he inserts an excursus on the

---

[44] *Phys.* Θ 255 b 29–31: ὅτι μὲν τοίνυν οὐδὲν τούτων αὐτὸ κινεῖ ἑαυτό, δῆλον· ἀλλὰ κινήσεως ἀρχὴν ἔχει, οὐ τοῦ κινεῖν οὐδὲ τοῦ ποιεῖν, ἀλλὰ τοῦ πάσχειν. For a similar stress on the internal principle of motion in these cases, see *Phys.* B 192 b 9–15, where the things that exist are divided into those that exist by nature and those that exist by virtue of other causes. An exclusive characteristic of the things that exist by nature—among which the elements are explicitly mentioned—is that they have an intrinsic principle of motion and rest. Admittedly the term *kinêsis*, which is used here refers to all kinds of change—they are in fact all enumerated—but locomotion is certainly among them: 'each of them [scil. of the things existing by nature] has within itself a principle of motion and of rest, in respect of place, or of growth and decrease, or by way of alteration' (192 b 14–16). It is clearly implied that the principle of natural motion and natural rest for the elements is intrinsic, and in what follows (192 b 20–22) this principle is identified as their nature. Not even a hint on the alleged role of natural place.

[45] *GC* 337 a 8 ff.

use of the term 'potentially'.[46] For the two different kinds of external cause are each linked to a different kind of potentiality that is being actualized. An example will make this clear. Water, being itself heavy, is potentially light, since it can change into air, which is light. This potentiality—let us call it 'first potentiality'—is actualized by the external cause (in this case: heat) which turns water into air. This generator—it is usually and conveniently referred to by the Latin scholastic term *generans*—is the first and most important external cause identified by Aristotle. Now as soon as water has become air, it may still remain potentially light in a different sense: it is no longer heavy, but it may be hindered from being actually light, i.e. from being in its own natural place.[47] Let us call this second way in which an element can be potentially light 'second potentiality'. Here the second kind of external cause comes in: that which removes the obstacle —the *removens impedimentum*, as it was conventionally called in medieval scholastic discussions—allows the element to actualize this second potentiality.[48]

We may conclude, accordingly, that there are three factors being introduced here, the thing's inner tendency to be moved, the generator and the *removens impedimentum*. I shall here ignore the difficult question of how these three factors interrelate,[49] and

---

[46] Cf. *Phys.* Θ 255 a 30–32: ἐπεὶ δὲ τὸ δυνάμει πλεοναχῶς λέγεται, τοῦτ' αἴτιον τοῦ μὴ φανερὸν εἶναι ὑπὸ τίνος τὰ τοιαῦτα κινεῖται, οἷον τὸ πῦρ ἄνω καὶ ἡ γῆ κάτω.

[47] Cf. *Phys.* Θ 255 b 11–12: ἐνέργεια δὲ τοῦ κούφου τὸ ποὺ εἶναι καὶ ἄνω ('the actuality of lightness consists in the light thing being in a certain place, namely high up'; note that the revised Oxford translation here misleadingly translates Aristotle's *energeia* as 'activity').

[48] We should note, however, that Aristotle makes it quite clear that this *removens impedimentum*, though in certain cases indispensable, is only *in a way* to be regarded as a cause, whereas the *generans* is more properly to be called the external cause of an element's natural motion.

[49] Weisheipl (1985a) 80–83, and (1985b) 101–108, argues that, properly speaking, only the *generans* and the *removens impedimentum* are true causes, and that the inner tendency of the elements is not. Sorabji (1988a) 220–221, on the other hand, speaks of the *generans* and the *removens impedimentum* as 'a pair of low-grade causes which do not meet his full requirements for efficient causation' because they are not in contact with what they move (on the contact requirement, however, see Weisheipl (1985b) *passim*), whereas he regards the internal nature of the elements as the proper cause of their natural motions. Interestingly, these two different interpretations echo the medieval controversy between the Thomist and the Scotist theories of 'gravity', on which see Maier ([2]1952) 156–158.

turn directly to what for our present purpose are the two most important questions: where, and how, does natural place come in, and how does Aristotle account for the teleological character of elementary motion? Two passages in particular will have to be taken into account, if we want to answer these questions. First, at 255 a 24–28 we find the remark that

> 'Similarly, what is potentially of a certain quality or of a certain quantity or *somewhere* (που) is naturally movable when it contains the corresponding principle in itself and not accidentally (for the same thing may be both of a certain quality and of a certain quantity, but the one is an accidental, not an essential property of the other). So when fire and earth are moved by something the motion is violent when it is unnatural, and natural when it brings to actuality what they potentially possess'.[50]

Three categories of natural motion—or rather: of natural change —are mentioned: change of quality, quantity or place. In all three cases the change at issue is to be described as a potentiality being actualized. The change is natural, moreover, when the principle of change is internal and not accidental. As a corollary, natural change may be regarded as the actualization of a potential location, quantity or quality, in those cases where the actualization of this potentiality belongs to the essence of the thing at issue. A few lines further on, at 255 b 11, this principle is applied to the case of light and heavy elements:

> 'The actuality of lightness consists in the light thing being somewhere (που), namely high up: when it is in the contrary place it is being impeded. The case is similar with regard to quantity and quality. But, be it noted, this is the question we are trying to answer: how can we account for the motion of light things and heavy things to their proper places? The reason for it is that they have a natural tendency to go in a certain direction (πέφυκέν ποι); and this is what it is to be light or heavy, the former being determined by an upward, the latter by a downward tendency'.[51]

---

50 *Phys.* Θ 255 a 24–28: Καὶ κινητὸν δ' ὡσαύτως φύσει τὸ δυνάμει ποιὸν ἢ ποσὸν ἢ *πού*, ὅταν ἔχῃ τὴν ἀρχὴν τὴν τοιαυτὴν ἐν αὑτῷ καὶ μὴ κατὰ συμβεβηκός (εἴη γὰρ ἂν τὸ αὐτὸ καὶ ποιὸν καὶ ποσόν, ἀλλὰ θατέρῳ θάτερον συμβέβηκεν καὶ οὐ καθ' αὑτὸ ὑπάρχει). τὸ δὴ πῦρ καὶ ἡ γῆ κινοῦνται ὑπό τινος βίᾳ μέν ὅταν παρὰ φύσιν , φύσει δ' ὅταν εἰς τὰς αὑτῶν ἐνεργείας δυνάμει ὄντα.

51 *Phys.* Θ 255 b 11–12: ἐνέργεια δὲ τοῦ κούφου *τὸ ποὺ εἶναι* καὶ ἄνω, κωλύεται δ', ὅταν ἐν τῷ ἐναντίῳ τόπῳ ᾖ. καὶ τοῦθ' ὁμοίως ἔχει καὶ ἐπὶ τοῦ ποσοῦ καὶ ἐπὶ τοῦ ποιοῦ. καίτοι τοῦτο ζητεῖται, διὰ τί ποτε κινεῖται εἰς τὸν

Apparently, each element has its own natural and intrinsic quality (*poion*, e.g. hot or cold) and quantity (*poson*; the idea is presumably that the same quantity of matter expands or contracts in the process of elemental motion), as well as its proper location (*pou*). Now it is this location which is by definition linked to lightness and weight: if a light element is to reach its full actuality *qua* lightness, it will have to come to be somewhere specifically, viz. 'up'. In other words—and here we reach an important conclusion—the internal source of motion, which was referred to earlier as one of the factors at work in the process of elementary motion, is now being described in more detail in terms of the three categories *poion, poson*, and *pou*. It is implied that, for an element, to have such an internal source of motion, is to be potentially of a certain quality, of a certain quantity, and in a certain location. What does this teach us about the role of natural place?

It should be observed that Aristotle is quite unambiguous in describing the goal of elementary motion, not as 'natural place' but as *being somewhere* or as the element's *being in a natural place*. This is by no means a trivial distinction, because 'being in a natural place' describes a property in the category *pou* ('where'). As such it is inseparable from the substance to which it belongs, in the sense that, like other properties, it cannot exist independently. As a potential property of the elements—in this case a property that belongs to them in virtue of their own essence, or in virtue of their own weight or lightness[52]—it will be actualized unless prevented. Place itself, however, as Aristotle had been stressing time and again in *Phys.* Δ, is something extrinsic to and independent of that which is being located.[53] But there is more to it. For even

---

αὐτῶν τόπον τὰ κοῦφα καὶ τὰ βαρέα. αἴτιον δ' ὅτι πέφυκέν ποι, καὶ τοῦτ' ἔστιν τὸ κούφῳ καὶ βαρεῖ εἶναι, τὸ μὲν τῷ ἄνω, τὸ δὲ τῷ κάτω διωρισμένον.

[52] We are dealing, in other words, with an accidental property that is intimately bound up with the element's substantial form. The question to which this would naturally give rise, viz. whether the *generans* confers weight or lightness directly or indirectly (viz. *via* substantial form), was one of the more frequently discussed questions in medieval scholastic treatments on the subject of natural motion. Cf. Maier ([2]1952) 159–170.

[53] We may recall that the reason for his choosing a vessel (ἀγγεῖον) as a model for place was that a vessel 'does not belong to the thing' (*Phys.* Δ 209 b 29–30: οὐδὲν τοῦ πράγματος). On the problems inherent in treating place itself as a property—as in one possible interpretation it is treated in the *Categories*—see chapter 4, 128–132. But, although it thus appears that place as such cannot consistently be regarded as a property of the emplaced body, there is, on the premises of Aristotle's system, nothing wrong with treating

the conclusion that the formal or final cause of natural motion is not simply *natural place*, but *being in a natural place*, should be further qualified. For the natural property of *being in its natural place* is concomitant to the full actualization of the form (*eidos*) of the element at issue, or at any rate of its form *qua* heaviness or lightness.[54] In other words, in so far as an element's being in its natural place is a formal or final cause, it is not an *independent* cause, but rather a factor in a more general cause, viz. the full actualization of the essence of the element. Just as fire, as soon as it is fire cannot help being hot, it cannot help moving upward (in fact, according to Aristotle it is precisely in virtue of this characteristic that the (natural) motion of inanimate substances differs from the motion of animate substances). So this is how Aristotle accounts for the goal-directed character of natural motion: the teleology is so to speak built into the internal source of motion. There is no hint at all that natural place as such is to be counted as an independent final cause, and hence as an independent fourth causal factor—next, that is, to the internal source of motion, the generator and the *removens impedimentum*.[55]

---

being in a natural place as a natural property as long as place itself is regarded as something in its own right (otherwise attempts to explain physical motion would end up in a hopeless muddle).

54 That we are dealing with the actualization of the potentialities of the elements *qua* heavy or light is a point that is particularly stressed by Alexander of Aphrodisias *ap*. Simp. *In Ph.* 1213, 3–8: ὁ δὲ Ἀλέξανδρος τοὺς κατὰ φύσιν τοῖς σώμασιν τόπους ἐνεργείας αὐτῶν εἰρῆσθαί φησι, τουτέστι τελειότητας. "τότε γὰρ αὐτῷ δοκεῖ, φησίν, ἕκαστον τῶν σωμάτων *ἐνεργείᾳ βαρὺ ἢ κοῦφον* εἶναι, ὅταν ἐν τῷ οἰκείῳ τόπῳ ᾖ. οὐ γὰρ *πῦρ ἢ ὕδωρ ἢ γῆ ἢ ἀὴρ δυνάμει* λέγεται εἶναι πρὸ τοῦ ἐν τοῖς οἰκείοις εἶναι τόποις· εἰ γὰρ δυνάμει τι τούτων ἦν, οὐκέτι *κινούμενα* ἦν, ἀλλὰ *γινόμενα*".

55 The idea that natural place is an independent cause, next to the internal source of motion of the elements, is an essential part of Sorabji's interpretation: cf. (1988a) 222. But Sorabji is fair enough to admit that this is a reconstruction of Aristotle's position, and that Aristotle himself never speaks of place as a final cause. Lang (1992) 63–84 goes far beyond Aristotle's text in arguing that natural place actually is the mover identified by Aristotle in this fourth chapter of *Phys.* Θ. Unfortunately her analysis confuses what Aristotle *does* say about the role of natural place, viz. that the goal of the natural motion of the elements is their 'being in a natural place', and what Aristotle does *not* say, viz. that natural place itself is the end of natural motion. In addition I fail to see how she makes sense of the role Aristotle himself ascribes to the generator at the end of this chapter: according to Aristotle it is primarily the generator that is 'that by which' the elements are moved. Lang (1992) 81 (with n. 102) quotes this conclusion which she calls 'brief and uncomplicated', but it is not at all clear how it fits in with her own interpretation according to which 'that by which' is natural

We shall now turn to the second piece of main evidence to be discussed here, the third section of the fourth book of the *De Caelo*. Aristotle there tackles the question of what constitutes the light and the heavy, and, in particular, why some bodies naturally move upward and others downward. Here again change of quantity, quality and place are treated on a par:

> The local movement of each body to its own place must be regarded as similar to what happens in connection with other forms of generation and change. There are, in fact, three kinds of movement (κινήσεις), viz. change of size, change of form and change of place, and in each one the change is observed to proceed from opposites to opposites (or intermediate states).[56]

He goes on to argue that the thing which produces change and the thing which changes are not related in a fortuitous way. In the case of natural motion that which produces the change is that which produces weight and lightness:

> Now that which produces upward and downward movement is that which produces weight and lightness, and that which is moved is that which is potentially heavy and light, and the movement of each body to its own place is motion towards its own form.[57]

The contents of this passage resemble the ideas expressed in *Phys.* Θ in two important respects. Firstly, that which produces natural motion is said to be that which produces the heavy and the light, which, we may presume, is to be identified with the *generans* mentioned in *Phys.* Θ. Secondly, that which is moved is claimed to be that which is potentially heavy or light, so that natural motion is here again represented as the actualization of a potentiality. Here Aristotle must be referring to the second actualization described in *Phys.* Θ, 4. In fact, the statement of *Phys.* Θ that the

---

place—unless she wants to argue, very implausibly, that what at 256 a 1 is called 'what generated and made it light or heavy' is natural place (yet, that this is what she has in mind, seems to be suggested by the sequence of her arguments on p. 81).

[56] *Cael.* Δ 310 a 20–25: Περὶ μὲν οὖν τοῦ φέρεσθαι εἰς τὸν αὑτοῦ τόπον ἕκαστον ὁμοίως ὑποληπτέον ὥσπερ καὶ περὶ τὰς ἄλλας γενέσεις καὶ μεταβολάς. ἐπεὶ γάρ εἰσι τρεῖς αἱ κινήσεις, ἡ μὲν κατὰ μέγεθος, ἡ δὲ κατ' εἶδος, ἡ δὲ κατὰ τόπον, ἐν ἑκάστῃ τούτων τὴν μεταβολὴν ὁρῶμεν γιγνομένην ἐκ τῶν ἐναντίων εἰς τὰ ἐναντία καὶ τὰ μεταξύ, κτλ.

[57] *Cael.* Δ 310 a 31–34: εἰ οὖν εἰς τὸ ἄνω καὶ τὸ κάτω κινητικὸν μὲν τὸ βαρυντικὸν καὶ τὸ κουφιστικόν, κινητὸν δὲ τὸ δυνάμει βαρὺ καὶ κοῦφον, τὸ δ' εἰς τὸν αὑτοῦ τόπον φέρεσθαι ἕκαστον τὸ εἰς τὸ αὑτοῦ εἶδός ἐστι φέρεσθαι κτλ.

actuality of lightness consists in the light thing being somewhere is here paralleled by the statement that for an element to move towards its own place is to move towards its own form. It thus becomes clear that what Aristotle means by the latter phrase is not simply that place itself is form—as he has indeed sometimes been supposed to have meant—but rather that for a body to be in its proper place is part and parcel of its attaining its proper form or actuality.[58]

Aristotle goes on to explain in what sense natural motion may be regarded as an instance of motion of 'like to like':

> But since the place of a thing is the boundary of that which contains it, and all things that move upward or downward are contained by the extremity and the centre, and this boundary comes to be, in a sense, the form of that which is contained, for something to move to its own place is for it to move to its like. For the successive members of the series are like one another: water, I mean, is like air, and air like fire, and between intermediates the relation may be converted, though not between them and the extremes; thus air is like water, but water is like earth, for the relation of each outer body to that which is next within it is that of form to matter.[59]

The passage starts with the definition of place that issued from the discussion of *Phys.* Δ, once more a sign that Aristotle did not think that the subjects of place on the one hand and of natural place on the other could be dealt with in complete separation. On this

---

[58] In connection with this passage also Furley (1976) 83–100, esp. 93 and n. 46, argues that place is not a cause in any of the four traditional categories, though it is a factor in the formal cause which is, in its turn, the *logos* or *eidos* of the primary bodies. Different interpretations are to be found in Ross (1936) 563, who claims that natural place is *like* form in that it exhibits an attractive influence, 'just as form which matter is destined to assume has attractive influence on the matter', and Machamer (1978) 377–387, esp. 380, who argues that we should not think that natural place is a form, but rather that an element attains its proper form when it is in a natural or organic unity with its like. The latter interpretation draws heavily on *Cael.* Δ 310 b 10–11, quoted in the text above (see text to next footnote), where it is said that the the surrounding body becomes in a sense the form of the surrounded thing. But Machamer's interpretation of this passage cannot stand: see below, n. 60, and the text thereto.

[59] *Cael.* Δ 310 b 8–15: ἐπεὶ δ' ὁ τόπος ἐστὶ τὸ τοῦ περιέχοντος πέρας, περιέχει δὲ πάντα τὰ κινούμενα ἄνω καὶ κάτω τό τε ἔσχατον καὶ τὸ μέσον, τοῦτο δὲ τρόπον τινὰ γίγνεται τὸ εἶδος τοῦ περιεχομένου, τὸ εἰς τὸν αὐτοῦ τόπον φέρεσθαι πρὸς τὸ ὅμοιόν ἐστι φέρεσθαι· τὰ γὰρ ἐφεξῆς ὅμοιά ἐστιν ἀλλήλοις, οἷον ὕδωρ ἀέρι καὶ ἀὴρ πυρί. ἀνάπαλιν δὲ λέγειν τοῖς μὲν μέσοις ἔστι, τοῖς δ' ἄκροις οὔ, οἷον ἀέρα μὲν ὕδατι, ὕδωρ δὲ γῇ· ἀεὶ γὰρ τὸ ἀνώτερον πρὸς τὸ ὑφ' αὐτό ὡς εἶδος πρὸς ὕλην.

definition of place, Aristotle argues, it can be understood that the natural motion of an element is motion towards its like. For the surrounding boundary is in some way the form of the surrounded thing. This sentence has sometimes been understood as implying that place *qua* surface is a kind of 'outward form' of the located body.[60] However the sequel makes clear that this is not what Aristotle means. Place is not a form *qua surrounding surface*, but it is as it were a form *qua* surface *of the surrounding body*. For to each element the surrounding element is said to be as form to matter. Water is potentially air, and as such it may be regarded as the matter of air. So if it finds itself surrounded by air, it in a way joins its like (its form). It does so even more clearly when it changes into air and moves into the appropriate natural place.[61]

The idea here appears to be the same one that was expressed rather briefly and enigmatically at the end of the dialectical discussion in *Phys.* Δ where Aristotle set out to show that his own definition of place allowed a meaningful use of the notions of natural place and natural motion.[62] Motion towards a thing's natural place is motion towards its like. Since that which is like is the surrounding body, a theory of place defining place as the limit *of the surrounding body* is according to Aristotle the only theory of place that fits this conception of natural motion.

5.1.4 If we combine our findings the following picture emerges. In *Phys* Δ Aristotle quite unambiguously denies that place is a cause. Moreover, nowhere else does he so much as imply that he has changed his mind on this subject: he never speaks of (natural) place as a cause. We may therefore safely assume that, whatever many of his medieval and modern commentators say, Aristotle was not committed to the idea that natural places figure as final causes of natural motion. Yet we also concluded that the statement that place has a certain *dunamis* should be taken to mean

---

[60] Cf. Machamer (1978) 377–387, esp. 380: '[...] the earth in its natural place is given form or shape by the surrounding water. Water is the form of the planet [sic] earth in that it provides the innermost containing boundary of the earth'. Incidentally, this rather uncharitable interpretation (note e.g. that it is simply not true that the earth is everywhere surrounded by water) is already to be found in Philoponus, 580, 7–16.

[61] On this Aristotelian interpretation of the 'like to like principle' see Solmsen (1960) 282–283.

[62] On which see above, 205–206.

that place is made to play at least some role in the explanation of natural motion. This left us with two questions. Firstly, how then could Aristotle's theory account for the clearly goal-directed character of natural motion? Secondly, what role then *did* he accord to natural places, and what was it in his own theory that made Aristotle think that it was better equipped to account for natural motion than the rival theory of void-space of the early atomists?

As to the former question, the relevant passages in *Phys* Θ, 4 and *Cael.* Δ, 3 make clear that the *telos* or goal of their natural motion is not simply *natural place*, but *being in a natural place*. As a natural property the element's *being in its natural place* is concomitant to the full actualization of its form (*eidos*) of the element at issue, or at any rate of its form *qua* heaviness or lightness. In other words, we are not dealing with an *independent* formal or final cause.[63] We are rather dealing with a cause that is, so to speak, programmed into the elements' own constitution and that, as such, belongs to the elements' inner capacity of being moved—as we saw, one of the three causal factors identified by Aristotle. The other two factors involved in the process, the *generans* and the *removens impedimentum* rather figure as efficient causes.

As to the second question (what role *did* Aristotle accord to place), we can now further confirm the provisional conclusion we reached at the end of our discussion of 5.1.2. The role of the concept of natural place itself, as it emerged from our study, appears to be this. It allows one to *specify* the 'being somewhere'

---

[63] I am being deliberately vague about whether we should speak of a formal or a final cause, and my interpretation certainly does not hinge on whether we opt for the former or the latter possibility. As I have indicated in the introduction to the present section, most later Aristotelians spoke of natural place as a final cause. Aristotle himself for sure only uses the word form (*eidos*) in the contexts we are here discussing, but of course in Aristotelian natural processes form and *telos* often coincide (on which see *Phys.* B 198 a 24–27, with the famous example of 'man generates man', where even the efficient cause coincides with the formal and final cause). Of course Aristotle also claims (at 198 b 8–9) that the final cause should be invoked to show that 'it is better thus' (διότι βέλτιον οὕτως), but that claim is immediately qualified by his remark that 'better' here should be understood 'not without qualification, but with reference to the substance in each case'. This, I think, leaves room for a fairly 'loose' use of the notion of a final cause. For a defence of a more restricted use of the term 'final cause' see Gotthelf (1987) esp. 214, n. 18, where it is argued that the natural motions of the elements are not 'for the sake of' anything, so that here the notion of a final cause should not be invoked.

that as a natural location constitutes the goal of the natural motion of the elements. In other words, however strongly this 'being somewhere' may be bound up with the (internal) nature of the elements, it cannot be specified independently of an extrinsic independent place. Precisely because Aristotle is able to give a meaningful specification of the 'being somewhere' that is the goal of this particular natural motion, viz. a specification in terms of his own conception of place as a surrounding surface, his theory is to be preferred, or so Aristotle himself believes, to, say, the conception of an isotropic void space endorsed by the atomists. The latter are unable to make sense of the notion of a natural place, and hence also unable to make sense of the notion of 'being naturally somewhere', since one part of the void is no different from another.[64]

In sections 5.3 and 5.4, I shall try to show that this interpretation squares with what remains of the early Peripatetic discussions of Aristotle's theory of place and natural place. For the moment we may just note that our reconstruction of Aristotle's position shows that, whenever early Peripatetics like Theophrastus were claiming that natural places were not a cause of natural motion, they were adopting rather than criticizing Aristotle's position. But on closer view even some of the later commentators who credit Aristotle with the view that natural place *is* a (final) cause of natural motion come close to the interpretation here put forward. Thus Simplicius explicitly states that, although we might be inclined to think natural place is a cause of elementary motion, the right thing to say is that 'to be in a place' is the *telos* of that motion.[65] When further on in his commentary the same Simplicius suggests that place can be regarded as a cause after all, he is clearly not representing Aristotle's position, but rather speaking from his own point of view.[66] As for Averroes, it is true that he does speak of

---

[64] It is perhaps legitimate to infer—although Aristotle never says so—that on his line of thought natural place figures as a necessary condition (not as a cause) of natural motion. We may compare what he says elsewhere (*Phys.* Γ 200 b 20 and *Cael.* A 275 b 17 ff.) about place as such being a necessary condition of motion.

[65] Simp. *In Ph.* 533, 22–25: ἀλλ' εἰ ἄλλο ὁ τόπος καὶ ἄλλο τὸ ἐν τόπῳ εἶναι, ὡς δηλοῖ καὶ ἡ ποῦ κατηγορία ἄλλη οὖσα παρὰ τὸν τόπον αὐτὸν ἐν τῷ ποσῷ τεταγμένον, τὸ δὲ τέλος τοῖς σώμασίν ἐστιν, εἴπερ ἄρα, τὸ ἐν τῷδε τῷ τόπῳ εἶναι, οὐκ ἂν εἴη ὁ τόπος τὸ τέλος.

[66] *In Ph.* 600, 30–34: πρὸς δὲ τὴν ἀπὸ τῶν αἰτίων ἀπορίαν ῥητέον, ὅτι εἴπερ

the elements reaching their entelechy or perfection in their natural places. But he also speaks of these places as 'ubi', i.e. as attributes of the elements belonging to the category *pou*.[67] Also Thomas, even if he does speak of *locus* as the goal of elementary motion, makes it very clear that he is taking *locus* to be equivalent to 'being in a place', for he explicitly regards it as an attribute consequent upon substantial form (in other words, he as well is thinking of the category *pou*).[68] For these thinkers the difference between 'place' and 'being in a place' was apparently not important enough to be consistently maintained, so that we should regard their talk about natural place being a final cause of natural motion as hardly more than a *façon de parler*. After all it should be recalled that Aristotle himself in his later work sometimes used the label '*topos*' to refer to the category *pou*.[69]

φυσικῶς ἐφίεται τὰ σώματα καὶ φέρεσθαι ἐπὶ τοὺς οἰκείους τόπους καὶ μένειν ἐν αὐτοῖς, τελικὸν ἂν εἴη αἴτιον ὁ τόπος ὡς *συνεκτικὸν καὶ τελεσιουργὸν καὶ περιποιητικὸν τῆς ἑκάστου τελειότητος*. The italicized terms show that Simplicius is here arguing from the point of view of his own theory of place—which is largely that of his master Damascius—according to which place is an active form-like principle that shapes things and keeps them together. See on this conception of place Sorabji (1988a) 207–211; Sambursky (1982) 21–22; and Algra (1992) 157–162. The term τελεσιουργὸν, for example is alien to Aristotle's theory of place, but does occur in Simplicius' own characterization of Damascius' conception of place at *In Ph.* 601, 18–19: οἱ δὲ τελεσιουργὸν τῶν σωμάτων [scil. εἶναι τὸν τόπον] ὡς ὁ ἡμέτερος Δαμάσκιος.

67 Cf. Averroes in *Aristotelis de physico auditu libri octo cum Averrois Cordubensis variis in eosdem commentariis*, Venezia 1562, 122 K (the immediate sequel of the passage quoted above, n. 14): 'et hoc intendebat cum dixit ipsum [scil. locum] habere potentiam, in quantum apparet ex motu terrae ad inferius, et ignis ad superius, quod querunt per istum motum *ubi* conveniens sibi, quae convenientia non est, nisi per potentiam naturalem existentem in illo *ubi*'.

68 Cf. *Comm. in Phys.* II, lectio 1: 'quia sicut alia accidentia consequuntur formam substantialem, ita et locus, et per consequens moveri ad locum'. To this we may add the evidence of lectio 8 of the commentary on *Phys.* IV, where Thomas discusses the passage in which Aristotle argues that his own concept of place allows a meaningful explanation of the phenomenon of natural motion (212 b 29–213 a 1). According to Thomas this explanation amounts to saying that it is the *proximitas naturae* between the successive elements which causes their natural motion and rest. No causal role for natural places seems to be implied: 'dicit ergo primo, quod si ponatur locus esse terminus continentis, rationabiliter assignari potest causa, quare unumquodque corpus feratur ad proprium locum: quia illud corpus continens ad quod consequenter se habet corpus contentum et locatum, et quod ab eo tangitur terminis simul existentibus, et hoc non per violentiam, est proximum ei secundum naturam. [...] sic igitur proximitas naturae, quae est inter corpus continens et contentum, est causa quare corpus naturaliter movetur ad suum locum'.

69 On this see above, ch. 4, n. 5.

For others, however, the distinction between 'natural place' and 'being in a natural place' may simply have been too subtle. But apart from that, also the desire to frame a more simple or *prima facie* more appealing model of the teleology of natural motion may have contributed to the endorsement of the view that natural place is a cause of natural motion. In this connection, we may now briefly return to what is no doubt the most curious interpretation defended by some later scholastics (Bonaventura, Roger Bacon) viz. that natural place figures as an *efficient* cause.

It can be argued that Aristotle's account as it stands has little explanatory force. After all he explains natural motion by merely asserting that it is part of the nature of each element to be in such-and-such a place, and that the process of natural motion is the actualization of this intrinsic form. That is as far as his account goes. Although, as we have seen, he does connect the notions of 'natural place' and 'being in a natural place' (in so far as the former is necessary to specify the latter), he makes no attempt to connect the *intrinsic* tendency of the elements to be in their natural places and these *extrinsic* places themselves. We would expect him to give an account of how and why each element strives to be in its natural place, i.e. how and why it tries to be surrounded by 'its like', as Aristotle puts it. In fact, the very use of the old saying 'like moves to like' would seem to imply an underlying view of the cosmos as some kind of organic unity. It has sometimes been argued that Theophrastus explicitly made this further step,[70] but Aristotle never says anything of the sort. For him a more 'phenomenological' approach apparently suffices: we see that it belongs to the nature of the elements to move to their natural places, and no further explanation at a 'deeper' level seems to be required. Now what thinkers like Bonaventura and Roger Bacon maybe tried to do, was to fill up this gap and to provide something like a proper explanation of the mechanism of natural motion. A more thorough investigation of the details of this particular part of the *Rezeptionsgeschichte* of Aristotelian physics certainly seems to be a *desideratum*. At the same time, it is also clear that such an investigation would go beyond the limits set to this study.

---

70 Van Raalte (1988) 189–215; Van Raalte's assessment has been criticized, however, by Ellis (1988) 216–223.

5.2 *The problem of the immobility of place*

5.2.1 In the previous section I have tried to show that, even if Aristotle's conception of the role of natural places in the explanation of natural motion was perhaps not *immediately* clear, it could nevertheless be plausibly and consistently reconstructed from a number of pointers in his own work. Moreover, I have suggested —and I shall adduce further evidence in the last two sections of this chapter—that divergent interpretations of Aristotle's theory of natural places came into vogue only at a fairly late date. The problem that will be discussed in the present section, the problem of the immobility of place, is different in both respects. Firstly, we are dealing with a problem which, as I shall argue, could not really be solved according to proper Aristotelian principles. Secondly, as a problem it has been recognized and discussed since the days of the early Peripatos. We find references to it in the remaining fragments of the works of Aristotle's pupils Theophrastus and Eudemus. It also occurs as one of the main objections raised against Aristotle's theory of place by Philoponus in his *Corollarium de Loco.*[71] Indeed the problem became so well known as to become one of the standard problems discussed in scholastic treatments of Aristotle's physics in the later Middle Ages.[72]

My own discussion of this problem in the present section is structured as follows. I shall first, in 5.2.2, discuss the passages where Aristotle formulates the requirement that place should be immobile and where he provides his own solution for the problems this requirement would seem to pose for his own theory in the case of things emplaced in a flowing, or otherwise moving, medium. We shall see that his own solution is far from clear and that it called forth various interpretations in the history of the Aristotelian tradition. Next, in 5.2.3, we shall study the problem in a more systematic way and ask whether Aristotle could have solved it at all. It will then appear that the problem of the immobility of place is closely linked with the problem of the ontological status of Aristotelian place as a surface, and that the only way of solving it would have been to stop regarding this Aristotelian place as an existent *sui generis.*

---

[71] Phlp. *In Ph.* 264, 14–33.
[72] Grant (1981b) 61–72.

5.2.2 One of the more salient features of the dialectical inquiry into the nature of place (*topos*) to be found in the fourth book of Aristotle's *Physics* is the complete rejection of the concept of place as an underlying self-subsistent three-dimensional extension. Aristotle nevertheless describes this concept as one which easily comes to mind because it is part and parcel of the way in which people ordinarily think and speak about *topos*. At *Phys.* Δ 212 a 7–14, a text quoted and discussed in the previous chapter,[73] he adds two factors which may be held responsible for this erroneous common conception. For one thing, the phenomenon of air seeming to be incorporeal apparently contributes to the conception that place is a three-dimensional incorporeal something in which things are located. The main reason, however, why people go astray in this respect is, according to Aristotle, that they draw the wrong conclusion from the fact that replacement presupposes fixed surroundings (apparently in the sense of 'a fixed frame of reference'). For they believe that this entails that what remains fixed is an underlying extension (διάστημα) which is different from the extension of the bodies replacing each other. Aristotle of course rejects this inference, but it should be noted that he nevertheless subscribes to the premise that replacement presupposes a fixed surrounding *something,* a vessel, a that-which-stays-behind. Even if, in other words, the location of a body is according to Aristotle to be defined in relation to its surroundings rather than in terms of an underlying extension, he does not go so far as to define place itself as a *mere relation.* Instead, he sticks to a more naive reified conception of place.

The requirement that this surrounding something should be immobile is explicitly put forward—and the word 'immobile' is added to Aristotle's provisional definition—in the following passage (212 a 14–20):

> And, just as a vessel is a mobile place, so place is an immobile vessel. Therefore, when an enclosed thing moves and changes within something that moves, like a ship in a river, it has the surrounding thing serving as a vessel rather than as a place. But place should be immobile. Therefore the whole river is rather the place, because the whole is immobile. So the first immobile surface of the surrounding thing, that is place.

---

[73] See above, ch. 4, 178–179.

The distinction between a mobile vessel and an immobile place is in itself quite clear. Problems arise, however, as soon as we realize that place, on Aristotle's line of thought, should be conceived of as the surface *of a surrounding body*. For often this surrounding body will move, or at least be mobile. Aristotle' solution to this problem as exemplified by the boat in a flowing river is not as clear-cut as he himself apparently wants us to believe. In the case of the boat the direct surroundings of the boat *qua flowing* are said to be only a vessel (ἀγγεῖον) for the boat, not a place. His use of the word 'vessel' indicates that he is probably thinking of a boat flowing along with the current—for in that case it is always the same surface that contains the boat (although the surface is itself moving along)—and that this is what constitutes the analogy with something which is being carried in a vessel. This is also how Alexander, Themistius, Simplicius and Philoponus interpret the example.[74] It is to be noted, however, that the problem is of a more general nature, and that it also arises in a form which would make it difficult even to maintain that the boat is in its direct surrounding surface as in a vessel (consider e.g. the case of a boat being moored in a river with the current flowing past). But even when we take the problem in the more restricted version probably envisaged by Aristotle, it is difficult to be satisfied with his answer that in the case of a boat floating down a river the place is rather 'the river as a whole' (ὁ πᾶς μᾶλλον ποταμὸς) because that is immobile.

Indeed Aristotle's answer seems to have already troubled the ancient commentators, and various attempts on their part to interpret it have been recorded. Alexander of Aphrodisias, for a start, offers a complicated solution which goes way beyond Aristotle's own wording. He argues that what is meant is not so much that the river is the place of the boat in the strict sense (πρώτως). The idea is rather that the boat is in the water as in a vessel, whereas in

[74] We find this interpretation *per implicationem*, in Alex. Aphr. *ap*. Simp. *In Ph*. 584, 3–5: ἀλλ' εἴη ἂν ὁ μὲν ποταμὸς πᾶς ἐν τόπῳ, ὥς φησιν ὁ 'Αλέξανδρος, τῷ *μὴ συγκινεῖσθαι* αὐτῷ τὸ πέρας τοῦ περιέχοντος αὐτόν, ἡ δὲ ναῦς ἐν ἀγγείῳ , and more explicitly in Them. *In Ph*. 118, 28: ... ἐπειδὰν ἐν κινουμένῳ τι *κινῆται* ὡς πλοῖον ἐν ποταμῷ κτλ., Simp. *In Ph*. 583, 20–22: καὶ λέγει ὅτι ὅταν κινούμενόν τι σῶμα ἐν κινουμένῳ τινὶ ἐνῇ ὥστε *συγκινεῖσθαι* αὐτῷ τὸ ἐν ᾧ ἐστιν ὥσπερ τὸ ἐν ποταμῷ πλοῖον ἢ ξύλον οὐ τέμνον τὸ ὕδωρ, and Phlp. *In Ph*. 586, 13–14: *συγκινεῖται* γὰρ τῷ ῥεύματι ὥστε τὸ περιέχον προσεχῶς τὸ πλοῖον ὕδωρ κυριώτερον ἀγγεῖον ἂν λεχθείη.

its turn the river, *qua* flowing water, is in its banks as in a place strictly speaking. In virtue of the fact that the relation of 'being in something' appears to be transitive, the boat will be in the river, *qua* banks, as in a place, albeit not when place is taken in the strict, primary, sense.[75] Simplicius presents this interpretation of Alexander as an alternative to his own rather helpless explanation which suggests that the boat for sure *is* in the river as in a place. Simplicius, like Alexander, takes the word 'river' in this case to refer to the banks, and he suggests that this river-*qua*-banks is immobile in the required sense.[76] He adds, however, that Aristotle deliberately used the word 'rather' (μᾶλλον) because the river banks cannot really serve as the place of the boat, since they do not fulfill the requirement of contiguity.[77]

In his commentary on the passage at issue, Philoponus suggests another line of defence which is also rather lame. Place as the hollow surface of the surrounding body, he argues, may be regarded as in itself immobile; it moves only accidentally by virtue of the fact that the body to which it belongs, is moving.[78] This 'defence' is of course of little value, since it merely makes the *ontological* point that a surface cannot move in its own right, so that in that sense it might be said to immobile. However, the immobility required by Aristotle's argument is a *physical* immobility, and it concerns place conceived precisely *as a thing in its own right*. In his *Corollarium on Place* Philoponus opts for a different explanation. He here more or less subscribes to Alexander's position in arguing that the 'river as a whole' of which Aristotle speaks, should be understood as the river *banks*. He adds that these are a place for the boat not in the primary sense, but only as a *koinos topos*.[79] Most modern commentators take over this interpretation.[80] On the other hand they also appear to be sensitive to the problems it involves.[81]

---

75 Alex. Aphr. *ap.* Simp. *In Ph.* 584, 5–14.

76 Simp. *In Ph.* 583, 30–33.

77 Simp. *In Ph.* 583, 33–584, 2.

78 Phlp. *In Ph.* 590, 17–18: ... κοίλη ἐπιφάνεια, ἥτις ἀκίνητός ἐστι καθ' αὑτήν, κινεῖται δὲ κατὰ συμβεβηκὸς τῷ τὸ σῶμα οὗ ἐστι πέρας κινεῖσθαι.

79 Phlp. *In Ph.* 586, 19–20.

80 Cf. Hussey (1983) and Wagner (1967) *ad loc.*, and Moreau (1965) 39.

81 Hussey (1983) *ad loc.* even proposes to regard the river example as an interpolation, though he does not provide any reason for doing so apart from the fact that it is difficult.

The most important—indeed crucial—objections would be, first, that it involves Aristotle neglecting the criterion that place has to be neither larger nor smaller than the located body, and, second, that it involves his abandoning the requirement that place should be in immediate contact with what is in place. But there is no hint whatsoever that Aristotle was willing to abandon this requirement. On the contrary, as Richard Sorabji has pointed out, he goes on to repeat it at least once in the immediate sequel of the passage we are here discussing.[82] But apart from being at odds with Aristotle's own claim that place should be an *immediate* instead of a *remote* container, the interpretation defended by Alexander and others does not even succeed in adequately serving the purpose for which it seems to have been designed. For it would certainly not leave Aristotle with a workable theory of place. To mention just one possible objection, it would according to this line of thought never be immediately clear what may count as the place of a given body. One would always have to determine first which of the successive surrounding bodies are moving. In the case of the boat on the river, for instance, we would be facing the rather counter-intuitive consequence that, as soon as the tide stops, the nearest immobile surface constituting the place of the boat will switch at once from being that of the banks to being that of the water surrounding the boat.[83]

An alternative interpretation which is *prima facie* more appealing—if only because it does not involve Aristotle's abandoning his own explicit requirement that place should be contiguous, although, as we shall see, it involves problems of its own—is to take Aristotle's words 'the river as a whole' to refer to the river taken *in abstracto*, or as a geographical entity.[84] Following this line of thought the boat's place will be the rim of a boat-sized hole in the river as a geographical entity. This, however, will not get Aristotle off the hook either. For it remains difficult to imagine how the surface of this boat-sized hole in the river can be

---

[82] *Phys.* Δ 212 a 29–30; also 212 b 19; see Sorabji (1988a) 188.

[83] For a fuller discussion of this example and similar ones see Sorabji (1988a) 188–189.

[84] This interpretation is put forward by Burnyeat (1984) 230, n. 15. I defended a similar view in Algra (1988b) 131, n. 127. As will be clear from the text I now defend a slightly more skeptical position, in so far as I do not want to go beyond the claim that this is what Aristotle *might* have said when pressed to clarify his remarks on the boat in the river.

immobile. Richard Sorabji has given a convenient overview of the problems Aristotle would run into following this interpretation.[85] If we consider a moving boat, what immobile rim can it be said to leave behind? It might be thought that in its course the boat is surrounded as it were by a series of instantaneous rims. There is nothing in the text, however, to make us think that when speaking of immobile places Aristotle had instantaneous entities in mind. Moreover instantaneous places would not seem to fit Aristotle's conception, referred to above, of places as 'things' that can be left behind. If we consider the case of a moored boat in a flowing river, we run into the same difficulty. In both cases we shall need some specification of the sense in which the successive surrounding surfaces of the flowing water are to be regarded as *one* and as *immobile.* Treating them as *one* would again seem to involve treating them *in abstracto,* and as independent of the water of which they are the successive surfaces. The surrounding surfaces would in that case have to be regarded as parts of a kind of overall grid of locations.[86] That, however, would also go way beyond Aristotle's actual wording. It would, moreover, entail a conception of place, viz. as an incorporeal entity in its own right, which is hard to fit into Aristotle's ontology. I shall presently revert to this.

Before we pursue these problems further, I would like to point out that both the traditional interpretation, defended by Alexander and others, and the alternative interpretation put forward by Burnyeat are charitable interpretations, in the sense that they presuppose that Aristotle actually *had* a more or less technical account of the immobility of place, to which the cryptic words 'the river as a whole' would supposedly refer. I am afraid, however, that we are hardly entitled to make this presupposition in this case. First of all, as we have just seen, Aristotle would on both accounts in the end run into difficulties anyway. In view of the serious nature of these difficulties it is hard to imagine that he

---

85 Sorabji (1988a) 189–190.

86 Aristotle may at times have gone quite a long way into this direction (it may e.g. be that also the difficult passage at 212 b 1–2 about the place of the parts of an unplaced body has to be interpreted in this way) but since he nevertheless stuck to talking about place as the limit of the surrounding body, his position remains ambiguous in this respect, an ambiguity of which —as Ross (1936) 576, noted—he does not seem to have been aware. On this ambiguity see also Moreau (1965) 42–43.

would have remained content with either of the two interpretations, *if* he had thought this problem through. The obvious answer, therefore would seem to be that he did not actually think over all problems connected with the immobility of place. Secondly, it is hard to imagine that the simple and unspecific words 'the whole river' were meant to refer to anything as specific as 'the bed and banks of the river' or 'the surface of the boat-sized hole regarded *in abstracto*'. It would therefore seem to be more palatable to suppose that the problems arose precisely because Aristotle simply added the immobility requirement without working out a more specific and technical account of immobility. Now it will be recalled that Aristotle envisaged the work of the philosopher of nature as a process starting out from the *phainomena* of ordinary experience ('what is more known to us') and of sifting and refining these *phainomena* into the more technical account constitutive of 'what is more known by nature'.[87] It appears, or so I would suggest, that Aristotle could in the present case be reproached, even on his own terms, for failing to complete this process, i.e. for failing to move beyond the rather inarticulate *phainomena* of everyday thinking and speaking, according to which we may indeed say that we determine the place of a boat in a flowing river not by reference to mobile, and actually moving, surface(s) of the surrounding portions of water, but rather by reference to 'the whole river', i.e. the river regarded as a geographical entity. He apparently did not see that it was impossible to refine this rough account into a more technical account that would fit into the overall system of his ontology and physics.

5.2.3 So much for Aristotle's own rather defective treatment of the question of the immobility of place. We may now conclude this section by considering which problems an ideal Aristotelian, or an ideal Aristotle, would have had to solve—*if*, that is, he would have cared to address them properly. A sharpened awareness of these problems may be of help later on, when I shall try to interpret the contributions of the early Peripatetics Eudemus and Theophrastus to the debate on this subject.

There appear to be basically two problems in connection with the immobility of Aristotelian place. First, a specification is needed

---

87 See on this the previous chapter, esp. 175.

of the sense in which place as a surrounding surface should be conceived as immobile. *With respect to what,* in other words, is it immobile? Here again Aristotle himself offers no clear and explicit answer.[88] In fact, one gets the impression that where he explicitly applies the immobility criterion he does not go beyond common experience, which usually just *assumes* that one body is moving while something else remains at rest. Here again, in other words, he fails to offer the technical account which his physics would actually seem to demand. Nevertheless his cosmology in general does provide a few clues which might be helpful in framing such a more technical account. After all, the Aristotelian cosmos contains fixed parts, such as the heavenly spheres, and fixed directions, such as 'up' (ἄνω) and 'down' (κάτω) which could in principle be used as reference points in a more technical explanation of the immobility of places. These clues were in fact taken up and worked out by later Aristotelians, beginning, as we shall see, with Eudemus of Rhodes.[89] They tried to give a more technical and more specific definition of the immobility of place in terms of a relation to fixed reference points, such as the centre of the cosmos and the outermost heavenly sphere. Thus many medieval commentators distinguish between what they call *formal* place on the one hand and *material* place on the other. In the case of a boat in a river the *material* place would be the ever different surrounding surfaces of the water, whereas the *formal* place would be identified as this surrounding surface *taken as immobile* in virtue of the introduction of the concept of fixed distances to e.g. the outer sphere of the heavens or the poles.[90]

This, however, immediately brings us to the second problem connected with the immobility of Aristotelian place. For even if the concept of immobility itself could in this way be given clear outlines, we would still be left with the problem of the immobility *of Aristotelian place.* For such a 'reified' place, i.e. place as the surface of the (mobile) surrounding substance, can not be regarded as immobile in the required sense. In fact one might say that

---

88 The issue is briefly discussed by Sorabji (1988a) 190, but seems to be in need of further study.

89 On Eudemus, see below, 252–254. On the medieval contributions see Grant (1981b) esp. 63–72 with n. 20 and 21.

90 On the concepts of formal and material place in medieval physics, see Grant (1981b) 63–72.

Aristotelian ontology, treating surfaces as inextricably bound up with the bodies to which they belong, does not allow a satisfactory solution to the problem of the immobility of place. For surfaces of mobile substances are not themselves immobile, whereas the idea of a self-subsistent surface as part of a kind of overall grid of (abstract) locations would be as alien to Aristotle's ontology as the idea of a self-subsistent three-dimensional extension. This actually means that one of the initial *aporiai* which according to Aristotle any theory of place will have to deal with, viz. the *aporia* what place exactly is—given that it is neither a body, nor an *asômaton*, and neither an element, nor a compound of elements[91]—does not receive a proper answer in Aristotle's own account.

In fact the only really satisfactory solution to the problem at issue here—at least the only satisfactory solution that preserves the morphology of Aristotelian place as a surrounding surface—is the 'nominalist' one proposed by Ockham in the later Middle Ages. In his treatment of the subject of the place of a body in a flowing medium, Ockham took over the by then familiar distinction between *material place* and *formal* place, but he made an important addition in further defining formal place (which alone combines the Aristotelian requirements of contiguity and immobility) as the successive material surfaces *considered as immobile* or *considered as if they were materially immobile.* [92] This of course goes way beyond anything that is in the text of *Phys.* Δ and, given Aristotle's overall realism, it could hardly count as a solution along true-blue Aristotelian lines. Nevertheless, for those who were willing to accept this, it did show a way to stick to the idea of place as a surrounder rather than an underlying three-dimensional extension, while evading the pitfalls connected with Aristotle's *Verdinglichung* of place. Another way to do so would have been to regard place no longer as a surrounding *thing*, but as a *relation* between a body and its surroundings. This, as we shall see, is a solution suggested by Theophrastus.

---

[91] *Phys.* Δ 209 a 13–18 (item A 4 c in the overview of section 4. 4. 4).

[92] Cf. William Ockham, ed. by Boehner (1944) 87: '...Philosophus intendit dicere, quod locus est immobilis per aequivalentiam, hoc est quod tantum valet locus ad salvandum omnia, quae ponuntur de loco, ac si realiter esset immobilis, quia ita possunt in eodem loco succedere sibi diversa corpora loco non mutato, sicut possent, si locus esset immobilis.'

5.3 *Theophrastus' critique of Aristotle's theory of place*

5.3.1 Theophrastus, Aristotle's pupil and successor as head of the Peripatos, continued working and writing on the whole range of subjects which had been covered by Aristotle. Scholarly opinions have differed as to the extent to which Theophrastus was an original thinker.[93] These different assessments were of course partly due to the fragmentary nature of much the evidence. Partly they were also called forth by the various conceptions people had of Aristotle's philosophy. We cannot here go into these matters at length. It will be sufficient to note that there seems to be a growing *communis opinio* according to which Theophrastus should be seen as someone further explaining and working out Aristotle's philosophy, while not hesitating to criticize his master on various points or to supply answers where Aristotle had left loose ends. Now one of the reasons why it is often so difficult to assess the extent to which Theophrastus actually took original stances, is that he availed himself of a rather modest and sometimes also rather diaporematic style. It should be realized, however, that in this respect he did not differ that much from the Aristotle we know from the school writings. Indeed, it may be salutary to think of Theophrastus not so much as someone who saw himself as a faithful follower of a dogmatic philosophy framed by Aristotle, but rather as someone continuing and refining Aristotle's work, while regarding this work as a kind of common project to be carried on, in a characteristically free spirit, by all members of the school. The fragments on place which have been preserved, deriving from Theophrastus' eight books of *Physics*, testify, or so it appears to me, to Theophrastus' free handling of the Aristotelian material.

Fr. 146 and 149 in the edition of Fortenbaugh, Huby, Sharples

[93] Different assessments in Steinmetz (1964) (defending the picture of Theophrastus as a fairly original thinker); J. B. McDiarmid (1953), 'Theophrastus on the Presocratic Causes', *HSPh* 61 (1953) 85–156 (Theophrastus as largely following Aristotle); and H. B. Gottschalk, review of P. Steinmetz, *Die Physik des Theophrasts*, *Gnomon* 39 (1967) 17–26 (Theophrastus as someone who introduced the results of empirical observations and *ad hoc* speculation into a basically Aristotelian framework). For similarly divergent opinions expressed by our ancient sources, see Thphr. fr. 72a FHSG (Boethius depicting Theophrastus as continuing and further developing the work and thought of Aristotle) and fr. 694 FHSG (Quintilianus depicting Theophrastus as someone who as a rule did not hesitate to dissent from Aristotle).

and Gutas (FHSG; fr. 21 and 22 in the old collection of Wimmer) —two passages in Simplicius' commentary on Aristotle's *Physics* —constitute virtually all the available information concerning Theophrastus' ideas about place.[94] Fr. 146 (Simp. *In Ph.* 604, 5–11) contains a relatively straightforward enumeration of what Simplicius describes as a set of *aporiai* put forward by Theophrastus in connection with Aristotle's famous final definition of place as the 'inner boundary of the surrounding body'. As to fr. 149, an allegedly *verbatim* quotation to be found in the final part of Simplicius' *Corollarium de loco* (Simp. *In Ph.* 639, 13–22), the situation is more complicated. In the first place, it is not immediately clear what exactly Theophrastus was trying to convey in these rather condensed phrases, nor what position Simplicius assigns to Theophrastus' conception of place in his *Corollarium de Loco.*[95] Accordingly, opinions differ as to how the contents of this fragment relate to the *aporiai* of fr. 146 and to Aristotle's theory of place. Secondly, it is not immediately obvious to what extent Theophrastus was himself positively committed to the ideas expressed in fr. 149.

All in all, we are facing a *status quaestionis* which is, roughly, as follows. According to what I shall refer to as the 'traditional' view—a view defended by Jammer and Sambursky[96]—fr. 149 testifies to Theophrastus having developed a 'relational' theory of place as a full-blown alternative to Aristotle's defective theory. Sambursky characteristically compared the view expressed in Thphr. fr. 149 with Leibniz' theory of place. Pierre Duhem, on the other hand, saw fr. 149 as dealing with the primacy of *natural* place and, more or less following Simplicius, assumed a close resemblance between this view and Damascius' theory of 'essential place' (*topos ousiôdês*).[97] Whereas these scholars offered little beyond a categorical statement of their position, a more detailed

---

94 I shall henceforth refer to these fragments by their number in the FHSG collection only. For some additional testimonia of minor importance see below, n. 102.

95 I have discussed Simplicius' view of the relation Theophrastus-Iamblichus-Simplicius in Algra (1992) 157–162; for a different view see Sorabji (1988a) 204–215.

96 Cf. Jammer (1954) 21; Sambursky (1962) 2; and *idem* (1982) 12–13; see also Steinmetz (1964) 156–157.

97 Duhem (1913) 349–350. On Damascius' conception of place see Sambursky (1982) 21–22; Sorabji (1988a) 207–210; Algra (1992) 157–162.

analysis was provided by Sorabji.[98] According to Sorabji fr. 149 should not be read as representing anything like a full scale concept of place. He argues that it may best be understood as an argument (or rather an objection) with a much more limited scope, specifically directed against Aristotle's conception of (the dynamic character of) *natural* place.[99]

In the present section I shall use the findings of the earlier sections of this chapter to determine what position should be assigned to Theophrastus' ideas about place in general, and to fr. 149 in particular, in the *Rezeptionsgeschichte* of Aristotelian physics. This will enable us to provide an overall interpretation of Theophrastus' position which differs both from the traditional one and from that put forward by Sorabji. I shall argue, against the 'traditional' view, that the evidence does not indicate that Theophrastus ever worked out the suggestions of fr. 149 into a detailed and coherent alternative theory of place. Even if the fragment represents ideas endorsed by Theophrastus *in propria persona*, as I indeed believe it does, we should take into account that its phrasing points to a dialectical context. At the same time I disagree with Sorabji's interpretation in that I do not believe that the argument has Aristotle's concept of *natural* place as its exclusive, or even primary, target. The present section should therefore be regarded as an attempt to defend a qualified version of the traditional view by means of a closer study of the evidence.

Its structure is as follows. We shall first, in 5.3.2, study Theophrastus' *aporiai* of fr. 146 FHSG and the way in which they relate to Aristotle's theory. Next, section 5.3.3 will deal with Thphr. fr.

---

98 Sorabji (1988b). Since the contents of this paper substantially reappear in the relevant chapters (viz. ch. 11 'The Immobility of Space', and 12 'Is Space Inert or Dynamic') of Sorabji (1988a) references will be to the book.

99 I have defended a rival interpretation in Algra (1992). In part the arguments of that paper and of the present section overlap, in part they are complementary. In particular the reader is referred to Algra (1992) for a fuller analysis of the way in which Simplicius' is dealing with Theophrastus than could be given within the limits of the present chapter. As indicated in the text, also Duhem, like Sorabji, believed fr. 149 to be primarily dealing with a conception of natural place opposed to Aristotle's. According to him, however, it is not so much the alleged dynamic character of Aristotelian natural place that Theophrastus is objecting to, but rather Aristotle's idea that natural place is logically posterior to place as such (a relation which, he claims, Theophrastus wanted to be reversed). For my view of the relation place-natural place in Aristotle and Theophrastus see below, 238–240.

149 FHSG and the ways in which it has been interpreted by scholars and commentators. We shall focus in particular on the question whether the fragment is directed against Aristotle's conception of natural place or against his conception of place as such. We shall also try to find out to what extent Theophrastus was himself committed to its contents: are we dealing with an objection, a full-blown alternative conception, or a mere suggestion? In our concluding subsection 5.4.4, I shall try to show that this approach allows us to develop a coherent interpretation of Theophrastus' position.

5.3.2 Thphr. fr. 146 (Simp. *In Ph.* 604, 5–11) consists of a list of *aporiai*, which runs as follows:

> One should know that in his *Physics* Theophrastus too raises difficulties (ἀπορεῖ) of the following sort against Aristotle's account of place:
> (1) a body will be in a surface;
> (2) place will be in motion;
> (3) not every body will be in a place; for the sphere of the fixed stars will not;
> (4) if the spheres are taken together, the heaven as a whole will also not be in a place;
> (5) things in a place will no longer be in a place if, without their changing themselves at all, their surroundings are removed.[100]

The first of these *aporiai* deals with a problem we have not yet discussed. Aristotle had claimed that a place should be equal in size to that which is contained by it. It is probably mainly against this background that Theophrastus now points out that the idea that a body is in a surface is odd. As Sorabji has pointed out, Aristotle might have replied that the sense in which a surrounding surface is equal to the surface of the emplaced body sufficiently accommodates the common sense notion that a thing is equal to its place and fits it exactly.[101] Nevertheless one might well ask whether this response would not overstretch the common sense notion of place as being equal to what is in it. Moreover, the

[100] Simp. *In Ph.* 604, 5–11: ἰστέον δὲ ὅτι καὶ ὁ Θεόφραστος ἐν τοῖς Φυσικοῖς ἀπορεῖ πρὸς τὸν ἀποδοθέντα τοῦ τόπου λόγον ὑπὸ Ἀριστοτέλους τοιαῦτα· ὅτι τὸ σῶμα ἔσται ἐν ἐπιφανείᾳ, ὅτι κινούμενος ἔσται ὁ τόπος, ὅτι οὐ πᾶν σῶμα ἐν τόπῳ (οὐδὲ γὰρ ἡ ἀπλανής), ὅτι ἐὰν συναχθῶσιν αἱ σφαῖραι, καὶ ὅλος ὁ οὐρανὸς οὐκ ἔσται ἐν τόπῳ, ὅτι τὰ ἐν τόπῳ ὄντα μηδὲν αὐτὰ μετακινηθέντα, ἐὰν ἀφαιρεθῇ τὰ περιέχοντα αὐτά, οὐκέτι ἐν τόπῳ.

[101] Sorabji (1988a) 192–193.

additional point could be raised that, according to Aristotelian ontology, the more obvious thing to say would be that surfaces are in bodies, rather than the other way round. Anyway, this *aporia* clearly casts doubt on the identification of place as a *surface*. It thus concerns what we might call the *morphology* of Aristotelian place. *Aporia* (2) focuses on the problem discussed above in section 5.2. There it was shown that the problem of the immobility (or rather: the lack of immobility) of Aristotelian place was closely connected with the ontological status of place as a surface. So, albeit in a less straightforward way, also this *aporia* can be said to have the morphology of Aristotelian place as its target. [102]

The next two *aporiai* (3) and (4), take up the problem of the 'disemplacement' of the (outer sphere) of the heavens. Aristotle had argued that, in the absence of a surrounding something, the outer sphere of the heavens, hence the heavens as a whole, and hence in a way even the cosmos as a whole, are not in a place. At the same time he had conceded that we do in fact use the notion of place in connection with the heavens, viz. in so far as they are rotating and hence in a way changing place. Aristotle had also made clear, however, that when speaking of place in the latter sense, people were actually not so much speaking of the place of the heavens as a whole, but rather of the place of its parts. This fairly ambiguous statement—what is supposed to be meant by 'the place of the parts' of the heavenly sphere ?—caused a lot of discussion among later commentators, and the whole problem of the emplacement of the heavens became one of the standard issues to be explained by ancient and medieval Aristotelians.[103] In section 5.4 I shall briefly discuss Eudemus' attempt to solve the

[102] We may here add that, to judge from a number of other minor testimonia concerning Theophrastus' conception of place, he was particularly keen to stress the requirement that place should be immobile. Cf. Simp. *In Ph.* 566, 18 (fr. 147 FHSG *in app.*); 583, 7 ff. (fr. 147 FHSG *in app.*); 606, 32 (= fr. 147 FHSG = Eudemus fr. 79 a, b and c Wehrli), and Procl. *ap.* Simp. *In Ph.* 612, 1–7 (= fr. 148 FHSG). See also below, the section on Eudemus, 252–255.

[103] See also the previous chapter, 185. For an excellent discussion of the way this topic was treated by the ancient commentators, see Sorabji (1988a) 194–196. For medieval discussions see Grant (1981b) 72–79. It may be that *aporia* (4) even refers to the whole cosmos not being in a place. Aristotle himself argues that according to his own theory there is a sense in which neither the heavenly sphere or spheres (212 b 8–10 and 11–13) nor the cosmos as a whole (212 b 14 τὸ δὲ πᾶν οὐ ποῦ) are in place, and he makes clear that the word οὐρανός can refer both to the heavens and to the cosmos as a whole (212 b 17–18: ὁ γὰρ οὐρανὸς τὸ πᾶν ἴσως).

problem. But we may here leave later discussions of these matters for what they are, and simply take notice of the fact that the background for Theophrastus' *aporiai* (3) and (4)—i.e. the reason why the outer sphere, paradoxically, is not in a place—is that the outer sphere of the cosmos has no 'surrounding something'.[104]

A connected, though less familiar, *aporia* is (5), which deals with a similar paradoxical disemplacement of intracosmic substances if the requirements of the thought experiment at issue were to be fulfilled, i.e. if the immediately surrounding substances were taken away.[105] I believe that the point Theophrastus wants to make here is the logical absurdity of the idea of a thing's not even simply *changing* its place, but altogether *losing* it, as soon as (counterfactually, of course) its immediate surroundings are taken away, while it does not change or move in any way itself.

---

[104] *Phys.* Δ 212 a 31: ᾧ μὲν οὖν σώματι ἔστι τι ἐκτὸς σῶμα περιέχον αὐτό, τοῦτο ἔστιν ἐν τόπῳ, ᾧ δὲ μή, οὔ. Cf. also 212 b 8: ὁ δ' οὐρανός, ὥσπερ εἴρηται, οὔ που ὅλος οὐδ' ἔν τινι τόπῳ ἐστίν, εἴ γε μηδὲν αὐτὸν περιέχει σῶμα.

[105] Sorabji (1988a) 197–198 believes that also this *aporia* concerns the immobility of place. He maintains that it must be about replacing rather than, as the text would at first sight seem to suggest, about removing altogether a thing's surroundings. He argues that if the paradox in this case consisted in the body at issue coming to have no place at all, Aristotle could have replied (1) that the experiment could never be performed, and (2) that even if it could, we ought not to be surprised that the removal of all surroundings should suffice to deprive a thing of location. 'It is likely, therefore, that the experiment Theophrastus intends is the more familiar one of replacing, not removing altogether, a thing's surroundings' (Sorabji (1988a) 198). Parallel examples are to be found, according to Sorabji, in such various authors as John of Ripa, Simplicius and Philoponus. I would object, firstly, that the latter examples, all dealing with the case of a person or thing being surrounded by a draught of air, and hence by a moving surface, are better taken as illustrating the second *aporia* ('place will be in motion'). Secondly, as to Aristotle's imaginary ripostes, one may doubt the relevance of (1), as we are obviously dealing with a thought experiment. Note, moreover, that at *Phys.* Δ 212 a 32–34 Aristotle himself envisages a similar thought experiment concerning a mass of water with no outer surroundings. Neither, to my mind, can (2) be used to prove that Theophrastus cannot have meant that the removal of its surroundings deprives a thing of location. First of all, the third *aporia* clearly shows that Theophrastus surely did regard the fact that the removal of all surroundings would derive a body of its place as a paradoxical consequence of Aristotle's theory. Secondly, moreover, Theophrastus is not even speaking of the removing of *all surroundings*. Our fragment merely speaks of the removing of the *immediate* surroundings (ἐὰν ἀφαιρεθῇ *τὰ περιέχοντα* αὐτά. It is of course in this sense that περιέχειν is used of Aristotelian places: see e.g. the passages quoted in the previous footnote. I therefore find myself unable to adopt Sorabji's interpretation, and I accordingly prefer to take this fifth *aporia* as showing one more absurdity following on the conception of place as a surrounding 'something'.

If we once again review this set of *aporiai*, we may conclude that they can all be read as attacking the *morphology* of Aristotelian place. *Aporia* (1) is an objection against conceiving place *as a surface tout court*; (2) objects against seeing place as the surface *of a mobile surrounding body*; (3), (4) and (5), also focus on Aristotle's *Verdinglichung* of place in showing the absurdities that follow if place is identified as an immediately surrounding *something*.

We may note that we thus have a fairly strong *prima facie* argument in favour of the 'traditional interpretation' of fr. 149, according to which this fragment, in so far as it suggests viewing place no longer as a 'thing' but as the order and position of individual bodies, contains Theophrastus' alternative conception of place. For it should be stressed that this alternative relational conception, viz. place as order and position, would enable him to solve all five *aporiai* of fr. 146.

5.3.3 It is now time to have a closer look at the text of fr. 149 FHSG (Simp. *In Ph.* 639, 13–22) itself, in order to find out whether and to what extent it allows of the interpretation that has just been suggested. After having set out the concept of place of his master Damascius, Simplicius goes on to show that this concept is, in his own words, 'neither entirely new, nor unknown to famous philosophers'. The first to be mentioned is Theophrastus, and this time we are presented with a *verbatim* quotation.

> Also Theophrastus seems to have subscribed to this view in his *Physics*, in a passage where he continues his account in an aporetic fashion and says: 'Perhaps place (*topos*) is not a substance (*ousia tis*) in its own right (*kath' heauton*), but we speak of it because bodies have an order (*taxis*) and position (*thesis*) in conformity with their natures and powers, and similarly also in the case of animals and plants and in general of things with a differentiated structure, whether animate or inanimate, if they have a nature that exhibits form. For in these too there is an order and position of parts in relation to the whole substance. And this is why each of them is said to be in its own place through having its own proper order. For each part of the body too would desire and demand its own place and position'.[106]

---

106 Καὶ γὰρ καὶ Θεόφραστος ἐν τοῖς Φυσικοῖς φαίνεται τὴν ἔννοιαν ταύτην ἐσχηκὼς περὶ τόπου ἐν οἷς φησιν ὡς ἐν ἀπορίᾳ προάγων τὸν λόγον· 'μήποτε οὐκ ἔστι καθ' αὑτὸν οὐσία τις ὁ τόπος, ἀλλὰ τῇ τάξει καὶ θέσει τῶν σωμάτων λέγεται κατὰ τὰς φύσεις καὶ δυνάμεις, ὁμοίως δὲ καὶ ἐπὶ τῶν ζῴων καὶ φυτῶν καὶ ὅλως τῶν ἀνομοιομερῶν εἴτε ἐμψύχων εἴτε ἀψύχων, ἔμμορφον δὲ τὴν φύσιν

There are two factors which might cast doubt on the 'traditional' interpretation of this fragment and of how it relates to fr. 146. First, it appears to speak of *natural* place rather than of place *tout court*, an aspect which was given much weight in the interpretations of Duhem and Sorabji. Secondly, it may be argued that it contains an *objection*, presumably against Aristotle's theory of (natural) place, rather than any positive statement and hence that it remains to be decided to what extent Theophrastus was himself committed to its contents. We shall here first concentrate on the former question, viz. whether the two conceptions opposed in fr. 149 are conceptions of place or of *natural* place.

As we noted, Richard Sorabji has defended the thesis that in this fragment Theophrastus is objecting to Aristotle's idea (expressed at *Phys.* Δ 208 b 10–11) of natural place having a kind of power (*dunamis*). As a result of Theophrastus' downgrading of the role of place in favour of the role of nature and form, Aristotle's habitual talk about natural places would turn out to be a mere *façon de parler*. According to Sorabji's line of thought this downgrading of talk about place is best understood as a challenge to natural place, not to place in general. For, he argues, Theophrastus' suggestion 'would not cover the ordinary sense in which we say that a rock rolled along the ground has left its former place. At best it would only allow us to say that a rock has left its place, when it was thrown upwards in the air, contrary to the natural arrangement dictated by form. It is therefore easiest to suppose that Theophrastus believes some further conception of place is in play when we talk of things changing their place'.[107]

I believe this interpretation meets with a number of serious objections. Firstly, the wording of the fragment itself (speaking of place not being an οὐσία) indicates that it contains an alternative conception of *topos* in general; the concept of natural place only comes in later. Our source Simplicius, moreover, quite clearly takes Theophrastus as advocating a different *morphology* of place, rather than as defending a different explanation of natural motion and natural place.

ἐχόντων. καὶ γὰρ τούτων τάξις τις καὶ θέσις τῶν μερῶν ἐστι πρὸς τὴν ὅλην οὐσίαν. διὸ καὶ ἕκαστον ἐν τῇ αὑτοῦ χώρᾳ λέγεται τῷ ἔχειν τὴν οἰκείαν τάξιν, ἐπεὶ καὶ τῶν τοῦ σώματος μερῶν ἕκαστον ἐπιποθήσειεν ἂν καὶ ἀπαιτήσειε τὴν ἑαυτοῦ χώραν καὶ θέσιν'.

[107] Sorabji (1988a) 203.

Secondly, talk about natural place would seem to presuppose a more general concept of place-as-such. In Aristotle's *Physics* the two subjects are closely connected, natural place being regarded as a species of the genus place.[108] Also the wording of fr. 149 points in this direction. The statement that the *proper* place (αὑτοῦ χώρα) is determined by the *proper* position (οἰκεῖα τάξις) only makes sense if it is assumed that there is also a *non-proper* place being determined by a *non-proper* position. Hence, as a kind of common denominator, a general concept of place seems to be involved, a concept defined by order and position—which happens to be the very concept with which Simplicius credits Theophrastus. The upshot, in other words, is that things have a place assigned to them corresponding with their position, and, more specifically, a natural place which corresponds with their natural position.[109] In that case there is no need to suppose that the conception of place outlined would not allow us to account for adventitious motion: we may surmise that according to Theophrastus a stone rolling down a slope changes place without *ipso facto* exchanging its natural place for an unnatural one.

Thirdly, even if it were supposed, for the sake of the argument, that Aristotle was committed to a dynamic conception of natural place, and that Theophrastus wanted to attack it, there was no need for the latter to change the whole outlook of Aristotelian place by introducing the idea of order and position. He might in that case just as well have stuck to Aristotle's own concept of place as the surface of the surrounding body, while adding the rider that this surface *as such* is fully inert and that natural motion is caused by the *phuseis* and *dunameis* of the bodies themselves. That, instead, Thphr. fr. 149 contains a more radically different conception of

---

[108] See above, n. 25 .

[109] Sorabji translates the words κατὰ τὰς φύσεις καὶ δυνάμεις as '*through* their natures and powers'. In that case the implication of Theophrastus' word would seem to be that the nature of bodies is the *sole* factor determining their place, so that for any body to change place would be to exchange its natural place for an unnatural one (or *vice versa*). The conception of place which then results would indeed not cover the adventitious motion of e.g. a stone rolled over the ground. At this point, however, we are in danger of reading too much into the text. If we take the phrase κατὰ ... δυνάμεις in the most natural way, i.e. as meaning 'in accordance with their natures and powers' the implication is merely that, according to the natural tendencies of 'things', we assign to them natural as well as non-natural places.

(natural) place indicates a more general dissatisfaction on Theophrastus' part with the ontology and morphology of Aristotelian place.

Finally, and most importantly, on the basis of the findings of our section 5.1 we may conclude that there never was any such thing as an Aristotelian theory of the dynamic character of natural place for Theophrastus to argue against. I have tried to show that according to Aristotle the *dunamis* with which (natural) place is commonly credited is not to be equated with any causal power. Rather, the statement that place has a *dunamis* appears to boil down to nothing more than the thesis that natural place is a necessary condition of natural motion and, hence, that the concept of natural place allows us to make use of the notion of natural motion in a meaningful way. In addition, I have argued in section 5.1 that Aristotle, like Theophrastus, regarded the natural motion of the elements as being determined by their essence or nature, and that he also spoke of natural motion as the actualization of a potentiality. Hence Theophrastus' talk about natural motion being determined by the natures (φύσεις) and potentiatlities or capacities (δυνάμεις) of the elements at issue does not appear to differ substantially from Aristotle's habitual description of the same phenomenon.[110]

The suggestion that fr. 149 betrays a more general dissatisfaction on Theophrastus' part with the ontology and morphology of Aristotelian place becomes even stronger, when we take account of yet another interesting element in this text. For in this fragment Theophrastus explicitly extends the idea of place to the parts of non-homogeneous substances. True, in his biological works also Aristotle repeatedly makes use of the idea of the parts of animals and plants having a proper and natural position. However, in the only technical discussion of place, in *Phys.* Δ, he

110 See e.g. *Cael.* A 276 b 4–5: φέρεσθαι *πέφυκε*, and 278 b 30–32 which speaks of the οἰκεῖοι τόποι of some elements in connection with the way they move κατὰ *φύσιν*. Note, in addition, that in his *Meteorologica* Theophrastus made use of the concept of natural place in a quite Aristotelian way. Of the vapors that ascend from the earth it is said that they move more quickly when they approach their natural places, whereas they move more slowly when they are distant from them. But here also, just as in in Aristotle, there is no implication that natural place figures as a cause. For a convenient translation of the relevant passages from the 10th century Arabic translation by Ibn al-Khammar see now Daiber (1992) 264.

argues that such parts are normally speaking not in a place of their own.[111] The reason is obvious. In such cases the continuity of the parts precludes our conceiving of the place of these parts in the Aristotelian fashion, i.e. as a surrounding surface. Accordingly, Theophrastus' explicit statement that we also speak of place in these cases (λέγεται ... ὁμοίως δὲ καὶ ἐπὶ τῶν ζῴων καὶ φυτῶν καὶ ὅλως τῶν ἀνομοιομερῶν εἴτε ἐμψύχων εἴτε ἀψύχων, ἔμμορφον δὲ τὴν φύσιν ἐχόντων) may be interpreted as another polemical stab, showing that this alternative conception of place can also account for the common idea that the parts of a body have a place of their own, and that it thus saves more *phainomena* than the conception of place which emerged from Aristotle's discussion in *Phys.* Δ.

If fr. 149 presents us with an alternative conception of place as such, the question nonetheless remains why it gives such prominence to the notions of the natural tendencies and natural position of things. Here, for lack of context, we can only speculate. Still, I believe there are some plausible answers which present themselves. In the first place, Theophrastus may well think it is the phenomenon of natural motion which is primarily responsible for our conception of place *tout court.* This appears to be a quite Aristotelian line of thought. Also at *Phys.* Δ 208 b 8 ff. the phenomenon of natural motion is adduced as one of the reasons why we think place exists. Hence we may probably say that although for Aristotle natural place is in the end logically posterior to place as such, it is prior in another sense. For the phenomenon of natural place is initially more obvious to us than the eventual definition of place as such. The fragment we are here discussing

---

[111] Cf. Phys. Δ 211 a 29: ὅταν μὲν οὖν μὴ διῃρημένον ᾖ τὸ περιέχον ἀλλὰ συνεχές, οὐχ ὡς ἐν τόπῳ λέγεται ἐν ἐκείνῳ, ἀλλ' ὡς μέρος ἐν ὅλῳ and 211 b 1: ἔτι ὅταν μὴ διῃρημένον ᾖ, ὡς μέρος ἐν ὅλῳ λέγεται, οἷον ἐν τῷ ὀφθαλμῷ ἡ ὄψις ἢ ἐν τῷ σώματι ἡ χείρ κτλ. At 212 b 3 Aristotle explicitly (ὥσπερ δ' ἐλέχθη) refers back to this distinction by his statement that 'some things are in place potentially, others actually'. It is not entirely clear when Aristotle thinks this potentiality is actualized. It is perhaps most plausible to suppose that he thought that when e.g. a hand starts waving, and hence starts moving on its own, it gets a place properly speaking. Alternatively, he may have thought that these parts are only in a place if they are actually separated from the body to which they belong. The former possibility seems to be suggested at 211 a 17 ff. where the parts of a body are also said to 'move accidentally' i.e. by being involved in the motion of the body as a whole and to be 'capable of moving on their own': Ἔστι δὲ κινούμενον τὸ μὲν καθ' αὑτὸ ἐνεργείᾳ, τὸ δὲ κατὰ συμβεβηκός· τοῦ δὲ κατὰ συμβεβηκὸς τὸ μὲν ἐνδεχόμενον κινεῖσθαι καθ' αὑτό, οἷον τὰ μόρια τοῦ σώματος καὶ ὁ ἐν τῷ πλοίῳ ἧλος κτλ.

shows that also Theophrastus appears to focus in particular on the way in which we commonly use the notion of place (note his stress on how we *speak* about place). That in such a context talk about natural place figures prominently should not surprise us.

But it is possible to think of yet another reason why the notion of natural place could be expected to play an important role in Theophrastus' theory of place. The fact that there are *natural* places provides also Theophrastus' relational account of place—tentative though this account may be—with a fixed frame of reference (or better: with a number of fixed relations), and hence saves it from being completely relativistic. Also in this connection we may compare Arist. *Phys.* Δ 208 b 8–26, where it is argued that the phenomenon of natural motion provides us with absolute and natural directions.[112]

We may conclude that the two conceptions opposed in fr. 149 are conceptions of place in general, not merely of *natural* place. In so far as this fragment also offers a conception of natural place different from Aristotle's, the evidence appears to indicate that it is primarily the *morphology*, rather than any alleged *dynamic character*, of Aristotelian natural place that Theophrastus is objecting to. In fact if we compare Thphr. fr. 149 FHSG with the texts in which Aristotle discusses the dynamics of natural motion, we shall have to conclude that there were no substantive disagreements on this subject between the two thinkers.

5.3.4 Thus far our story may be regarded as a reasoned defence of the interpretation put forward—albeit *ex cathedra*—by scholars such as Jammer and Sambursky. It remains to be decided, however, to what extent Theophrastus was himself actually committed to the alternative conception of place outlined in fr. 149. With regard to this question scholars have reached various conclusions. Jammer and Sambursky do not express any doubts and plainly assert that fr. 149 contains the outlines of an alternative theory of place defended by Theophrastus *in propria persona.*[113] Also Duhem

---

[112] Cf. esp. *Phys.* Δ 208 b 15 ff.: ἡμῖν μὲν γὰρ οὐκ ἀεὶ τὸ αὐτό [scil. δέξιον καὶ ἀριστερόν], ἀλλὰ κατὰ τὴν θέσιν ... ἐν δὲ τῇ φύσει διώρισται χωρὶς ἕκαστον. οὐ γὰρ ὅ τι ἔτυχέν ἐστι τὸ ἄνω , ἀλλ' ὅπου φέρεται τὸ πῦρ καὶ τὸ κοῦφον κτλ. Here also, in other words, the implication is that mere localization by *thesis* without fixed determinations by *phusis* would amount to a too relativistic account of place.

[113] Jammer (1954) 21: 'Theophrastus ... comes to the conclusion that space

and Sorabji, however much their interpretations may in other respects differ from those of Jammer and Sambursky, appear to assume that fr. 149 represents Theophrastus' own positive views, viz. his alternative conception of natural place.[114] More reservations were expressed by Zeller and Regenbogen, who regarded the contents of fr. 149 as a cautious suggestion on Theophrastus' part.[115] An even more sceptical stand has been taken by David Sedley who, in a footnote to a study on Philoponus' concept of space, points out that Simplicius claims that Aristotle's theory of place was shared by the 'whole Peripatos'. Sedley thinks 'the whole Peripatos' should include at least Eudemus and Theophrastus. Accordingly he regards the *aporiai* of fr. 146 and fr. 149 as 'merely exploratory'.[116]

In the present section I want to argue that the position taken by Zeller and Regenbogen is, in the light of our (scanty) evidence, the most plausible one. To this purpose I shall first discuss the question of the meaning of Simplicius' words ὡς ἐν ἀπορίᾳ προάγων τὸν λόγον. Next, I shall discuss the implications of Simplicius' statement about 'Aristotle and the whole Peripatos' sharing one and the same conception of place.

It is perhaps tempting to take the phrase by which Simplicius introduces Thphr. fr. 149, ὡς ἐν ἀπορίᾳ προάγων τὸν λόγον, as an indication that we are here dealing with an *aporia* of the same kind as those of fr. 146, viz. an *objection*. This indeed seems to be the way in which both Sorabji and Sedley interpreted Simplicius' words.[117] Of course the term *aporia* (as well as the verb *aporein*) is

---

is no entity in itself but only an ordering relation that holds between bodies and determines their relative positions'; Sambursky (1982) 12: 'For Theophrastus place is not a reality in itself, but is defined by the juxtaposition and coexistence of bodies and their parts'.

114 Duhem (1913) 350: 'Théophraste modifie la définition du lieu naturel de telle sorte qu' elle ne suppose pas l' éclaircissement préalable de la notion du lieu'; Sorabji (1988a) 204, concludes, in spite of his deference to Sedley's more sceptical approach, that it is plausible that Theophrastus put forward the suggestion of fr. 149 'not as a puzzle to be refuted, but as something deserving serious consideration'.

115 Zeller–Mondolfo (1966) 372: 'Teofrasto *tendeva*, per parte sua, a ricondurre il concetto di spazio all' ordine' [my italics]; and *ibid.* n. 110: 'Teofrasto dice, anche se *in modo di dubbio* ...' [again my italics]. Regenbogen (1940) 1549, speaks of fr. 149 as containing Theophrastus' 'eigene, vorsichtig wie gewöhnlich, vorgetragene Auffassung'.

116 Sedley (1987) 140, n. 1.

117 Sorabji (1988a) 202, speaks of 'Theophrastus' second fragment, in

often used to denote a particular puzzle, or a particular objection. However, it should be recalled that the word may also bear what seems to be the more basic connotation of 'the mental state of being unable to decide between two or more conflicting views'.[118] I would suggest taking the word *aporia* in Simplicius' introductory phrase in the latter sense, particularly because Simplicius speaks of ἐν ἀπορίᾳ προάγων *τὸν λόγον*, thus implying that it is Theophrastus' *logos* which is in a state of *aporia*. Accordingly, we may translate Simplicius' introductory phrase as: 'continuing his account in an aporetic fashion' (the implication being that Theophrastus is unable or unwilling to draw any definite conclusions),[119] rather than as raising a puzzle' or 'continuing his account in the form of a puzzle' (in which case the implication would be that we are dealing with just one more specific puzzle or objection). In that case the upshot of these introductory words is that Theophrastus confronted Aristotle's theory as a whole (as well as the *aporiai* it involved) with hardly more than his own cautious suggestion (place as a relation, not as an entity in its own right), although the first word of the actual quotation (μήποτε) may be taken to suggest that he tended to favour the proposed alternative. Students of Theophrastus, especially those of the so-called 'metaphysical fragment', are familiar with this unobtrusive mode of presentation. In general Theophrastus appears to have refrained from claiming definite truths.[120]

---

which he switches from raising difficulties for Aristotle's definition of place to raising difficulties which may be taken to concern his idea of natural place'. Sedley (1987) 140, n. 1, speaks of both fr. 146 and fr. 149 FHSG as 'Theophrastus' *aporiai*—puzzles—about Aristotelian place'.

118 Cf. Arist. *Top*. 145 b 1: ἡ ἀπορία ἰσότης ἐναντίων λογισμῶν. And *ibid*. 145 b 16–20: ὁμοίως δὲ καὶ τῆς ἀπορίας δόξειεν ἂν ποιητικὸν εἶναι ἡ τῶν ἐναντίων ἰσότης λογισμῶν· ὅταν γὰρ ἐπ' ἀμφότερα λογιζομένοις ἡμῖν ὁμοίως ἅπαντα φαίνηται καθ' ἑκάτερον γίνεσθαι, ἀποροῦμεν ὁπότερον πράξωμεν. For a similar use of ἀπορεῖν in Simp. cf. *In Cat*. 118, 9–10: ἀπορεῖ μήποτε ... καὶ λύει τὴν ἀπορίαν.

119 The new edition of *FHSG* (vol. 1, 305) translates ὡς ἐν ἀπορίᾳ προάγων τὸν λόγον along similar lines: 'as one who in an impasse tries to advance the argument'.

120 He characteristically inserts words like ἴσως (*caute adseverantis*, as the index verborum in Ross' edition of Theophrastus' *Metaphysics* comments); see also Regenbogen (1940) 1552 Theophrastus' 'methodische Vorsicht') 1556 (Theophrastus' 'skeptische Vorsicht, die im Ganzen doch wohl das Grundelement seines Denkens ausmacht').

We may next turn to the second problem which is of relevance to our attempts to determine the extent of Theophrastus' commitment to the contents of fr. 149, viz. the curious juxtaposition in Simplicius' *Corollarium* of the claim that Aristotle's conception of place was shared by *the whole Peripatos* (viz. in the initial diaeresis, at *In Ph.* 601, 20),[121] and the claim, made with respect to fr. 149, that Theophrastus to some extent foreshadowed Damascius' relational theory of place (viz. in the final part of the *Corollarium*, at *In Ph.* 639, 13–15; 642, 17–18). How should these apparently incompatible claims be weighed against each other?

We may recall, to begin with, that Simplicius seems to have been well acquainted with Theophrastus' *Physics*. Indeed he may well have had first hand knowledge of it.[122] At any rate he is our main witness for it, providing a number of direct quotations and numerous paraphrases. We should therefore not too easily dispose of his words when, in what he himself regards as the final and most important diaeresis of all possible conceptions of place, he claims, on the basis of our fr. 149, that Theophrastus suggested a concept of place more like Damascius' than like Aristotle's.

---

[121] At the beginning of the *Corollarium* Simplicius professes that he will add some contributions of his own to the articulation (*diarthrôsis*) of the possible views of place, since in this respect Aristotle's contributions (especially his fourfold diaeresis of the possible conceptions of place to be found at *Phys.* Δ 211 b 5 ff.) are a bit outdated, if only because interesting new views have been defended by later philosophers. He then provides a more refined initial diaeresis (*In Ph.* 601, 14–25), which divides theories of place as follows. Some argued that it was (A) corporeal (*sôma*; only Proclus), others that it was (B) incorporeal. Among the latter class there were those who considered it (B1) unextended. This group can be further divided, for some thought (B1a) that unextended place was merely a substrate (*hupokeimenon*; example: Plato), whereas others assumed (B1b) that it had some effective power (*telesiourgos*; example: Damascius). Those who (B2) thought of place as extended endowed it either (B2a) with two dimensions (example: Aristotle 'and the whole Peripatos'), or (B2b) with three dimensions. The latter group in turn should be divided into (B2b–α) those who admitted empty space or void (examples: Democritus *cum suis*, Epicurus), and (B2b–β) those who did not (examples: the more famous among the Platonists, and Strato).

[122] Note that the author of the commentary on the *De Anima*, which bears Simplicius' name (Simp. *In De An.* 136, 29) claims to have written an *epitomê* of Theophrastus' *Physics*. The authorship of Simplicius is disputed and some scholars prefer to attribute the treatise to Priscianus Lydus; see Bossier–Steel (1972). See also Hadot (1978) 194–202. Yet even if the commentary is to be attributed to Priscianus Lydus, who, like Simplicius, was a pupil of Damascius, the passage referred to testifies to the fact that in the philosophical circles in which Simplicius was active Theophrastus' *Physics* was still read and commented upon.

Neither should we forget that elsewhere (i.e. in our fr. 146) Simplicius exclusively records Theophrastus' *doubts* about Aristotle's conception of place and that he *nowhere* implies that Theophrastus positively tried to defend Aristotle in this respect, whereas he does refer to such attempts on the part of Eudemus and Alexander.

At the same time, it seems significant that Simplicius does not provide more evidence for Theophrastus' heterodox view: he merely comes up with a single quotation (and a rather diaporematic one at that) and there is no suggestion that he could have added numerous parallel passages. It seems a fair guess that there was nothing more to be quoted as evidence. This plausibly explains the absence of the name Theophrastus from Simplicius' initial diaeresis of the different theories of place. In the eyes of Simplicius, apparently, Theophrastus' unaristotelian suggestion of fr. 149 was not enough to make him deserve a separate label in the initial diaeresis.[123] And it is obvious that it is a separate label that would have been required if Simplicius had wanted to include Theophrastus' name. He could not—as he did in the final diaeresis—simply append Theophrastus' name to the view ascribed to Damascius. For whereas in the final diaeresis the *differentia specifica* distinguishing Damascius' conception from that of others is merely what we may call its *morphology* (i.e. the fact that Damascius defines place in terms of *taxis* and *thesis*, in which respect Theophrastus, to judge from fr. 149, may indeed count as his predecessor),[124] the initial diaeresis stresses the *dynamic character* as the *proprium* of Damascian place;[125] and this dynamic character is precisely what Theophrastus appears to have denied to place in fr. 149.

---

[123] Note that also in the final part of the *Corollarium* Theophrastus' view is not so much included as something worth mentioning in its own right, but rather as a witness to the more elaborate theory of Damascius.

[124] On the structure of Simplicius' own final diaeresis and on the way in which this diaeresis connects Theophrastus, Iamblichus and Damascius, see Algra (1992) 157–162. There I have tried to show that Simplicius groups together Theophrastus, Iamblichus and Damascius on the basis of a rather limited common ground between their theories (viz. the fact that they all define place in terms of 'order and position'). Note that, since Simplicius apparently had first-hand knowledge of Theophrastus' *Physics*, and since on the other hand there are hardly any reasons to assume that he misrepresents or misunderstands Theophrastus' position, the way he interprets fr. 149 himself is of great interest.

[125] On the structure of Simplicius' initial diaeresis see above, n. 121.

We may conclude therefore that, whereas we cannot explain away Theophrastus' figuring as a heterodox Aristotelian in the final part of Simplicius' *Corollarium*, the absence of his name from the initial diaeresis is, in the light of the contents of fr. 149, not too surprising. Hence, if the claims (1) that Aristotle's conception of place was shared by the whole Peripatos (initial diaeresis of the *Corollarium*) and (2) that in so far as his conception of place is concerned Theophrastus should be grouped with Damascius rather than with Aristotle (final part of the *Corollarium*) contradict each other and, hence, cannot both be taken at face value, it clearly seems to be (1) which should be taken *cum grano salis* (note, incidentally that (1) is inexact anyway, Strato figuring as a prominent counterexample in the same diaeresis). We had best take the rather cavalier 'the whole Peripatos' as primarily meant to refer to 'orthodox Aristotelians' like Eudemus and Alexander of Aphrodisias, whose attempts to rescue Aristotle's concept of place are referred to elsewhere in Simplicius' commentary.

5.3.5 We may now draw some threads together. We have concentrated on different aspects of Thphr. fr. 146 and 149. It appeared that, from a systematic point of view, both could very well be explained as a reaction against Aristotle's 'reified' conception of place, fr. 146 pointing to a number of difficulties that would follow from such a conception, fr. 149 suggesting a relational conception of place as a plausible alternative. This charitable interpretation is not contradicted by the evidence itself. Contrary to what has been supposed by others, we have found no indications that fr. 149 merely deals with *natural* place, nor that it reacts to an allegedly Aristotelian theory of dynamic natural place. Moreover—as I have argued elsewhere in more detail—our interpretation of fr. 149 appears to be supported by the way in which Simplicius arranges his material and presents Theophrastus' position in the final part of his *Corollarium de Loco*.[126] Still it must be admitted that the first sentence of fr. 149, as well as the way it is introduced by Simplicius, point to an originally diaporematic context: Theophrastus probably did not go beyond a mere suggestion.

All this results in the following picture of the place of Theophrastus in the *Rezeptionsgeschichte* of (this aspect) of Aristotelian

---

[126] See Algra (1992) 157–162.

physics. Theophrastus was sensitive to the weaknesses of Aristotle's theory of place, which he criticized for good reasons. His adducing *aporiai* while at the same time putting forward a tentative alternative solution appears to be quite in line with what is otherwise known of his style and method. It should be pointed out, however, that in this respect Theophrastus was more or less continuing Aristotle's own dialectical approach. I believe, accordingly, that it is potentially misleading to judge his contributions solely or even predominantly in terms of orthodoxy *versus* heterodoxy—or, for that matter, in terms of faithfulness *versus* originality—where tenets are concerned. We probably do him more justice by regarding him as someone who saw Aristotelian philosophy as a kind of common enterprise and who in that spirit, and while remaining faithful to Aristotle's own critical and dialectical approach, tried to refine and enlarge the Aristotelian corpus of knowledge. In the present case this approach apparently resulted in the following position. In spite of his criticisms he did not confront the Aristotelian theory of place with a fully worked out alternative. In other words, in so far as there was any apparent doctrinal heterodoxy on Theophrastus' part, it consisted merely in his leaving the *aporiai* of fr. 146 unanswered, while presenting our fr. 149 as hardly more than a good *suggestion*. It is perhaps for this reason that his ideas about place did not become widely known.

5.4 *Aristotle's theory of Place and Eudemus of Rhodes*

5.4.1 Another pupil of Aristotle's, Eudemus of Rhodes, also wrote a *Physics*. In general Eudemus appears to have stayed closer to Aristotle's original text and thought than his fellow-pupil Theophrastus, but here of course the defective state of the evidence concerning both thinkers should make us cautious in drawing definite conclusions. Be that as it may, the fragments of Eudemus' *Physics* do suggest that, like Theophrastus, he aimed at providing something more than a merely exegetical commentary. The overall structure of the work appears to have been that of a shortened paraphrase of Aristotle's *Physics*, but where necessary it contained additions and clarifications.[127] Here again it is Simplicius

[127] As for the abbreviated character of Eudemus' work, we know from Simp. *In Ph.* 1036, 13–15 (fr. 109 Wehrli), that Eudemus, presumably in view of the overlap between the seventh and eighth book of Aristotle's *Physics*, did

who in his commentary on Aristotle's *Physics* has included a number of references to and descriptions of Eudemus' work.

In the edition of Wehrli's *Die Schule des Aristoteles* the fragments of Eudemus that discuss Aristotle's theory of place are the nrs. 72 to 80. In some of these cases we are dealing with hardly more than concise paraphrases, small corrections, or rearrangements of the Aristotelian material. Thus, fr. 72 contains an observation on the primacy of locomotion as compared to other kinds of κίνησις. Fr. 74 elaborates on Aristotle's critique of the conception of place as a three-dimensional extension (*Phys.* Δ 209 a 4 ff.) by specifying that it is not only impossible for two bodies to completely interpenetrate when they are *physical* bodies, but also in case one of them is a *physical* and the other a *mathematical* body. Fr. 76 explains, in a somewhat more explicit fashion than Aristotle's own text, why place cannot be matter (viz. because matter is mobile). In fr. 77 we find Eudemus refining Aristotle's overview of the senses in which something can be said to be 'in' something else. Where Aristotle in *Phys.* Δ 210 a 20–21 had spoken of one rather general sense in which ἡ ὑγίεια ἐν θερμοῖς καὶ ψυχροῖς καὶ ὅλως τὸ εἶδος ἐν τῇ ὕλῃ, Eudemus specifies that it remains to be discovered whether πάθη and ἕξεις are in οὐσίαι in the same way in which μορφή is in ὕλη. Finally, fr. 88 presents us with a clarification of the gist of Zeno's paradox of place and of the way in which Aristotle's theory of place is able to dodge its vicious *regressus*.

More interesting for our present purpose are those fragments on place (frs. 73, 75, 79 and 80 Wehrli) that deal with some of the more problematic aspects of Aristotle's theory. These fragments show that at least a number of the problems signalled by Theophrastus and discussed in the previous section were also treated by Eudemus. This need not really surprise us. For one thing, as we have noticed, Aristotle's theory involved some loose ends as well as a number of consequences that were sometimes as obvious as they were embarrassing, and we may assume that these were discussed within the Peripatos at a fairly early date. The fragments of Theophrastus and Eudemus on place may thus be seen to reflect a recent or even still ongoing discussion within their

---

not discuss the contents of the seventh book, but went straight on to the eighth. For a survey of the main characteristics of Eudemus' approach to Aristotelian philosophy see Wehrli (^{2}1983) 530–534.

school. But also after Eudemus had left Athens to settle down in Rhodes, he remained in touch with Theophrastus, so that even at that stage the two thinkers may well have influenced each other.[128]

In the present section we shall focus on the fragments of Eudemus in which he tackles some of the problems that had also puzzled Theophrastus. First, in 5.4.2, I shall discuss his ideas about the role of Aristotelian place in the explanation of natural motion (fr. 75 Wehrli). In 5.4.3 we shall turn to his ideas about the immobility of place and about the related issue of the ontological status of Aristotelian place (frs. 73, 79, and 80 Wehrli). In 5.4.4, finally, I shall discuss Eudemus' solution to the problem of the (dis)emplacement of the heavens (fr. 80 Wehrli).

5.4.2 One of Eudemus' fragments on place, fr. 75 Wehrli (Simp. *In Ph.* 533, 14–19), provides some interesting additional information about the way in which the early Peripatetics interpreted Aristotle's slightly ambiguous stance on the dynamic character or inertness of place. In his commentary on *Phys.* Δ 209 a 18–22, the very passage in which Aristotle denies that place could serve as one of the four causes, Simplicius quotes some polemical lines from the third book of Eudemus' *Physics* in which Democritus is attacked.

> 'Is it then the motive cause? Or is that impossible as well, Democritus? For it needs to be able to move and to have some power'.[129]

Although these lines are in themselves not very informative, we are given some clues as to their purport by the context in Simplicius and by some other texts in which Aristotle attacks the early atomist conception of motion in a void. First of all, Simplicius goes on to point out that Democritus identified place as void, and that void 'in its own nature would be powerless and inactive'. This suggests that the upshot of Eudemus' argument was that the Democritean void ought to have some motive power but did not

[128] Eudemus (in Rhodes) and Theophrastus (in Athens) exchanged letters on the text of Aristotle's Physics; on which see Simp. *In Ph.* 801, 13 ff. (= Eudemus fr. 6 Wehrli).

[129] Simp. *In Ph.* 533, 16–17: ἀλλ' ἆρά γε, φησίν, ὡς τὸ κινῆσαν; ἢ οὐδὲ οὕτως ἐνδέχεται, ὦ Δημόκριτε· δεῖ γὰρ κινητικὸν εἶναι καὶ ἔχειν τινὰ δύναμιν. For the way in which Simplicius introduces this quotation see below, n. 132.

have it. This tallies with Aristotle's own polemical strategy against the atomists which transformed their belief that void was a necessary condition for motion into the thesis that void was actually a cause of motion, only to turn the tables on them by showing that void, far from causing or even allowing motion, would actually rather preclude it.[130] Puzzling though this Aristotelian representation of the early atomist position may be—Aristotle elsewhere seems to have taken a fairer stance by pointing out that the early atomists did not provide a causal explanation for atomic motion at all and that they even denied that it made sense to look for a cause in things eternal[131]—it apparently appealed to Eudemus.

Now what is important for our present purpose is whether the words 'for it needs to be able to move and to have some power' should be taken to imply that Eudemus himself subscribed to the view that place should have a causal status. At this point we are fortunate to have further information about the context of our quotation. For Simplicius introduces it by the statement that Eudemus took it as agreed among all (ὁμολογούμενον) that place is not a cause.[132] We must conclude, therefore, that the upshot of Eudemus' polemics was not that Democritus failed to see something which Aristotle did see, viz. that a proper explanation of locomotion requires a causal role to be played by place. His point in saying that 'it needs to be able to move and to have some power' was rather the familiar polemical one that the physical system of the atomists themselves required place, or rather void, to play a causal role, whereas they were forced at the same time to agree with Eudemus and all others that place cannot play it.[133]

---

[130] See above, n. 38; see also the appendix to chapter 6, below, 336–339. Another text in which Aristotle represents the atomists as having maintained that the atoms move *because* of the void is *Phys.* Θ 265 b 24–27.

[131] Cf. *GA* 742 b 17 ff.: where Aristotle ascribes the following position to Democritus: τοῦ μὲν ἀεὶ καὶ ἀπείρου οὐκ ἔστιν ἀρχή, τὸ δὲ διὰ τί ἀρχή, τὸ δ' ἀεὶ ἄπειρον, ὥστε τὸ ἐρωτᾶν τὸ διὰ τί περὶ τῶν τοιούτων τινὸς τὸ ζητεῖν εἶναί φησι τοῦ ἀπείρου ἀρχήν. At *Metaph.* A 985 b 19, moreover, he even reproaches the atomists for not having identified the source or cause of atomic motion.

[132] The *verbatim* quotation is introduced in the following words (Simp. *In Ph.* 533, 14–17): διὸ καὶ ὁ Εὔδημος ἐν τρίτῳ τῶν Φυσικῶν παρακολουθῶν τοῖς ἐνταῦθα λεγομένοις καὶ *ὡς ὁμολογούμενον ἕκαστον τῶν αἰτίων ἀφαιρῶν τοῦ τόπου* (follows the quotation printed above, n. 129).

[133] I defended a slightly different interpretation of these lines in Algra (1992) 151, n. 26.

There is no implication, in other words, that Eudemus himself was willing to credit place with any properly dynamic character. In fact we may note that saying that it is 'agreed among all' or 'generally agreed' (ὁμολογούμενον) that place is none of the four causes would be a very odd thing to do for an early Peripatetic who did not notably dissent from Aristotle's theory of place, if Aristotle himself had in fact not subscribed to this 'common conviction'. Thus we may regard this testimony as additional evidence supporting the reconstruction of the Aristotelian conception of the inertness of place which we provided in the first section of this chapter. If we are correct, both Theophrastus (as suggested in the previous section) and Eudemus, however critical their approach towards Aristotle's theory of place may have been in other respects, took over and defended Aristotle's position on this particular issue.

5.4.3 The fragments of Eudemus also provide interesting information on the way in which he tried to cope with the problem of the immobility of place. In the first place Simplicius informs us of the fact that both Theophrastus and Eudemus assumed the immobility of place as one of the *axiômata* from which the inquiry of the physicist should start.[134] Although Aristotle had not included the immobility of place among the 'apparent data' (ὅσα δοκεῖ ἀληθῶς καθ'αὑτὸ ὑπάρχειν αὐτῷ) about place, he did of course add the requirement that place should be immobile later on in the course of his dialectical procedure.[135] We may accordingly suppose that by including the immobility of place among the *axiômata* Eudemus and Theophrastus were drawing the full consequences from Aristotle's own position rather than criticizing that position. Nevertheless their move may point to grown awareness of the importance of the fact that place should be immobile. In section 5.2.3, I have tried to show that the required immobility of place involved two problems for anyone defending the Aristotelian position. Firstly, *with respect to what* is place supposed to be

---

[134] Eudemus fr. 79 a, b, and c. See also above, the section on Theophrastus, n. 102.

[135] These apparent data are ticked off at *Phys.* Δ 210 b 32–211 a 7 (item D 1 in the overview of section 4. 4. 4 in the previous chapter); the requirement that place should be immobile is added at 212 a 14–21 (item D 5 in the overview of 4. 4. 4).

immobile? Secondly, how can Aristotelian place, as a surrounding surface, be immobile in actual practice? On the basis of the evidence we know that Eudemus had an answer to the first problem, while it is at least probable that he was aware of the second.

In a context dealing with the emplacement of the outer sphere of the heavens, Simplicius shows us how Eudemus specified the immobility of ordinary Aristotelian places (fr. 80 Wehrli, partly):

> Having said that place must be the limit, in so far as it surrounds, of the surrounding body which was immobile he [i.e. Eudemus] added: 'For that which moves is like a vessel, and that is why we determine places in relation to the heavens. For they do not change place, except in their parts'.[136]

Whereas Aristotle in *Phys.* Δ failed to offer a more technical account of immobility, Eudemus apparently drew the consequences of the fact that the Aristotelian cosmos contained fixed parts and fixed directions which could in principle be used as reference points in such a more technical explanation. As we see, he claimed that the immobility of place was to be determined ultimately with reference to the sphere of the heavens, a surrounding substance which itself no longer changes place, even though it rotates (which means that in a sense its parts change place, on which see the next subsection). Here again, in other words, we see Eudemus working out a view which, although not put forward explicitly by Aristotle himself, was already present *in nuce* in the cosmological system as exposed in the *Physics.*

But perhaps there is more to it. The sentence by which Simplicius introduces our fragment contains a puzzling qualification: place is said to be the limit of the immobile surrounding body *in so far as it [i.e. this limit] surrounds* or *taken qua surrounding* (τοῦ περιέχοντος *ᾗ περιέχον* πέρας). Although the lack of further context precludes our drawing definite conclusions, it is tempting to take these words as implying that Eudemus in some sense prefigured the later medieval distinction, discussed above in 5.2.3, between formal place and material place. For, just as in medieval physics material place was defined as the (mobile) surface of the

---

136 Simp. *In Ph.* 595, 5–8 (part of fr. 80 Wehrli): εἰπὼν ὅτι τοῦ περιέχοντος ᾗ περιέχον πέρας ὁ τόπος ἂν εἴη ἀκινήτου ὄντος ἐπήγαγε· 'τὸ γὰρ κινούμενον ἀγγειῶδες, καὶ διὰ τοῦτο τῶν τόπων τὴν ἀναφορὰν πρὸς τὸν οὐρανὸν ποιούμεθα. οὗτος γὰρ οὐ μεταλλάσσει τόπον ἄλλον κτλ.'.

actually surrounding (mobile) body, whereas formal place was defined as this surrounding surface *taken as immobile* in virtue of the introduction of the concept of fixed distances, so here Eudemus may have tried to save the immobility of place by regarding place *qua surrounding* surface instead of *qua* surface *of a* surrounding (mobile) *body*, and by further specifying the immobility of this surrounding surface by reference to the fixed sphere of the heavens.

Whether he did so or not, however, sooner or later anyone taking the position defended by Eudemus or by the later medieval scholastic commentators would have to face the question of whether Aristotelian ontology allows anything like an immobile surrounding surface. This takes us back to the second problem outlined above. Elsewhere in this chapter I have suggested that this basically ontological problem may have been at the background of both Theophrastus' suggestion that perhaps place is not a thing in its own right and Ockham's later 'nominalist' conception of the immobility of place. Although there is no evidence that Eudemus was equally prepared to consider such essentially radical solutions to the problem, we do have another passage in Simplicius which shows that Eudemus was at least aware of it. At *In Ph.* 523, 22–28, as part of his commentary on Aristotle's own avowal, at *Phys.* Δ 208 a 32, that the subject of place is a difficult one, Simplicius adds an extra reason adduced by Eudemus (fr. 73 Wehrli):

> Eudemus says that a further cause of the difficulty of the problem of place is that [the notion of] place is not easy to grasp, because it altogether escapes us when the body in it is removed, and it is not possible to apprehend it in itself, but, if at all, in combination with something else, like the sounds of the so-called consonants. For with 'a' added the sound of 'b' and 'c' becomes clear.[137]

Here it is argued that one of the inherent difficulties of the subject consists in the fact that place escapes us once the emplaced body is taken (or thought) away. Place cannot be conceived in its own right, but merely together with something else (presumably: the

---

[137] Simp. *In Ph.* 523, 22–28 (fr. 73 Wehrli): ὁ δὲ Εὔδημος τῆς δυσκολίας τοῦ προβλήματος καὶ τοῦτο αἴτιον εἶναί φησι τὸ μὴ ῥᾴδιον εἶναι ἐπιλαβέσθαι τοῦ τόπου, διότι ὑποφεύγει πάντως ἐξαιρουμένου τοῦ ἐν τόπῳ σώματος, καὶ καθ' αὑτὸ μὲν οὐκ ἔστιν αὐτὸν νοῆσαι, εἴπερ δὲ ἄρα, μεθ' ἑτέρου, ὥσπερ τοὺς μὲν ἀφώνων καλουμένους φθόγγους· μετὰ γὰρ τοῦ α ὁ τοῦ β καὶ ὁ τοῦ γ δῆλος.

substances which are in place), just as consonants can only be conceived (or: only occur) in combination with vowels. True, as our description shows, it is not quite clear whether Eudemus focuses on the possibility or impossibility of our having a *conception* of place in its own right or on the possibility or impossibility of place itself *being* something in its own right. But then of course, following an Aristotelian line of thought, the idea that we cannot *think of* place in its own right would certainly imply that place cannot *exist* in its own right.

Hence also Eudemus testifies to what I called the ontological problems involved in Aristotle's theory of place. Like Theophrastus in fr. 149, he appears to have doubted whether place is an *ousia* in its own right. We do not know, however, what consequences he drew from that doubt. At any rate, he kept close to Aristotle's own theory, in sticking to the conception of place as a surrounding surface. It is hard to imagine that he defended a non-reified version of this conception of place-as-a-surface, a solution along the lines preferred later on by Ockham. Most probably his attitude *vis-à-vis* a problem of this kind was the quite Aristotelian one of supposing that even if it could not be solved, 'to have considered the problem well, is enough for the lover of understanding'.[138]

5.4.4 We may finally briefly return to fr. 80 Wehrli. It shows that Eudemus also addressed the problem of the disemplacement of the (outer sphere of) the heavens, the problem discussed by Theophrastus in the third and fourth *aporia* of fr. 146. Because Aristotle arguably used the word οὐρανός to refer both to the outer sphere, i.e. the sphere of the fixed stars, and to the heavenly region as a whole, whereas he also mentioned the fact that the universe as a whole (τὸ πᾶν) was not in a place, there were in fact three problems: the emplacement of the outer sphere, the emplacement of the heavenly region as a whole, and the emplacement of the whole cosmos. As we have seen, the first two of these were treated as separate *aporiai* by Theophrastus. According to Simplicius, Eudemus limited his attention to the problem of the emplacement

138 This Aristotelian way out is referred to by Simplicius in the passage immediately following on Eudemus fr. 73 Wehrli, at *In Ph.* 523, 28–30: κἂν γὰρ μηδὲν ἄλλο ἐπὶ τῶν δυσλήπτων, ἀλλὰ τό γε ἀπορῆσαι περὶ αὐτοῦ καλῶς ἱκανὸν τῷ φιλομαθεῖ, ὡς ἐν ἄλλοις αὐτὸς ὁ Ἀριστοτέλης διδάσκει.

of the heavenly region as a whole.[139] Simplicius goes on to provide a purportedly *verbatim* quotation, the first part of which has already been discussed above. The part that is relevant in the present connection shows us Eudemus adhering rather closely to the contents of *Phys.* Δ:

> For what moves is like a vessel, and that is why we determine places in relation to the heavens. For they do not change place, except in their parts; for they revolve in the same place. But are the heavens themselves in a place or not, and how, in either case, could this be so? For they are not in a place as a whole, unless there is something outside; in that way they would be contained; we shall look into this. But the stars and everything within the outermost body are within its limit in so far as it surrounds them. Things thus in something are said to be in a place; and that in which parts are, is also that in which the whole is; so in that way it will be in a place. But it is also somewhere in another way; for the whole is in its parts, since '"somewhere" is said in many ways.[140]

On closer view Eudemus appears to come forward with two solutions. The first one in fact merely fleshes out the solution suggested rather cursorily by Aristotle at *Phys.* Δ 212 a 31–b 22, viz. that the heavenly sphere is in a place in one sense (viz. in so far as its parts are in a place), though not in a place in another sense (viz. in so far as there is nothing outside it). The second, put forward in the last line of the fragment, uses a distinction made elsewhere in *Phys.* Δ between various ways in which something may be said to be 'in' something else. Aristotle used the fact that 'being in something' was 'said in various ways' (πολλαχῶς λέγεσθαι) to circumvent the infinite regress of Zeno's paradox of place (on which see the previous chapter, section, 4.4.4, item C3). As we know from fr. 78 Wehrli (briefly discussed above), Eudemus rephrased both Zeno's paradox and Aristotle's answer to it in terms

---

139 Simp. *In Ph.* 595, 3: ὁ δὲ Εὔδημος τὸ μὲν πᾶν οὐδὲ ὀνομάζει· ἐπὶ δὲ τοῦ οὐρανοῦ τὴν ζήτησιν ποιεῖται τοῦ ὅλου.

140 Simp. *In Ph.* 595, 5–15: εἰπὼν ὅτι τοῦ περιέχοντος ᾗ περιέχον πέρας ὁ τόπος ἂν εἴη ἀκινήτου ὄντος ἐπήγαγε· 'τὸ γὰρ κινούμενον ἀγγειῶδες, καὶ διὰ τοῦτο τῶν τόπων τὴν ἀναφορὰν πρὸς τὸν οὐρανὸν ποιούμεθα. οὗτος γὰρ οὐ μεταλλάσσει τόπον ἄλλον, ἀλλ' εἴπερ, κατὰ μόρια· ἐν τῷ αὐτῷ γὰρ ἡ παριφορά. αὐτὸς δὲ πότερόν ἐστιν ἐν τόπῳ ἢ οὐκ ἔστιν ἢ πῶς ἑκάτερον; ὅλος μὲν γὰρ οὐκ ἔστιν ἐν τόπῳ, εἰ μὴ ἔστι τι ἐκτός· οὕτω γὰρ ἂν περιέχοιτο· τοῦτο δὲ ἐπισκεψόμεθα. τὰ δὲ ἄστρα καὶ ὅσα ἔντος τοῦ ἐξωτάτω σώματος ἐν τῷ ἐκείνου πέρατί ἐστιν, ᾗ περιέχον ἐστί. τὰ δὲ οὕτως ἔν τινι ἐν τόπῳ λέγεται, ἐν ᾧ δὲ τὰ μόρια, τὸ ὅλον λέγομεν εἶναι. οὕτω μὲν οὖν εἴη ἂν ἐν τόπῳ. ἔστι δὲ καὶ ἄλλως που· ἐν τοῖς μορίοις γάρ ἐστι τὸ ὅλον, πολλαχῶς δὲ τὸ που'.

of 'being somewhere' (*pou*). According to Eudemus, what Zeno missed and what Aristotle saw, was that 'being somewhere' is said in many ways. Quite in line with all this, the fragment we are here discussing shows Eudemus introducing one of Aristotle's senses in which something can be 'in' something else as a special sense in which something can be 'somewhere': viz. in the way in which the whole is in its parts. Also in this sense the heavenly sphere as a whole can be said to be somewhere. For it can be said to be in its parts.

In this fragment Eudemus' purpose is clearly to save on the one hand the common-sense notion that everything is somewhere—and the more technically Aristotelian point that, in so far as the various parts of heavens rotate, they will need to be in a place in some sense[141]—while on the other hand maintaining the Aristotelian position that the heavens as a whole are not in a place in the particular sense required by locomotion.

As a final observation we may add that the two arguments adduced to prove that the heavenly sphere as a whole may still be said to be somewhere, are on closer view not all that successful. The first argument clearly makes use of the idea that within the heavenly region each inner sphere, in so far as it is surrounded by another, is in a place. But apart from the fact that, as Richard Sorabji aptly points out, such places would not be immobile in the required sense, this defence is also unsuccessful in that it still leaves the outermost sphere, the sphere of the fixed stars, without a place.[142] The second argument is of no avail in this respect, for it only provides a sense in which the heavenly region, *taken as a whole* may be said to be somewhere. In addition, this second argument appears to be of little help anyway. For although it presents us with a sense in which the heavens may be said to be 'somewhere', this is not a physically meaningful sense. Being somewhere in the sense in which a whole is in its parts has nothing to do with the problems of the location and locomotion of physical bodies—it cannot, for instance, throw any light on the rotation of the heavenly body—and as such it is not really at home in the present context dealing with physical place.[143]

---

141 Cf. *Phys.* Δ 212 a 10: ἐφ' ᾧ δὲ κινεῖται ταύτῃ καὶ τόπος ἔστι τοῖς μορίοις.
142 Sorabji (1988a) 194–195.
143 See on this also Sorabji (1988a) 195.

5.4.5 Our conclusions can be brief. From our discussion of the Eudemian fragments on place the following picture emerges. Eudemus was well aware of some of the main problems involved in Aristotle's theory of place. In particular, he addressed the subjects of the immobility of place, its ontological status, and the emplacement of the heaven. In that respect his fragments on place and the two Theophrastean fragments discussed in the previous section appear to have the same background. At the same time Eudemus' strategy in coping with these problems in Aristotle's theory appears to have been mainly conservative. In the extant fragments we witness him making particular passages in Aristotle less hard to swallow by on the one hand smoothing Aristotle's own wordings and making them more explicit, while on the other hand adducing extra material from elsewhere in Aristotle to explain the issues at hand. He thus appears to have been intent on remaining within the general framework of Aristotle's theory, and there are no indications that he seriously considered potentially radical solutions of the kind suggested by Theophrastus in fr. 149.

Conclusions

In the present chapter I have discussed several early Peripatetic attempts to come to terms with Aristotle's theory of place. These were studied against the background of Aristotle's theory of place itself and the obscurities and problems it involved. As was already noted in the previous chapter, Aristotle's dialectical discussion in *Phys.* Δ exhibited a number of rough edges and loose ends. Thus he nowhere explicitly reconciled his own two claims that place should have some power and that it should not be counted as one of the four familiar causes. In section 5.1 it was shown, or so I hope, that it is possible to reconstruct his position by a closer study of the dialectical structure of the discussion of *topos* in *Phys.* Δ and by adducing a number of other relevant passages from elsewhere in the *Physics* and the *De Caelo*. In the course of this chapter it became clear that the resulting picture of the non-dynamic character of Aristotelian place was confirmed by the few remarks on this issue that have come down to us from Theophrastus and Eudemus.

Two other problems left open by Aristotle, viz. the interrelated problems of the immobility of place and its ontological status, seem to have been more difficult to solve, as I have tried to show in section 5.2. They were discussed—together with a number of other problems, such as the problem of the emplacement of the heavens—by both Eudemus and Theophrastus

It appears—if we are allowed to draw some general conclusions from the scanty fragments that have come down to us—that each of these two pupils of Aristotle's continued his master's work in his own way: Theophrastus by continuing Aristotle's critical dialectical approach, which involved his feeling free to sometimes add some rather radically alternative suggestions, Eudemus by mainly filling out Aristotle's own suggestions by adducing material from elsewhere in his work or by rephrasing Aristotle's arguments in clearer terms. But even if Eudemus appears to have been the more 'orthodox' of the two, we should not overestimate the strength and the extent of Theophrastus' dissent from Aristotle. It appears to have consisted mainly in his leaving the *aporia* of fr. 146 unanswered, while putting forward the contents of fr. 149 as hardly more than a suggested alternative. Moreover, it is worth noting that in a way also the conception of place as a relation between bodies, suggested in fr. 149, may be regarded as constituting a sensible elaboration rather than a complete rejection of the Aristotelian position. For in so far as it still defines the place of a thing in terms of its surroundings rather than in terms of a διάστημα (whether in the Platonic or in the atomist sense), it remains on the Aristotelian side of the line drawn by Aristotle himself at *Phys.* Δ 209 b 1–7.[144] And unlike the alternative proposed by Strato, this conception of place could in principle be taken over *ceteris paribus*, leaving the rest of the system of Aristotelian physics intact.

At the same time, it should be clear that Theophrastus' solution, however hesitantly put forward, is far superior from a systematic point of view. It might even be claimed that it transforms Aristotle's (and Eudemus') rather naive theory of *place* (focusing on the location of individual substances) into what we might call a theory of *space* (in principle allowing an account of the sum total

---

144 Anticipating his later fourfold division (211 b 6 ff.) of possible candidates for *topos* Aristotle there distinguishes only two ways of looking at place: one might regard it as a surrounder (in which case one might wrongly believe it to be form) or as the διάστημα τοῦ μεγέθους (in which case one might be tempted to mistakenly identify place as matter).

of spatial relations within the cosmos).[145] This brings us to the curious fact that this novel conception of place did not have a wider appeal. As we saw, we actually have to wait for Damascius to take up Theophrastus' suggestion. This is probably partly due to the fact that Theophrastus omitted to elaborate his point and that, as a consequence, it did not become widely known. In addition, the relational conception of place suggested by Theophrastus, if worked out properly, was much more technical and much farther removed from everyday usage and ordinary experience than its contemporary rivals. We need only look at Aristotle's theory of *topos* and the way in which it was taken seriously in antiquity (and beyond) to see to what extent lack of technicality and closeness to common thinking and speaking were commonly counted as virtues.

This in its turn leads us to the question of the influence of (Eudemus and) Theophrastus in general. To some extent the doubts, criticisms and refinements of Aristotle's theory put forward by Eudemus and Theophrastus may have proved seminal. At any rate later critics of the Aristotelian position, such as Simplicius, found it worthwhile to refer to their ideas or to add quotations from their work. And the mere fact that Aristotle's theory of place had come under attack within the Peripatos and that even a relatively faithful pupil like Eudemus had felt obliged to advocate some changes, may have encouraged the much bolder dissent of a philosopher like Strato of Lampsacus. Yet, it should be stressed that the precise extent of the influence of these early Peripatetics is impossible to determine.[146] At any rate there is no positive evidence that any of the later critics of Aristotle was directly influenced by Theophrastus or Eudemus, and it should be kept in mind that these critics probably did not even need their examples. Indeed, Aristotle himself provided enough ammunition, for example by failing to answer the question of the ontological status of place, by failing to provide a more technical account of immobility,[147] and by attacking the most obvious rival view (place as a three-dimensional extension) with very unsatisfactory arguments.

---

[145] On this use of the denominations 'place' and 'space' see chapter 1, 20.

[146] For a different assessment, stressing the influence of Theophrastus' position on thinkers like Philoponus, Strato and the Stoics, see Sorabji (1988a) 211–214.

[147] On Aristotle's rather untechnical notion of immobility see above, 227–228.

CHAPTER SIX

PLACE, SPACE, AND VOID IN STOIC THOUGHT

Introduction

One of the problems confronting the student of the philosophy of space of the early Stoics is the fragmentary nature of the evidence. Apart from two quotations from a work of Chrysippus we are dependent for our information on testimonia of a doxographical character (Arius Didymus, Aëtius, Sextus) and on the work of the Stoic astronomer Cleomedes who probably post-dates the early Stoics by some five centuries. In practice this means that the early Stoic theory of space has to be reconstructed and that, consequently, the philological component of any historical study in this field will be quite considerable. As another consequence, the conclusions will in a number of cases inevitably be of a rather provisional and hypothetical character.

At the same time the subject is of considerable interest. From a systematic point of view, the consequent materialism of the Stoics, in connection with their no less consequent rejection of the atomist hypothesis, arouses curiosity about the way in which they gave place and space a niche in their physical system. From a more purely historical point of view it might be asked whether in this respect early Stoic physics shows any traces of previous discussions, in particular whether the Stoics knew and reacted to Aristotle's theory of *topos* as expounded in the esoteric works. The latter question has lately aroused renewed interest as a result of the diametrically opposed conclusions of D. Hahm's *The Origins of Stoic Cosmology* on the one hand, and F. H. Sandbach's *Aristotle and the Stoics* on the other. According to Hahm the influence of the Aristotelian *pragmateiai* on early Stoic physics—including the theory of space—has been considerable. Sandbach, on the other hand, defended the 'traditional' thesis about the fate of the school writings. In his view they more or less disappeared from the philosophical stage during the first centuries following on Aristotle's death. Sandbach argued that early Stoic philosophy—

indeed early Hellenistic philosophy in general—showed no traces of acquaintance with these works.

In the present chapter I shall, in section 6.1, first deal with the preserved Stoic definitions of spatial concepts from what might be called a philological perspective. I shall try to determine how they are related and whether anything can be said about their provenance. Special attention will be paid to the concept of *chôra*. In section 6.2 I shall discuss the difficult text of Plutarch, *SR* 1054b ff., a text which contains two *verbatim* quotations from Chrysippus' *Peri dunatôn*. These two fragments are, as will appear, in more than one respect remarkable and interesting. Plutarch introduces the quotations in order to charge Chrysippus with grave inconsistency, and as yet, scholars have not succeeded in providing an interpretation which proved Plutarch wrong. However, I shall try to show that these two fragments can be made sense of, provided that we take proper account of the fact that they originally figured in a context dealing with modal logic. The interpretation which is to result can be connected with section 6.1 in that, together with the findings of that section, it allows us to reconstruct hypothetically Chrysippus' peculiar concept of *chôra*. It also prepares the way for section 6.3, in so far as it suggests that the early Stoic—at least the Chrysippean—theory of space was dominated by two different perspectives which cannot easily be reconciled.

Section 6.3 will deal with the ontological and epistemological aspects of the Stoic theory of place and of the void. I shall argue that in particular the theory of the void exhibited a fundamental ambiguity, in so far as the nature and ontological status of space was concerned. Section 6.4 discusses the doxographical evidence for Posidonius' allegedly heterodox position on the subject of the extent of the extra-cosmic void. The possble philosophical merits or demerits of that position will be assessed against the background of the analysis of the early Stoic theory of the void which was carried out in sections 6.2 and 6.3. The conclusions are followed by an appendix which focuses on a question to which I already referred, viz. whether the early Stoic theory of place and void shows any traces of Aristotelian influence.

6.1. Topos, chôra, *and* kenon: *early Stoic technical terms and definitions*

6.1.1 As *SVF* II, 503 ff., in a section entitled *De loco et inani,*[1] von Arnim has listed a number of fragments which provide us with definitions, ascribed to Chrysippus or 'the Stoics' (οἱ Στωικοί), of the concepts of *topos, chôra* and *kenon.* For lack of much further evidence, these definitions inevitably constitute the core of any discussion of what we might call the early Stoic theory of space. Though some scholars have treated this subject more or less *in extenso,*[2] it still calls for our attention, if only because as yet no consensus has been reached about the interpretations of the definitions, or about the way they fit in with their philosophical background. Moreover, one gets the impression that the (fragmentary) textual evidence is nowhere fully and adequately accounted for.

In the present section I shall first, in 6.1.2, present the surviving Stoic definitions of *topos, chôra* and *kenon* in full. In order to be able to shed more light on the nature and origin of the several definitions, I shall next, in 6.1.3, provide some information about the sources concerned and their reliability. In 6.1.4 I shall classify the preserved definitions and provide a tentative interpretation. It will appear that some are to be ascribed to Chrysippus, and others to what might be called the 'common Stoic orthodoxy'. Among the latter some substantially resemble the definitions provided by Chrysippus, whereas the definitions of *chôra* seem to differ to such an extent as to betray different underlying concepts. In 6.1.5, I shall provide a critical survey of the several ways in which these definitions of *chôra* have been interpreted in the scholarly literature, and I shall subsequently propose a provisional alternative interpretation of the Chrysippean concept of *chôra.*

---

[1] As to the texts printed by von Arnim under this heading, it should be noticed that *SVF* II, 501 (S. E. *M.* 10, 7), in spite of its apparent Stoic argumentational structure, does not reflect original Stoic arguments. Most of its contents derive from Aristotle *Phys.* Δ, so that there is no reason to print it in *SVF.* On the other hand there are some other texts, absent from the section *de loco et inani* (though partly printed elsewhere in *SVF*), which definitely belong here: the two Chrysippean fragments of Plu. *SR* 1054b–c and large scraps of the first chapter of Cleomedes' *Caelestia.*

[2] Cf. Bréhier (1907) 37–60; Goldschmidt (1953) 26–30; Sambursky (1959) 108–115; Pohlenz (1948–49) I, 65 and II, 37; Hahm (1977) 103–126; Christensen (1962) 14–25; Todd (1982) (for my criticisms of Todd's interesting but historically untenable interpretation see infra, n. 142); Algra (1988b),143–184 (an earlier version of part of the present chapter); Inwood (1991) (for my critique of Inwood's analysis, see below n. 168 and 181).

6.1.2 Stobaeus presents us with a scrap of Arius Didymus (fr. 25 Diels) which contains definitions of the spatial terms *topos*, *chôra*, and *kenon*. I here give a translation, following the text as printed by Diels (*DG*, 460). For the sake of convenience I have divided the text into four sections:

text 1

Arius Didymus fr. 25 = SVF *II, 503* = *LS 49 A (partly)*
(1) [Of Chrysippus]. Chrysippus declared *topos* to be that which is fully occupied by being (τὸ κατεχόμενον δι' ὅλου ὑπὸ ὄντος) or that which is able to be occupied by being and is *de facto* fully occupied whether by one thing or by several things (ἢ τὸ οἷόν ⟨τε⟩[3] κατέχεσθαι ὑπὸ ὄντος καὶ δι' ὅλου κατεχόμενον εἴτε ὑπὸ τινὸς ⟨εἴτε⟩[4] ὑπὸ τινῶν).
(2) If, of that which is able to be occupied by being, part is occupied and part not, the whole will be neither[5] void (*kenon*) nor *topos*, but a different something which has no name (ἕτερον δέ τι οὐκ ὠνομασμένον). For we speak of void (*kenon*) on the analogy of empty vessels (τοῖς κενοῖς ἀγγείοις παραπλησίως) and of *topos* on the analogy of full ones.
(3) *Chôra* is either[6] that which is larger and can be occupied by being, like a larger vessel of a body,[7] (τὸ μεῖζον οἷόν τε κατέχεσθαι ὑπὸ ὄντος καὶ οἷον μεῖζον ἀγγεῖον σώματος) or that which can contain a larger body (ἢ τὸ χωροῦν μεῖζον σῶμα).
(4) The void (*kenon*) is said to be infinite (ἄπειρον) for that which is outside the cosmos is suchlike, but *topos* is finite because no body is infinite. Just as the corporeal is finite, so the incorporeal is infinite; for time and the void are infinite. For just as the nothing constitutes no limit, so also is there no limit to the nothing, e.g. to the void. For by its own nature (κατὰ γὰρ τὴν αὑτοῦ ὑπόστασιν) it is infinite; but it is being limited when it is filled up (περατοῦται δ' αὖ τοῦτο ἐκπληρούμενον); but when that which fills it is taken away, you cannot conceive of its boundary.[8]

---

3 ⟨τε⟩ add. Diels.
4 ⟨εἴτε⟩ add. Heeren.
5 ⟨οὔτε⟩ κενὸν ἔσεσθαι οὔτε τόπον; ⟨οὔτε⟩ add. Canter.
6 For this use of πότερον ...ἢ ... see *LSJ s.v.* πότερον, II, 5.
7 Or: 'like a vessel which is larger than the body contained'.
8 Ar. Did. fr. 25: [Χρυσίππου]. Τόπον δ'εἶναι ὁ Χρύσιππος ἀπεφαίνετο τὸ κατεχόμενον δι' ὅλου ὑπὸ ὄντος ἢ τὸ οἷόν ⟨τε⟩ κατέχεσθαι ὑπὸ ὄντος καὶ δι' ὅλου κατεχόμενον εἴτε ὑπὸ τινὸς ⟨εἴτε⟩ ὑπὸ τινῶν. ἐὰν δὲ τοῦ οἵου τε κατέχεσθαι ὑπὸ ὄντος τὶ μὲν κατέχηται, τὶ δὲ μή, τὸ ὅλον ⟨οὔτε⟩ κενὸν ἔσεσθαι οὔτε τόπον, ἕτερον δέ τι οὐκ ὠνομασμένον· τὸ μὲν γὰρ κενὸν τοῖς κενοῖς ἀγγείοις λέγεσθαι παραπλησίως, τὸν δὲ τόπον τοῖς πλήρεσι· χώραν δὲ πότερον τὸ μεῖζον οἷόν τε κατέχεσθαι ὑπὸ ὄντος καὶ οἷον μεῖζον ἀγγεῖον σώματος ἢ τὸ χωροῦν μεῖζον σῶμα. τὸ μὲν οὖν κενὸν ἄπειρον εἶναι λέγεσθαι· τὸ γὰρ ἐκτὸς τοῦ κόσμου τοιοῦτ' εἶναι· τὸν δὲ τόπον πεπερασμένον διὰ τὸ μηδὲν σῶμα ἄπειρον εἶναι. καθάπερ δὲ τὸ

At first sight there seem to be good reasons for comparing this part of the text of Arius Didymus with another doxographical testimonium, viz. Aëtius I, 20, 1, which runs:

text 2

Aëtius I, 20, 1 = SVF *II, 504 = part of* SVF *I, 95, not in LS.*
[Zeno and his school argue that....] void (*kenon*), *topos* and *chôra* differ. The void (*kenon*) is vacancy of body (ἐρημίαν σώματος) whereas *topos* is that which is occupied by body (τὸ κατεχόμενον ὑπὸ σώματος), *chôra* being that which is partly occupied (τὸ ἐκ μέρους ἐπεχόμενον) like in the case of a wine jar.[9]

and with the evidence adduced by Sextus Empiricus, *M.* 10, 3:

text 3

Sextus Empiricus, M. *10, 3–4* = SVF *II, 505 (1st part) = LS 49 B.*
(1) The Stoics argue that void (*kenon*) is that which is able to be occupied by being but is *de facto* not occupied (τὸ οἷόν τε ὑπὸ ὄντος κατέχεσθαι, μὴ κατεχόμενον δέ), or an extension empty of body (διάστημα ἔρημον σώματος) or an extension unoccupied by body (ἢ διάστημα ἀκαθεκτούμενον ὑπὸ σώματος), whereas
(2) *topos* is that which (or: the place which) is occupied by being and is made equal in size to its occupant (τὸν ὑπὸ ὄντος κατεχόμενον καὶ ἐξισαζόμενον τῷ κατέχοντι αὐτὸν) ...
(3) and they say *chôra* is an extension partly occupied by body and partly unoccupied (διάστημα κατὰ μέν τι κατεχόμενον ὑπὸ σώματος κατὰ δέ τι ἀκαθεκτούμενον).[10]
(4) Some say that *chôra* is the place of the larger body (τὸν τοῦ μείζονος σώματος τόπον), implying that *topos* and *chôra* differ in so far as <using> the former <term> says nothing about the size of the body contained—for even when it contains a body of minimal size, it is still called *topos*—whereas <using> the latter implies that the size of the body contained is considerable (ἀξιόλογον ... μέγεθος τοῦ ἐν αὐτῇ σώματος).[11]

σωματικὸν πεπερασμένον εἶναι, οὕτως τὸ ἀσώματον ἄπειρον· ὅ τε γὰρ χρόνος ἄπειρος καὶ τὸ κενόν. ὥσπερ γὰρ τὸ μηδὲν οὐδέν ἐστι πέρας, οὕτως οὐδὲ τοῦ μηδενός, οἷόν ἐστι τὸ κενόν. κατὰ γὰρ τὴν αὑτοῦ ὑπόστασιν ἄπειρόν ἐστι, περατοῦται δ' αὖ τοῦτο ἐκπληρούμενον· τοῦ δὲ πληροῦντος ἀρθέντος οὐκ ἔστιν αὐτοῦ νοῆσαι πέρας.

[9] Aët. I, 20, 1 (in Stobaeus' version): (Ζήνων καὶ οἱ ἀπ' αὐτοῦ ....) διαφέρειν κενὸν τόπον χώραν· καὶ τὸ μὲν κενὸν εἶναι ἐρημίαν σώματος, τὸν δὲ τόπον τὸ ἐπεχόμενον ὑπὸ σώματος, τὴν δὲ χώραν τὸ ἐκ μέρους ἐπεχόμενον, ὥσπερ ἐπὶ τῆς τοῦ οἴνου πιθάκνης.

[10] Cf. also the parallel passage in S. E. *P.* 3, 124.

[11] S. E. *M.* 10, 3–4: καὶ οἱ Στωικοὶ δὲ κενὸν μὲν εἶναί φασι τὸ οἷόν τε ὑπὸ ὄντος κατέχεσθαι, μὴ κατεχόμενον δέ, ἢ διάστημα ἔρημον σώματος ἢ διάστημα ἀκαθεκτούμενον ὑπὸ σώματος, τόπον δὲ τὸν ὑπὸ ὄντος κατεχόμενον καὶ ἐξισαζόμενον τῷ κατέχοντι αὐτόν, νῦν ὂν καλοῦντες τὸ σῶμα, καθὼς καὶ ἐκ τῆς

These definitions are furthermore reflected by some lines in Cleomedes:

text 4

Cleomedes Cael *I, 1, 17–19 and I, 1, 20–25 Todd; not in* SVF *or LS.*
(1) The void (*kenon*) outside it (scil. the cosmos) stretches without end to all sides; of this the part which is occupied by body is called *topos* and that which is not occupied will be (mere) void ...
(2) every body has to be in something. That in which it is must be different from that which occupies and fills it, since it is incorporeal (ἀσώματον) and, so to speak, intangible (οἷον ἀναφές). Such a subsistent (ὑπόστασιν) which is able to receive body and to be occupied (οἵαν τ' ἂν οὖσαν δέχεσθαι σῶμα καὶ κατέχεσθαι) we call 'void' (*kenon*).[12]

6.1.3 Before discussing these various definitions in more detail, we should first see whether it is possible to shed some light on the character of the sources and whether they allow us to draw any definite conclusions as to the possible origins of the definitions.

The main features of Diels' reconstruction of part of the doxographical work of Arius Didymus, from the large scraps which are to be found in the work of the 5th century CE compiler Ioannes Stobaeus, have as yet not been seriously challenged.[13] It would appear, then, that the 1st century BCE doxographer Arius Didymus—the philosophical teacher of the emperor Augustus—wrote a book called *Peri haireseôn* ('On the Sects')[14] which contained more or less continuous expositions of at least the physics

---

μεταλήψεως τῶν ὀνομάτων ἐστὶ συμφανές. χώραν δὲ φασιν εἶναι διάστημα κατὰ μέν τι κατεχόμενον ὑπὸ σώματος κατὰ δέ τι ἀκαθεκτούμενον. ἔνιοι δὲ χώραν ἔλεξαν ὑπάρχειν τὸν τοῦ μείζονος σώματος τόπον, ὡς ταύτῃ διαφέρειν τοῦ τόπου* τῷ ἐκεῖνον μὲν μὴ ἐμφαίνειν μέγεθος τοῦ ἐμπεριεχομένου σώματος—κἂν γὰρ ἐλάχιστον περιέχῃ σῶμα, οὐδὲν ἧττον τόπος προσαγορεύεται—τὴν δ' ἀξιόλογον ἐμφαίνειν μέγεθος τοῦ ἐν αὐτῷ σώματος. (*τοῦ τόπου τὸν τόπον mss.: τοῦ τόπου ⟨τὴν χώραν⟩ Bake, followed by von Arnim).

[12] Cleom. *Cael.* I, 1, 17–19 Todd: τὸ δὲ ἐκτὸς αὐτοῦ κενόν ἐστιν ἀπὸ παντὸς μέρους εἰς ἄπειρον διῆκον. τούτου δὲ τὸ μὲν ὑπὸ σώματος κατεχόμενον τόπος καλεῖται, τὸ δὲ μὴ κατεχόμενον κενὸν ἂν εἴη. *ibidem* I, 1, 20–24: πᾶν σῶμα ἔν τινι εἶναι ἀναγκαῖον, τοῦτο δὲ, ἐν ᾧ ἐστι, τοῦ κατέχοντος αὐτὸ καὶ πεπληρωκότος ἕτερον εἶναι δεῖ, ἀσώματον ὄν καὶ οἷον ἀναφές. Τὴν οὖν τοιαύτην ὑπόστασιν, οἵαν τ' ἂν οὖσαν δέχεσθαι σῶμα καὶ κατέχεσθαι, κενὸν εἶναί φαμεν.

[13] See on Arius Didymus in general and on Diels' reconstruction: Moraux (1973–84) I, 259 ff.

[14] Cf. Stob. *Ecl.* II, p. 6, 13 Wachsmuth, where a *testimonium* on Xenophanes has the lemma Διδύμου ἐκ τοῦ Περὶ αἱρέσεων. On this title see also Moraux (1973–84) I, 270–71.

and ethics of the Platonic, Peripatetic, and Stoic school.[15] Most probably his treatment was organized according to schools, not according to subject.[16] In that case there must have been a more or less continuous part concerned with Stoic physics. From what remains, it appears that Arius Didymus specified the views of individual Stoic philosophers, whenever they put forward heterodox thoughts or definitions, or arguments of specific interest.[17] Such explicit ascriptions to individual philosophers are usually part of Arius' running text, i.e. not added as *lemmata*. Where there is no such explicit differentiation between the views of several Stoics, we must regard the text as giving a common Stoic view, even when there is a *lemma*. As was already remarked by Diels,[18] at least a number of the *lemmata* (usually genitives like Ζήνωνος or Χρυσίππου) are added by Stobaeus himself, who needed them because he cut apart the original running *exposé* of Arius.[19]

Fr. 25 explicitly credits Chrysippus with the invention of the definitions provided. Since this ascription occurs within the actual running text of Arius we may, according to the criteria just outlined, regard it as fairly secure. Moreover, Chrysippus was of course in general responsible for much of the systematization of early Stoic thought, whereas the fact that he is said to have written a book called 'On the Void' (*Peri tou Kenou*) and is reported also to have discussed the problems concerning the void in the first book of his 'Arts of Physics' (*Phusikai Technai*), points to a more than occasional interest in problems concerning space and void on his part.[20] All this lends much *prima facie* plausibility to the

---

[15] Stobaeus must have cut apart parts of a continuous discourse. This is apparent e.g. in the case of Ar. Did. fr. 2 and 3 Diels, where fr. 3 is obviously the sequel of fr. 2.

[16] Cf. Diels' introduction to his *Doxographi Graeci* (1879) 72. At least in so far as the ethical doxographies are concerned this view has been challenged by M. Giusta, *I dossografi di etica*, 2 vols. (Torino 1964—1967); however, see Moraux (1973–84) I, 264 ff.

[17] Therefore Moraux' remark ((1973–84) I, 262) that 'auf die individuellen Unterschiede zwischen den einzelnen Denkern ging Areios nicht ein' needs to be qualified.

[18] Diels (1879) 74: 'lemmata anepigraphis capitibus suo more margini adscripsit genetivo usus Stobaeus'.

[19] This is notably apparent where the *lemma* does not correspond with the text (as in fr. 40 or in fr. 23) or where it is superfluous (as in fr. 22 and our fr. 25).

[20] For these references see D. L. VII, 140 (= *SVF* II, 543) and Plu. *Comm. Not.* 1081 f (= *SVF* II, 518).

supposition that it was primarily Chrysippus who was the creative genius behind most of what we know of the early Stoic theory of space and of many of the surviving definitions. There is, in short, no reason to doubt the Chrysippean origin of the definitions provided by Arius Didymus (text 1).

As to Aëtius' account, the version of Stobaeus here has the lemma Ζήνων καὶ οἱ ἀπ' αὐτοῦ ('Zeno and his followers') which, as should be noted, in doxographical texts of this kind often means nothing more than 'the Stoics'. Ps.-Plutarch's version prefaced the corresponding section by the lemma 'the Stoics and Epicurus'.[21] Since, therefore, this text offers no clues for a more specific ascription we had best treat its contents as 'common Stoic'.

The same goes for the account in Sextus. About Sextus' sources virtually nothing is known with certainty. He probably took most of his information from the *peri haireseôn* tradition, doxographical works, in other words, like that of Arius Didymus. That may account for the fact that, like Arius (and like previous sceptics such as Aenesidemus), he is able to present heterodox definitions alongside those of the common Stoic orthodoxy. In the present case most of the definitions provided are ascribed to 'the Stoics'. Only the alternative definition of *chôra* is introduced by the neutral 'some' (ἔνιοι).

We now come to Cleomedes. At an uncertain date,[22] somewhere between the 1st the 3rd century CE, he wrote an astronomical handbook which was prefaced by an introduction to cosmology written from a Stoic point of view. It is tempting to assume that his being a Stoic himself makes him rather reliable as a source of information on the details and purport of certain Stoic arguments he refers to. As he himself confesses (II, 7, 11–14 Todd) he uses 'older and more recent sources', but most of all Posidonius. Yet the extent to which his treatise is dependent on Posidonius has often been overestimated. At any rate all *explicit* references to Posidonius in his *Caelestia* concern more specialized astronomical or meteorological issues. With meteorology and

---

[21] On the occurrence of Epicurus in this connection see below, n. 26. On the interrelation between the various witnesses for Aëtius see now the convenient survey in Runia (1989) 245 ff.

[22] On various attempts to fix the date of Cleomedes, see Schumacher (1975); Goulet (1980) 6 ff.; see also my remarks in Algra (1988a) 170, esp. n. 31.

astronomy constituting the subject proper of Cleomedes' treatise, the prominent place explicitly accorded at the end of the work to Posidonius as a source is therefore understandable. There are no indications, however, that Cleomedes was acquainted with all aspects of Posidonius' work, nor that he specifically used Posidonius when discussing the Stoic *principia physica* at the outset of the first book. There he does not mention Posidonius, and he implies that what he presents is the common school doctrine,[23] and on such an important physical issue as the nature and extent of the extracosmic void he is seen to represent the more orthodox Stoic position (infinite void) and not to follow Posidonius (finite void).[24] On the other hand this should not lead us to think that in these cases he had the original early Stoic texts at his disposal. In fact he never mentions any of the early Stoics by name. Thus, in so far as he renders the more orthodox Stoic positions, he is most probably using later Stoic scholastic literature—containing what might be called 'common Stoic' views[25]—or perhaps even common doxographical sources. His account, therefore, need not always present an untainted Stoic view.[26]

---

23 At *Cael.* I, 1, 43–46 Todd (a passage printed as *SVF* II, 537) he refers to the theory of the conflagration and the related theory of an extra-cosmic void as being supported by 'the most distinguished among the ancient physicists.'

24 Cf. Cleom. *Cael.* I, 1, 17 and I, 1, 112–150 Todd. On Posidonius' conception of the extra-cosmic void see below section 6.4.

25 Cf. Rehm (1922) 684, who offers some possible answers to the question how Cleomedes could have got his information on the Stoics. Although I agree with Rehm that it is significant that Cleomedes never mentions Chrysippus and that he probably got his information on orthodox Stoic physics from a later Stoic school book, this does not mean that he is merely a copyist. He knows how to argue for his (the Stoic) case, as is also shown by his engaged polemics against the Epicureans (cf. *Cael.* II, 1, 1 ff.;. II, 1, 404 ff.; II, 1, 467; II, 1, 523 ff. Todd) and the Peripatetics (esp. *Cael.* I, 1, 81 ff. Todd; in Algra (1988a) esp. 169–171, I have argued that the latter passage had the anti-Stoic arguments of Alexander of Aphrodisias as its primary target—*contra* Theiler (1982) 157, who thinks Cleomedes records Posidonius's answers to early Peripatetic objections, and Hahm (1977) 119, who suggests that Cleomedes reports *Zeno's* answers to Aristotle.

26 This may be the reason why Cleomedes' testimony differs in one curious respect from those of Arius, Aëtius and Sextus. In all these texts the concept of void which is presented is clearly of the (b)-kind outlined in chapter 1. The term *kenon* does not function as a synonym for 'space'. Rather we are told to apply it exclusively to *unoccupied* place. If we want to look for a concept of space we had better locate it in the notion of τὸ οἷόν τε κατέχεσθαι ὑπὸ ὄντος ('that which can be occupied by (a) body'). Cleomedes, however, also uses the term 'void' (*kenon*) as a substitute for 'space'. See on this my

All this allows us to draw the following conclusions concerning the probable provenance of the various preserved definitions. The definitions in Arius may fairly securely be attributed to Chrysippus. Those found in Aëtius and Cleomedes, as well as the main definitions provided by Sextus, may be labelled 'common Stoic' definitions. They appear to reflect what at a certain point became the orthodox Stoic spatial concepts. In principle, of course, this does not exclude the possibility that all or some of them are in the end also to be traced back to Chrysippus. Whether or not this is so, will have to be borne out by our further inquiries. Finally, in addition to the common Stoic definition of *chôra*, Sextus propounds what may probably be regarded as a 'heterodox' definition of *chôra*, apparently endorsed only by 'some Stoics', not by the majority.[27]

6.1.4 If next we try to determine the spatial concepts underlying each of the preserved definitions, the following picture presents itself. As a preliminary note I would like to point out that the definitions preserved by Arius and Sextus show that there was a kind of general concept of what we might call 'space' underlying the various concepts of *topos*, *chôra* and *kenon*. To designate this concept they used the expression 'that which is capable of being occupied by being' (τὸ οἷόν τε ὑπὸ ὄντος κατέχεσθαι—an expression found in Arius and Sextus) or they simply used the word *diastêma* ('extension'; cf. Sextus in part (3) of our text 3).[28] The definitions of *topos* provided by Aëtius and Sextus show no material departure from the Chrysippean point of view: though each time the wording is slightly different,[29] they can easily be

remarks above, chapter 2, 69–70. Note that there even seem to be traces of (probably indirect) Epicurean influence: especially the word *anaphes* ('intangible') is an Epicurean term. For confusion between the two different concepts of *kenon* in Epicurus see Long/ Sedley (1987) I, 30. For an example of the tendency in some doxographical sources to put the views of Epicurus and the Stoics on a par see Aët. I, 20 (especially the version of ps.-Plutarch).

[27] I here argue on the plausible assumption that Sextus' ἔνιοι refers to 'some *Stoics*', but in principle the possibility cannot be excluded that he just means 'some people'; we cannot be absolutely sure, therefore, that part (4) of our text 3 contains genuinely Stoic material.

[28] Note that this means that the Stoics differed from Aristotle in so far as they no longer defined the void as a *kind* of place (viz. as *unoccupied* place), but instead regarded both place and void as species of a more general 'entity', viz. what we might call 'space'.

[29] Chrysippus' first definition runs τὸ κατεχόμενον δι' ὅλου ὑπὸ ὄντος;

squared with the definitions ascribed to Chrysippus.[30] The underlying concept of *topos*, in other words, seems to be identical: place is 'that which can be occupied by being and is *de facto* occupied'; or, in short: place is occupied space.

Though the Chrysippean account in Arius Didymus does not offer a definition of *kenon*, it may plausibly be extrapolated from his definition of *topos* in Ar. Did. fr. 25 (τὸ οἷόν ⟨τε⟩ κατέχεσθαι ὑπὸ ὄντος καὶ δι' ὅλου κατεχόμενον). It must have been something like τὸ οἷόν τε ὑπὸ ὄντος κατέχεσθαι, μὴ κατεχόμενον δέ, which is in fact the common Stoic definition found in Sextus (text 3).[31] Anyway, from the last part of Arius fr. 22 (part (4) of text 1) we may infer that Chrysippus' concept of void was no different from that which underlies the surviving definitions, so that we may conclude that the early Stoics were also committed to one and the same concept of *kenon*.

In the case of the various definitions of *chôra* presented by our sources, the situation is significantly different. The common Stoic

---

Aëtius only leaves out the—redundant—qualification δι' ὅλου, which in Sextus' account is replaced by the words ἐξισαζόμενον τῷ κατέχοντι αὐτὸν; moreover Aëtius has ὑπὸ σώματος instead of Chrysippus' ὑπὸ ὄντος—not surprisingly since ὄν and σῶμα are well-nigh synonymous in Stoic ontology—whereas Sextus' definitions use both terms *promiscue*.

30 Inwood (1991) 247–248 is unnecessarily sceptical about the Chrysippean (or in general: early Stoic) credentials of the term 'extension' (διάστημα) in the definitions provided by Sextus. He believes that Sextus is here using the *Aristotelian* vocabulary and he adds, with a reference to the οὐκ ὠνομασμένον in the account in Arius, that 'it is not clear to me that Chrysippus used this term [scil. διάστημα], since [...] when he speaks of extension he says that it does not have a name'. However, apart from the fact that διάστημα is certainly not an exclusively Aristotelian term (it also occurs, e.g. in early Stoic definitions of time), Sextus is here quite obviously not so much paraphrasing the Stoic theory in his own terms, but rather rendering complete Stoic definitions. Moreover, the term which some of the definitions in Sextus use instead of διάστημα is τὸ οἷόν τε κατέχεσθαι. This suggests that διάστημα is *not* the equivalent of the οὐκ ὠνομασμένον in Arius Didymus, for the latter is only a *species* of τὸ οἷόν τε κατέχεσθαι (viz. τὸ οἷόν τε κατέχεσθαι *if* partly occupied, partly empty).

31 It may even be that this definition originally figured in Arius' text, but that, either Stobaeus or the later mss. tradition mistakenly left it out. In that case the words κενὸν ἔσεσθαι οὔτε τόπον at the end of the definition of the οὐκ ὠνομασμένον (which Canter corrected into ⟨οὔτε⟩ κενὸν ἔσεσθαι οὔτε τόπον) could be regarded as a contamination, caused by *saut du même au même*, of the final parts of the definitions of *kenon* and the οὐκ ὠνομασμένον respectively. Admittedly, from a textcritical point of view this is much less elegant than Canter's simple addition, but it remains odd that the concept of κενόν which in this text is referred to several times, is not properly defined.

definitions in Aëtius and Sextus seem to relate *chôra* and the other two spatial terms in such a way as to end up—in spite of the slight differences in wording—with an exact parallelism in definitions: *topos* is a *diastêma* which is *de facto* occupied by body, *kenon* is a *diastêma* (and as such capable of being occupied) which is *de facto* not occupied, and *chôra* is a *diastêma* which is *partly* occupied, *partly* unoccupied.[32] Given the common Stoic assumption that the cosmos is a *plenum*, so that it can have no empty space inside it,[33] the *kenon* has to be located outside the cosmos. Accordingly, *chôra* is at first blush best explained as (partly) occupied space *plus* (part of the) void. At its very largest *chôra* would then coincide with the (infinite) spatial aspect of that which some Stoics called *to pan* (the sum of the cosmos and the extracosmic void).[34]

Now the two Chrysippean definitions occurring in Arius Didymus as well as the 'heterodox' definition provided by Sextus, seem to differ substantially from the common Stoic definitions. First of all, the common Stoic definitions in Aëtius and Sextus define *chôra* primarily in terms of 'that which is capable of being occupied', 'that which is occupied' or 'an extension' (διάστημα), in other words, in terms of space. I already pointed out that, by thus linking *chôra* not to a particular body but to space, they leave open the possibility of a *chôra* that encompasses the sum total of (occupied and unoccupied) space and that, hence, is infinite in extent. Now in the Chrysippean account this notion of a (possibly infinite) space which is partly occupied, partly empty recurs, but

---

32 I take κατεχόμενον ὑπὸ σώματος to mean 'occupied by body', not just occupied by *a particular* body; similarly I take ἀκαθεκτούμενον to mean (absolutely) unoccupied, i.e. unoccupied by *any* body, not simply unoccupied by a particular body. Accordingly, on this reading of the definitions the concept of *chôra* here outlined only applies at what we might call the 'macro' level, i.e. in a cosmic setting, where there is space occupied by the cosmic body next to space that is absolutely unoccupied (i.e. the void).

33 See the texts printed at *SVF* II, 542 ff.

34 Cf. Aët. II, 1, 7 (= *SVF* II, 522): ... πᾶν μὲν εἶναι τὸ σὺν τῷ κενῷ τῷ ἀπείρῳ, ὅλον δὲ χωρὶς τοῦ κενοῦ τὸν κόσμον. See also the texts printed by von Arnim as *SVF* II, 523, 524, and 525. Though this terminological distinction is nowhere explicitly attributed to Chrysippus, I think there are good reasons for assuming that it was he who first used it. At least it may plausibly be inferred from Ar. Did. fr. 38 (= *SVF* I, 497) that the distinction postdates Cleanthes, whereas the *terminus ante quem* for the introduction of this distinction is provided by *SVF* III, 9, where it appears in a fragment from Apollodorus of Seleucia, a pupil of Diogenes of Babylon, who in his turn was a pupil of Chrysippus.

only as a spatial concept for which Chrysippus obviously did not bother to invent a name: οὐκ ὠνομασμένον. We may therefore assume that the *chôra* of which the Chrysippean account also gives two definitions amounts to a *different* spatial concept. This is also suggested by the fact that both definitions speak of *chôra* as just 'larger' (by contrast, in this same Chrysippean account the void is explicitly said to be *infinite*). Moreover, by identifying it as τὸ χωροῦν μεῖζον σῶμα or as a μεῖζον ἀγγεῖον σώματος they appear to define *chôra* not as a third coordinate species (i.e. next to *topos* and *kenon*) of τὸ οἷόν τε ὑπὸ ὄντος κατέχεσθαι, but rather as a relational concept. *Chôra*, in other words, is essentially defined by reference to a body which is its (potential) occupant—in particular it seems to be implied that *chôra* is that which provides room for more than, or more of, a particular body which it contains. Hence, it seems to be implied that *chôra* is always the *chôra of some particular body*[35]. Sextus' heterodox definition also links *chôra* with (a) body, but it does so in a different way. For it simply treats *chôra* as a *kind of topos*—hence as space fully occupied by a body. This trait sets this definition apart not only from the common Stoic definition of *chôra*, but also from the Chrysippean definitions as presented by Arius Didymus.

All this strongly suggests that whereas Chrysippus' concepts of *topos* and *kenon* found their way into the Stoic orthodoxy—the differences between the various surviving definitions being merely verbal[36]—his idiosyncratic definition of *chôra*—for reasons which at the present moment remain to be discovered—did not appeal to later Stoics. It remains for us to try and specify the

[35] Inwood (1991) 248 rightly draws attention to this point: 'The definition of room itself is a bit obscure: it seems to be a bit of extension which still has room for a larger body [...] and strictly speaking this makes "room" a merely relative term, unlike place and void'. I want to add that this view of *chôra* as being not just the sum total of *topos* and *kenon*, nor in some other way a third species of space, but rather a section of space *when regarded as the room for a particular body*, would account for the fact that *chôra* (unlike *topos* and *kenon*) does not occur in any of the existing lists of 'canonical' incorporeals (on which see below section 6.3).

[36] This is not to say that the possibility of these terminological distinctions preceding Chrysippus is to be excluded, although I think, for reasons outlined above, that Chrysippus is the most likely candidate for the authorship. What is important, though, is that there seems to be no break between Chrysippus and the later Stoic orthodoxy in this respect, whereas in the case of the various concepts of χώρα reflected by our definitions there are good reasons for assuming such a break.

nature and background of this typically Chrysippean concept of *chôra* and to find the reasons which may subsequently have induced other Stoics to abandon it. As far as I know two different interpretations of the Chrysippean definition of χώρα can be found in the scholarly literature.

Pohlenz, followed by Christensen and by Long and Sedley, chose to interpret Chrysippus' concept of *chôra* as just 'space', taken as the sum total of *topos* (being in fact the extension occupied by corporeal being) and *kenon* (the infinite extracosmic extension).[37] On this view the Chrysippean *chôra*—being, so to speak, the spatial aspect of *to pan*—would be no different from the common Stoic *chôra* of Aëtius and Sextus. Precisely this constitutes a fatal drawback for this interpretation. For it involves equalling—without any apparent reason—the *chôra* in the third part of Arius Didymus fr. 25, (our text 1a) with what in the second part of the very same fragment is *explicitly* referred to as οὐκ ὠνομασμένον. Furthermore, it does not really make sense of the two Chrysippean definitions of *chôra* which Arius Didymus does provide—nor, for that matter, of the heterodox definition provided by Sextus (which, in spite of the differences, to some extent resembles the Chrysippean definitions of Arius, but which does not appear to fit Pohlenz's interpretation).

A second, and to my mind most unlikely, interpretation has been put forward by Émile Bréhier.[38] He takes *chôra* in the first definition in the third part of Arius Didymus fr. 25 (τὸ μεῖζον οἷόν τε κατέχεσθαι ὑπὸ ὄντος etc.) to refer to infinite space as being partly occupied and partly unoccupied. To this extent his interpretation runs parallel to those of Pohlenz and Christensen. Otherwise, however, he concentrates on the alternative definition presented by Arius, τὸ χωροῦν μεῖζον σῶμα, which he translates 'le plus grand corps qui contient'. He does not bother to explain the possible relation between this definition, and the first one, but

---

[37] Pohlenz (1948–9) II, 37, where it is argued that Chrysippus (!) distinguished from τόπος and κενόν 'ein beides umfassendes Drittes, das nur zum Teil von Körperlichem ausgefüllt sei (er dachte an das All das Kosmos und das Leere umfasst ...), und nannte es, aber zweifelnd, χώρα.' Cf. also Christensen (1962) 24–25; and Long/ Sedley (1987) I, 296: 'The Stoics probably used the term 'room' to denote space which combines place and void (i.e. the 'all')'. On the interpretation of Long and Sedley see further below, 279.

[38] Bréhier (1907) 52–53.

defends his rather surprising translation by a reference to the Stoic theory of total blending (*krasis di' holou*): 'Ainsi quand un corps pénètre par mélange à travers un autre, et qu'il occupe seulement une partie du lieu du second, ce second corps est appelé, suivant Stobée, la place du premier'.

Apart from the fact that the way Bréhier constructs the Greek is odd,[39] this interpretation might at first glance seem not too implausible to be defended. For why couldn't we assume, in the case of the famous Chrysippean wine drop blending with the sea, that the sea is the *chôra* of the wine drop? In that case the qualification μεῖζον would have to refer to the situation of the two bodies before the total blending takes place, the two bodies being in the end of course fully coextensive (δι' ὅλων ἀντιπαρέκτασις).[40] Yet also this interpretation entails several difficulties. A first objection concerns Bréhier's examples. He only refers to cases where the χωροῦν σῶμα remains larger, even after the blending.[41] However, examples like 'l'âme du monde ... qui pénètre toutes les parties de l'univers n'est pénetrée par chacune d'elles que dans une de ses parties' are not supported by any evidence concerning the Stoic theory of *krasis*.[42] A second and more general objection is that

---

[39] If χωροῦν and μεῖζον were both meant to be nominatives qualifying σῶμα they would according to normal Greek usage have been connected by καὶ or τε καί. For this rule, and for possible exceptions—to which the present case would certainly not seem to belong—see R. Kühner/ B. Gerth, *Ausführliche Grammatik der Griechischen Sprache*, II (Hannover/ Leipzig 1898–1904) 277. It is far more natural indeed to take μεῖζον σῶμα together as the object of χωροῦν.

[40] Cf. Plu. *Comm. Not.* 1078c (= *SVF* II, 465): ... ἀμφότερα τῆς κράσεως δι' ἀλλήλων διιέναι καὶ μηδὲν ἐπιλείπεσθαι μηδενὸς μόριον ἀλλὰ ⟨πᾶν⟩ παντὸς ἀναπίμπλασθαι (accepting Madvig's ⟨πᾶν⟩).

[41] This is probably because otherwise there would be no reason why the two bodies could not be *each other's* χώρα.

[42] Of course 'l'âme du monde', being the active principle, permeates all passive matter, thus fashioning all kinds of 'substances'. These substances themselves (e.g. wine and water), it would seem, can in their turn completely blend. It makes no sense, however, to say that the cosmic *pneuma*, being itself so to speak a constituent of the individual substances, is 'penetrated' by each of them. Incidentally, it should be noted that Bréhier's wording is particularly unclear. On the one hand he argues that when two bodies are blended and one is larger then the other the larger one is called 'la place (= χώρα, k.a.) du premier'. About the cosmic *pneuma* being only partly penetrated by each particular part of the cosmos, he states, on the other hand, that 'on ne parle pas alors du lieu de ces parties, mais de la *place* qu'elles occupent *dans* l'âme du monde' [my italics]. It remains unclear whether χώρα is the larger body itself or just a place *in* the larger body.

there is no evidence that Chrysippus, or other Stoics thought that in the act of *krasis* one body served as the *chôra* or the place of the other. Rather this is a conclusion drawn by way of a *reductio ad absurdum* by opponents of the Stoics.[43] As a third objection it should be noted that Bréhier's interpretation, if correct, would imply that Chrysippus invested *chôra* with a status different from *topos* and *kenon*. For while the latter two are incorporeals (*asômata*), *chôra* would on this line of thought have to be a corporeal being (*on*). The way in which the definitions of the spatial terms are juxtaposed in Aëtius, Arius Didymus, and Sextus does not suggest a different ontological status for *chôra*. Finally, the only *verbatim* fragment I know in which Chrysippus uses the word *chôra*—a fragment to which I shall revert in the next section—uses the word in a sense which certainly cannot be squared with the interpretation proposed by Bréhier.[44] It appears, then, that his argument cannot be accepted.

If neither of these interpretations is successful, it seems worthwhile to look for an alternative interpretation—one, moreover, which takes account of the differences between the Chrysippean conception of *chôra* as rendered by Arius Didymus and what seems to have become the common Stoic conception. Let us first take the first definition of *chôra* to be found in Arius Didymus (text 1). It runs: τὸ μεῖζον οἷόν τε κατέχεσθαι ὑπὸ ὄντος καὶ οἷον μεῖζον ἀγγεῖον σώματος. The last words are somewhat ambiguous and the whole formula has accordingly been translated in various ways: 'the larger thing capable of being occupied by being and a kind of larger vessel *of* a body',[45] or 'the larger thing capable of being occupied by being and a kind of vessel which is larger *than* a body'.[46] The former translation is perhaps preferable, since we know from Cleomedes *Cael.* I, 1, 86–88 Todd, that the combination ἀγγεῖον σώματος could be used to refer to a vessel of a body, without implying that the vessel is *de facto* filled up. Cleomedes

---

43 Cf. Plu. *Comm. Not.* 1077e (= *SVF* II, 465): παρὰ τὴν ἔννοιάν ἐστι σῶμα σώματος εἶναι τόπον etc. I take it that Plutarch's polemical point is only the stronger because also on Stoic principles it is impossible for a body to be place (or for place to be a body).

44 Plu. *SR* 1054c, to be discussed below, section 6.2.

45 Hahm (1977) 130 n. 37, my italics.

46 Bréhier (1907) 52: 'ce qui est capable d'être occupé par un être et ce qui est plus grand *que* cet être, comme un vase plus grand qu'un corps' (my italics).

even argues that the assumption that there is an infinite void as an ἀγγεῖον σώματος does not commit one to the belief that there must also be an (equally infinite) body occupying it. According to Cleomedes such an ἀγγεῖον σώματος might be taken as simply τὸ οἷόν τε δέξασθαι σῶμα.[47] Similarly, in Arius Didymus (our text 1) the words τὸ μεῖζον ἀγγεῖον σώματος may be taken to refer to that which is in the same sentence also called τὸ μεῖζον οἷόν τε κατέχεσθαι ὑπὸ ὄντος. This *diastêma* may still be either actually filled up or actually (partly) empty.[48] Yet also on this reading of the definition the presence of the relative term μεῖζον indicates that *chôra* is conceived primarily in relation to something smaller than itself, presumably a smaller body which is contained.

The second definition to be found in Arius, viz. τὸ χωροῦν μεῖζον σῶμα, is less difficult. It can be translated as 'that which provides room for a larger body'. Note that this definition contains an etymologization in that it connects *chôra* with *chôrein*.[49]This verb usually means 'to provide room for <something>' and as such it is used to denote the capacity of a vessel.[50] Thus also in this case the definition seems to refer to a kind of larger space which *can* contain a larger body, but which need not necessarily contain that larger body *de facto*.

According to the heterodox definition in Sextus *chôra* is 'the place (*topos*) of the larger body'. Note that the use of the word *topos* in this connection—*topos* being always connected with body and hence finite—excludes the possibility that what is meant here is

---

47 Cleom. *Cael.* I, 1, 86–88 Todd: εἰδέναι οὖν χρή, ὅτι τὸ ἀγγεῖον σώματος διχῶς λέγεται, τὸ μὲν ἔχον σῶμα καὶ πεληρωμένον ὑπ' αὐτοῦ, τὸ δ' οἷόν τε δέξασθαι σῶμα. For a useful discussion of the arguments involved see now Sorabji (1988b) 132–134.

48 See also the definition of τόπος which is to be found in the same fragment. That definition qualifies the words τὸ οἷόν τε κατέχεσθαι ὑπὸ ὄντος by the addition καὶ δι' ὅλου κατεχόμενον; see also the (contrary) addition in the case of the definition of κενόν in S. E. *M.* 10, 3 (= our text 3).

49 That Chrysippus was particularly interested in etymology and etymologizations may appear from the fact that, according to the catalogue in D. L. VII, 200, he wrote two large monographs on the subject, viz. Περὶ τῶν ἐτυμολογικῶν πρὸς Διοκλέα ξ' and Ἐτυμολογικῶν πρὸς Διοκλέα δ'. The methodological relevance which Chrysippus accorded to etymology may be gauged from the remaining fragments of his *On the Soul* preserved by Galen. On this and on the importance of etymology and etymological explanation in Stoic thought in general, see now T.L. Tieleman, *Galen and Chrysippus*: Argument and Refutation in the *De Placitis* Books II–III (diss. Utrecht 1992) 182–189.

50 Cf., e.g., Hdt. I, 51: (κρητῆρ) ... χωρέων ἀμφορέας ἑξακοσίους.

the infinite sum of extracosmic and intracosmic space. Moreover, if *topos* is taken in the strict sense, it appears that according to this definition *chôra* is fully occupied by the body at issue. In this respect the definition seems to differ from both the Chrysippean definitions in Arius and the common Stoic definitions in Sextus and Aëtius. What are we to make of all this?

In so far as they consider *chôra* to be either larger than *topos* or a relatively large *topos*, these definitions may be regarded as simply elaborating on the way these terms were used in ordinary language (on which see above chapter 2, section 2.1). This, to judge from the preserved definitions, gave rise to three different conceptions:

(1) *chôra* as simply *topos* plus *kenon* (and hence as possibly infinite): a probably un-Chrysippean but apparently widely accepted Stoic view (attested in Sextus and Aëtius);

(2) *chôra* as a (presumably finite) larger space which provides room for a larger body; *de facto* this space may or may not be *fully* occupied: Chrysippus' view (as attested in Arius Didymus);

(3) *chôra* as just a *topos* of a larger body: the view of 'some Stoics' or perhaps even just 'some people' (attested in Sextus).

It may be that (3) is just a version of (2).[51] It may also be non-Chrysippean in origin. In fact, as I indicated earlier, it is even possible that it is not Stoic at all. In view of this *non liquet* I propose that from now on we concentrate on (2), which is certainly Chrysippean.

In the analysis carried out above, I have already tried to draw

---

[51] This, apparently was Hahm's view who implicitly suggested (1977) 130, n. 137, that for Chrysippus *chôra* and *topos* were more or less identical. He explains this by noting that 'Chrysippus seems closer to Aristotle and Epicurus who considered place to be synonymous with space'. This remark is misleading for the following reasons: (1) it is based on an inveterate but nevertheless wrong automatic translation of χώρα and τόπος (on which see my remarks in chapter 2) which, even if Hahm's interpretation were correct, would be pointless in the present context, since after all Chrysippus took the trouble to provide *separate* definitions for *topos* and *chôra*; (2) it is certainly wrong in the case of Aristotle, and at least in need of further qualification in so far as Epicurus is concerned (on which, again, see above, chapter 2, 52–58). Nevertheless, it cannot be excluded that Hahm is right in assuming that for Chrysippus *chôra* was, rather trivially, just a larger *topos*. In that case the interpretation of Chrysippus' *chôra* here advocated does not hold good. See however, my interpretation of Plu. *SR* 1054c–d in the next section.

out the full implications of the wording of the two Chrysippean definitions preserved by Arius (unfortunately this kind of close reading is all we have to go on). It appeared that both definitions depicted *chôra* as a (part of) space larger than a particular body contained, and as, so to speak, that which can contain *more of* that particular body (τὸ μεῖζον ἀγγεῖον σώματος, τὸ χωροῦν μεῖζον σῶμα). A further question, however, concerns the context in which these definitions might have been useful. One obvious answer to this question is provided by Long and Sedley who claim that Chrysippus 'recognized the less technical point that a spatial container within the world, for instance a half-filled wine jar, can accommodate more of the body which partly fills it'.[52] The implication is that the Chrysippean definition of *chôra* was intended to make a 'less technical' point than what I called the 'common Stoic' definitions, in that the latter were concerned with space in a 'cosmic' setting, whereas the former was only dealing with a common way of speaking about individual things and their places.

I doubt whether this interpretation gives an exhaustive picture of the physical importance of the distinction. For, although, of course also separate entities in the Stoic physical world do have their *topos*, and although it is conceivable that the term *chôra* could also figure in a description of a larger particular place or the place a place which can contain more of a particular body,[53] it should be noted that—perhaps due to the fact that according to the Stoics

---

52 Long/ Sedley (1987) I, 296. Long and Sedley, however, seem to ascribe these definitions to the Stoics in general.

53 Note that Arius' definition of *topos* also leaves room for a *topos* containing several bodies instead of just one. For it speaks of τὸ ... κατεχόμενον εἴτε ὑπὸ τινὸς ⟨εἴτε⟩ *ὑπὸ τινῶν* According to Brunschwig (1988) 95 the words ὑπὸ τινῶν should be taken to refer to the possibility of two co–extensive bodies occupying the same place by complete interpenetration (*krasis*). But, I find it hard to believe that we are here dealing with a reference to such a very specific and technical element of Stoic physics. On the contrary, the quite untechnical use of the term τι in these definitions (meaning here just 'a (corporeal) thing') rather seems to indicate that we should not read too much specific infomation into this fairly simple definition. I think it is much more probable that Chrysippus is here merely taking account of the fact that in some cases it makes sense to speak of one place being occupied by several smaller bodies, each occupying part of it. If this is correct, the difference between such a *topos* containing several smaller bodies, and *chôra* as a place which can contain *more than* or *more of* a particular body (but which is nevertheless not partly empty, but partly occupied by another body or bodies) appears to be merely one of perspective.

there is only a large-scale void—the whole context of the definitions at issue seems to be primarily concerned with what we might call 'large-scale *topos*'. Thus, as we saw (text 4), Cleomedes is able to state that 'outside the cosmos there is the void which stretches without end to all sides; of this the part which is occupied by body (τὸ μὲν ὑπὸ σώματος κατεχόμενον) is called *topos* and that which is not occupied will be mere void'. In the Chrysippean description of the οὐκ ὠνομασμένον in Arius the word σῶμα also seems to refer to the body *of the cosmos*. It is not unreasonable to expect that, accordingly, also the concept of *chôra* could meaningfully figure in a 'cosmic' context. In that case we may take it to be implied that the *chôra* of the cosmos is a (part of) space which is larger *than the cosmos* and which can contain *more of* the cosmos, but which presumably is still of finite extent. This strongly suggests that in what I called a 'cosmic' setting, Chrysippus used his idiosyncratic concept of *chôra* as τὸ μεῖζον χωροῦν σῶμα to denote the larger but still finite space which surrounds and encompasses the cosmos, and which provides room for the cosmic body when it becomes larger, i.e. during the conflagration.[54] Such an application of the concept of χώρα at a 'cosmic' scale would seem to make sense in view of the Chrysippean definitions which speak of a μεῖζον ἀγγεῖον (but not an *infinite* ἀγγεῖον) or a χωροῦν ('something which provides room).[55]

This brings us to the question whether Chrysippus had a reason to single out this particular spatial *diastêma* for special mention. We know that in general the *ekpurôsis* played an important role in the ancient Stoics' theorizing about space. The occurrence of an *ekpurôsis* was used to prove that 'there is' an extracosmic void into

[54] Cf. Plu. *Comm. Not.* 1077b (= *SVF* II, 618), where it is said that according to the Stoics at an *ekpurôsis* τὸν κόσμον .... τοῦ κενοῦ προσεπιλαμβάνοντα *χώραν* ἄπλετον.

[55] It is perhaps tempting to assume that additional evidence for this interpretation is constituted by the way the term χωρῆσαι is used in a testimony in Achilles *Isag.* 8 (= *SVF* II, 610; wrongly included in *SVF*, presumably because of the ascription to οἱ Στωικοί; I discuss this text below, in section 6.4), who ascribed to the Stoics in general the Posidonian view that the void is 'not infinite, but just large enough to provide room for the cosmos if it is dissolved' (τοσοῦτον ὅσον *χωρῆσαι* λυθὲν τὸ πᾶν). But apart from the fact that we are here dealing with the view *of Posidonius*, we are also dealing with a testimony deriving from the doxographical *placita* tradition (on the doxographical evidence for Posidonius' view of the extra-cosmic void, see below, section 6.4, and Algra (1993) 476–490), and there is little reason to trust its actual wording to be original.

which the cosmos could occasionally expand.[56] Particularly in view of the strict dichotomy in the Stoic system between body (*sôma*), being finite and determined, and the incorporeal void, being infinite and having no spatial determinations of itself,[57] it would seem to have made sense for Chrysippus to distinguish between the space which *can* be occupied by the cosmos ('can' here being taken in the specific non-physical sense it is invested with in Stoic modal logic, on which see the next section) and the space which *will* be actually occupied. Or, to put it differently, it would seem to make sense to distinguish between void *qua* empty space *as such* and void *qua* room for the cosmos. The former is infinite; the latter is only considerably *larger* than the cosmos. In fact, though most Stoics faithfully adhered to the dogma that the extracosmic void was infinite—for reasons which are to be discussed later—this thesis was, as we shall see in section 3, not unproblematic within the context of their ontology. In addition, the available evidence suggests that Posidonius—probably partly under the influence of Platonic and Aristotelian physics—advocated a return to a finitist cosmology in postulating a void which was not infinite but just large enough to contain the cosmos during conflagration.[58] Accordingly, it is not too implausible to suggest that Chrysippus may have introduced his curious concept of *chôra* in an attempt to bridge, so to speak, the gulf between his finitist theory of the cosmos on the one hand, and his theory of the infinity of the void and of the universe (*to pan*) on the other.

If our interpretation is correct, it would seem that the remark 'it is being limited when it is filled up' (περατοῦται δ' αὖ τοῦτο ἐκπληρούμενον) at the end of the last part of Arius Didymus fr. 25 would not only mean that empty space is provided with a limit (πέρας) in so far as it is *de facto* occupied, but also that in Chrysippus' cosmology space is limited (πεπερασμένον) in so far as it is *at times* occupied. In the next section we shall see that there are strong indications that Chrysippus endorsed this or a very similar view.

56 See Cleom. *Cael.* I, 1, 43 ff. Todd (partly printed in *SVF* II, 537) and Aët. II, 9, 2 (= *SVF* II, 609).
57 On which see the last part of Ar. Did. fr. 25 (printed above as text 1).
58 On which see below, section 6.4.

6.2 Chôra *and the interpretation of Plutarch* SR *1054b–1055a*

6.2.1 Plutarch *SR* 1054b–1055a has received little attention from scholars discussing the Stoic theory of space and has as a rule been neglected by students of Stoic modal logic. However, it should figure in both contexts. In the former, because, as an allegedly literal quotation, it can provide us with useful information concerning Chrysippus' use of spatial concepts, in particular of the concept of *chôra*. In the latter, because the argument it contains must have been part of a general discussion of the Chrysippean modal operators, as is to be inferred from the title of the work from which Plutarch, or his source,[59] took his quotation: *Peri dunatôn* ('On possibles').

The fact that von Arnim printed the text as three separate fragments (*SVF* II, 539, 550, and 551) in chapters dealing with the unity and coherence of the cosmos may have been partly responsible for this lack of attention. Furthermore, the text is very difficult to understand and it has often not been taken seriously, because it charges Chrysippus with grave inconsistency.[60]

Subsection 6.2.2 will provide text and translation of the two quotations from Chrysippus, and it will outline the two main inconsistencies he was accused of by Plutarch. Next, in 6. 2 3, I shall discuss Hahm's attempt to show that the two fragments do not contain Chrysippus' own positive views. In the rest of this section I shall try to show that, contrary to what Plutarch suggests, Chrysippus' view can be shown to be consistent. In 6.2.4 and 6.2.5 I shall try to show that one of the two alleged self-contradictions attributed to Chrysippus does not follow if proper attention is paid to the context from which the two fragments were lifted, viz. a context concerned with modal logic. In 6.2.6 I shall argue that Chrysippus can be defended against the other charge too, if our interpretation of *chôra* advocated in the previous section is accepted. The concluding subsection will be 6.2.7.

---

[59] That in this particular case Plutarch had recourse to Chrysippus' original text has been plausibly argued by Babut (1969) 229.

[60] Cf. Pohlenz (1948–49) I, 77; Sambursky (1959) 112; Goldschmidt (1953) 29–30 and 43; Furley (1966) 20 ff. An extensive discussion of the text is to be found as Appendix V in Hahm (1977) 260–266. Hahm's views will be critically discussed in the next subsection Note, incidentally, that all these studies discuss the text at issue merely as evidence of the Stoic theory of the coherence of the cosmos, not as evidence of the Stoic theory of space.

6.2.2 Plutarch's argument in *SR* 1054b–1055a may be summarized as follows. Chrysippus has time and again stated that there is no up, down, or centre in the infinite void. He has used this thesis, which we might call the thesis of the isotropy of the void,[61] as an argument against the Epicurean assumption of a downward movement of the atoms through the void. At the same time he argued—in the *Peri dunatôn*, from which Plutarch offers two allegedly *verbatim* quotations—that there is a centre in the void and that the fact that the cosmos happens to occupy that centre is partly responsible for its 'virtual indestructibility'. The first of the two quotations runs:

> Consequently, even in the case of the cosmos the question whether it should be said to be destructible requires deliberation, I think. All the same, the case seems rather to be as follows. For instance, to its virtual indestructibility a good deal is contributed even by the part of *chôra* [62] that it has occupied, that is to say through its being in the middle (ἡ τῆς χώρας κατάληψις, οἷον διὰ τὸ ἐν μέσῳ εἶναι), since, if it should be imagined to be elsewhere, destruction would most certainly attach to it.[63]

The second quotation, which, according to Plutarch, occurred almost immediately after the first one (μετὰ μικρὸν αὖθις), is as follows:

> For it has also in some such way been a concomitant of substance, as a consequence of the very fact that it is thus qualified (εὐθὺς τοιάδε τις οὖσα), to have occupied everlastingly the middle place (τὸν μέσον τόπον), so that for another reason, but also accidentally it does not admit of destruction <and> is in this very way everlasting.[64]

---

61 *SR* 1054e: Οὐ γὰρ ὑπάρχειν ἐν τῷ κενῷ διαφορὰν ᾗ τὰ σώματα δευρὶ μᾶλλον ἢ δευρὶ προσάγεται.

62 On this translation of the words ἡ τῆς χώρας κατάληψις see below, 305–306. Cherniss (1976) translates 'by the position that it has occupied in space'. The possibility cannot be entirely excluded that ἡ τῆς χώρας κατάληψις here means something like 'the concept of *chôra*', as has been suggested to me *per litteram* by David Furley. But for all I know the word κατάληψις usually means 'apprehension' rather than 'concept'; moreover, there is an, admittedly late but nevertheless neat, parallel for the use of the expression χώρας κατάληψις in the sense of 'the occupying of a certain space', in Simp. *In Phys.* 533, 20.

63 *SR* 1054c: διὸ καὶ ἐπὶ τοῦ κόσμου εἰ ῥητέον φθαρτὸν εἶναι αὐτὸν λόγου οἴομαι δεῖσθαι. οὐ μὴν ἀλλὰ μᾶλλον ἐμοὶ φαίνεται οὕτως ἔχειν· οἷον* εἰς τὴν ὥσπερ ἀφθαρσίαν πολύ τι αὐτῷ συνεργεῖ ἡ τῆς χώρας κατάληψις, οἷον διὰ τὸ ἐν μέσῳ εἶναι, ἐπεί, εἰ ἀλλαχῆ νοηθείη ὤν, καὶ παντελῶς ἂν αὐτῷ συνάπτοι ἡ φθορά. (* *emendavi ex* οἱονέι δ'; on this reading see below, n. 70).

64 *SR* 1054d: οὕτω γάρ πως καὶ ἡ οὐσία συντέτευχεν ἀιδίως τὸν μέσον

According to Plutarch, this constitutes a 'clear and obvious contradiction'.[65] Chrysippus, he argues, contradicts himself in two important respects:

(1) he postulates a centre in the infinite void, which contradicts his theorem that the void is isotropic; and

(2) he argues that the position of the cosmos is to a degree responsible for its indestructibility, for it is clear that Chrysippus is afraid that, if the cosmos were not placed in the centre of the void, its parts would naturally move towards that centre and thus cause the destruction of the cosmos; this contradicts the Stoic theorem that the void is inert, i.e. that it is not the structure of space, but the structure of the cosmos itself (1054e: τὴν τοῦ κόσμου σύνταξιν) which is responsible for the motion and position of its parts and thus for its coherence.

It will by now be clear that this Chrysippean account is of particular interest to us in that it uses the term *chôra* and deals with the structure and physical role of space in general. Our interpretation will in the end of course depend on whether we will follow Plutarch as to the contradiction or whether we shall be able to find another solution.

Generally speaking, we have to be careful not to accept the 'contradictions' detected by Plutarch too readily. To be sure, the Stoics, Chrysippus in particular,[66] often expressed themselves rather obscurely and sometimes rather carelessly. On closer scrutiny, however, by no means all the contradictions pointed out by Plutarch reflect a real conceptual ambiguity. In many cases they turn out to be merely verbal or not even that.[67] As Cherniss rightly remarked, Plutarch's paraphrases and interpretations, as well as the actual quotations, must each be judged for themselves, both in the context of Plutarch's own purpose and in comparison with all other available and relevant evidence.[68] This is what I intend to do in the present section.

κατειληφυῖα τόπον, εὐθὺς τοιάδε τις οὖσα, ὥστε καθ' ἕτερον τρόπον ἀλλὰ καὶ διὰ τὴν συντυχίαν μὴ ἐπιδέχεσθαι αὐτὴν φθορὰν ⟨καὶ⟩ κατ' αὐτὸ τοῦτ' εἶναι ἀίδιον (⟨καὶ⟩ add. Reiske).

[65] *SR* 1054d: καταφανῆ καὶ βλεπομένην ἐναντίωσιν.

[66] About Chrysippus' crabbed and obscure style see the texts printed as *SVF* II, 26, 28, 29, 288, and 902.

[67] I have presented a 'case study' of one of the 'contradictions' ascribed to Chrysippus by Plutarch in Algra (1990).

[68] Cherniss (1976) 405.

6.2.3 Before attempting to evaluate Plutarch's interpretation of the two Chrysippean quotations, we shall have to determine the precise status of these quotations themselves. For it has been extensively argued by Hahm[69] that they do not represent Chrysippus' own positive views, but that, on the contrary, they have been lifted from a Chrysippean *refutation* of the thesis that the position of the cosmos contributes to its indestructibility.

Hahm's interpretation mainly rests on three arguments. The first one is of a philological nature. It concerns the interpretation of the first two sentences of the first quotation: ' ... even in the case of the cosmos the question whether it should be said to be destructible requires deliberation, I think. All the same the case seems to be as follows (οὕτως)'. According to Hahm the second sentence should be read as stating that the cosmos *is* destructible, the word οὕτως being taken to refer back to φθαρτὸς [...] ὁ κόσμος (which is, of course, grammatically possible, though in the present context not the most natural way to read the sentence). This interpretation forces him to translate the word οἱονεί at the beginning of the next (third) sentence by 'as if ...' and thus to take it to introduce a more or less ridiculing description of the position Chrysippus wants to attack.

Hahm's second argument concerns the contents of the quotation. He argues that the eternity of the cosmos, which is at issue in the two quotations would be irrelevant for the cosmos' being indestructible, given the strict connotation of the term 'impossible' in Stoic modal logic. He stresses that the Stoics thought that many things which will never happen—such as the destruction of the cosmos—are still 'possible', in the sense of 'not being logically excluded'. Hence Chrysippus *cannot* have introduced the eternity of the cosmos to prove its indestructibility.

As a third, and related, argument Hahm adds that even if we grant that Chrysippus could have meant to defend the thesis that the cosmos is indestructible, the argument as it stands would have been of little help. For it could only concern the cosmos in one particular sense of the word, viz. 'the cosmos *qua ousia*' i.e. *qua* material substance. For the cosmos in its other attested sense, i.e. the cosmos as the particular *diakosmêsis* occurring between two conflagrations, *is* of course destructible. It only lasts until the next

---

[69] Hahm (1977) 260–266.

periodical conflagration. All in all, according to Hahm, we need not bother about how to square the two quotations with what is otherwise known about Chrysippus' views, since they do not reflect his own position.

As to the first point, the interpretation of the sentence starting—in the mss.—with οἱονεί, it should be noted that most scholars here saw a textual problem. The mss. reading of the third sentence of the first quotation is as follows: οἱονεὶ δ' εἰς τὴν ὥσπερ ἀφθαρσίαν ... etc. This gives awkward Greek and a very harsh transition from the preceding sentence. Consequently this passage has been variously emended by scholars like Wyttenbach, Pohlenz, and Cherniss.[70] Hahm, on the other hand, accepts the mss. reading and thus tries to make sense of it. To my mind he has not succeeded. In the first place the interpretation of the second sentence, which translates οἱονεὶ by 'as if' and takes it as qualifying συνεργεῖ, still leaves us with an awkward transition from the preceding sentence ('The cosmos is destructible. As if for the seeming incorruptibility ... etc.'). If one wants to retain the reading of the mss. by all means, it is preferable to take οἱονεὶ as qualifying ἀφθαρσίαν, as Pohlenz already remarked in his critical apparatus,[71] and to regard ὥσπερ as redundant (or as an

[70] Particularly problematic are the first word οἱονεί, and the resulting *doublure* with ὥσπερ. Wyttenbach proposed to read: εἰς δὲ τὴν οἱονεὶ ἀφθαρσίαν ... etc; Cherniss suggested [οἱονεὶ δ'] εἰς τὴν οἱονεὶ ἀφθαρσίαν. Both Cherniss and Wyttenbach argued for the substitution of οἱονεί for the mss. ὥσπερ by reference to *Def. Orac.* 425d–e which speaks of οἱονεὶ ἀφθαρσίαν. This, however, is not sufficient to warrant the proposed emendation: it merely shows that οἱονεί and ὥσπερ could be used synonymously. Pohlenz' οἷά τε δ' εἰς τὴν ὥσπερ ἀφθαρσίαν .... συνεργεῖν (scil. δοκεῖ ἡ ... κατάληψις) seems to require too many changes in the mss. reading, does not offer a natural explanation for the corruptions, and doesn't make much sense anyway, since the point does not seem to be that ἡ τῆς χώρας κατάληψις *can* contribute to the indestructibility of the cosmos, but rather that it actually *does.* I tentatively propose to read οἷον εἰς τὴν ὥσπερ etc., assuming an original dittography of ει- (or a dittography of εις, subsequently changed into -ει δ) and taking οἷον as 'for example', on which see *LSJ, s.v.* V, 2, b. This would accord well with the fact that, as I shall argue below, the argument Plutarch is here providing is only *one out of two* (or more). Even if, however, οἱονεί is left standing and taken—as Hahm would have it—as qualifying the whole sentence (or the verb συνεργεῖ), it seems preferable to translate it as 'in a way', 'so to speak' (which is how οἱονεί is used at *Def. Orac.* 425d), rather than as 'as if' (which is what Hahm proposes; but note that in that case one would expect an *irrealis* instead of the simple συνεργεῖ).

[71] M. Pohlenz (ed.), *Plutarchi Moralia*, vol. 6, 2 (Leipzig 1959^2) 51 *in app.*: 'οἱονεὶ certe non ad συνεργεῖν sed ad ἀφθαρσίαν quadraret'.

intrusion). Moreover, Hahm's interpretation fails to account for the occurrence of the particle δέ. A more serious objection, however, concerns the logic of Hahm's argument. It seems very unlikely that Plutarch would quote an argument as authentically Chrysippean if it immediately betrayed itself as actually being a ridiculing refutation of the arguments of an opponent, starting with 'as if ...'. Although, as I have already noted, Plutarch at times may be guilty of misrepresentations and misquotations, it would certainly not be in his own interest to do it in such an obvious way. It is difficult to see why what is clear to Hahm would not have been clear to Plutarch's readers.

Hahm's second argument, which concerns the correct interpretation of the Chrysippean modalities, will be extensively dealt with in the next subsections. Be it sufficient here to notice that, as it would appear, the Chrysippean modal terms could be used differently in different contexts and that the statement that the cosmos *qua* material substance (οὐσία) *cannot* be destroyed—in the sense that its destruction is 'physically' excluded—is not novel or unheard of in a Stoic context. In this connection it should be observed that the present quotations from Chrysippus are neatly paralleled by the *testimonium* of Arius Didymus fr. 37 (= *SVF* II, 599 *in fine*) where it is argued that the material substance (οὐσία) of the cosmos as well as its 'governing force' (διοικῶν) are indestructible, because there is nothing to cause such a destruction. This takes us to Hahm's third main point, the fact that the Chrysippean argument—if taken seriously at all—could only be sufficient to prove that the cosmos is indestructible in one particular sense of 'cosmos'.

True, we know that the Stoics could use the word *kosmos* in different senses. It may be used to denote God or matter, which are both eternal, or to denote each particular world order (*diakosmêsis*), which is not eternal.[72] That the two quotations from the *Peri dunatôn* do not deal with the cosmos *qua diakosmêsis* is clear. They speak about the eternity of the cosmos *qua* material substance (οὐσία, note that this word is explicitly used). But then there is no

---

[72] Cf. D. L. VII, 137 (= *SVF* II, 526): λέγουσι δὲ κόσμον τριχῶς etc.; Cleom. *Cael.* I, 1, 1 ff. Todd (= *SVF* II, 529): τοῦ κόσμου πολλαχῶς λεγομένου ὁ νῦν ἡμῖν λόγος etc.; Ar. Did. fr. 29 (= *SVF* II, 528): .....διὸ κατὰ μὲν τὴν προτέραν ἀπόδοσιν ἀΐδιον τὸν κόσμον εἶναί φασι, κατὰ δὲ τὴν διακόσμησιν γενητὸν καὶ μεταβλητὸν κατὰ περιόδους κτλ.

need to suppose that any argument on Chrysippus' part that the cosmos is indestructible should have to apply to *all* senses of the word 'cosmos' (or to all senses of the word 'destructible').[73]

We may conclude, then, that Hahm does not provide sufficient arguments to assume that the two quotations provided by Plutarch do not reflect Chrysippus' own positive views. That, however, is not to say that these quotations present no problems. On the contrary, how precisely they fit into Chrysippus theory of modalities and what role is played by spatial factors in securing the imperishability of the cosmos are questions which are not easy to answer in the absence of any further information about the context. Accordingly, our attempts in the next two subsections to reconstruct the modal context of the fragments as well as the underlying physical theory concerning the role of space, are inevitably to a large extent hypothetical.

6.2.4 Hahm's claim that the fragment's alleged *contradiction* of what is otherwise known of Stoic modal logic counted as an argument against the authentically Chrysippean character of its contents, confronts us with the question what exactly *is* known about early Stoic modal logic. I am afraid that Hahm's own, implicit, answer was rather bold, especially since in this field there is very little firm ground to stand on.

Attempts to reconstruct the Stoic theory of modalities have usually been based on two lists of definitions, to be found in Diogenes Laërtius and Boethius.[74] In addition, there are a number of texts—like our quotations from *Peri dunatôn*—which contain less technical information about the Stoic use of terms such as

---

[73] Note that also the word φθορά could be used in more than one sense, viz. to denote either the natural elementary change (κατὰ φύσιν μεταβολή)—in which sense the term was also applicable to the *ekpurôsis*—or destruction *stricto sensu* (κυρίως). Cf. Ar. Did. fr. 36 (= *SVF* II, 596): ....οὐ γὰρ ἐπὶ τῆς τοῦ κόσμου κατὰ περιόδους τὰς μεγίστας γινομένης φθορᾶς κυρίως παραλαμβάνουσι τὴν φθορὰν .... ἀλλ' ἀντὶ τῆς κατὰ φύσιν μεταβολῆς χρῶνται τῇ προσηγορίᾳ τῆς φθορᾶς. Our two Chrysippean quotations seem to deal with φθορά and ἀφθαρσία in the strict sense of the word and with the cosmos *qua* οὐσία (although the alleged ἀφθαρσία of the cosmos in this sense pertains both to the cosmos *qua* material substance, and to each particular *diakosmêsis*, for at a conflagration the *diakosmêsis* is not destroyed in the strict sense of the word, but it *changes* into fire).

[74] D. L. VII, 75 (= *SVF* II, 201, 1st part = *FDS* 914, partly), Boethius *In Arist. de Interpr.* III, 9 (= *SVF* II, 201, 2nd part = *FDS* 985).

'possible' and 'necessary'. It has proved extremely difficult—not to say positively impossible—to fit all these texts (lists of definitions and additional evidence) into a coherent interpretation.

In their preserved form the two lists of definitions provide difficulties of their own. The correct constitution of the text is disputed. Attempts to interconnect the several definitions in a coherent way seem to involve textual changes but there is no consensus about which changes are necessary and why.[75] The

[75] An attempt which by now many scholars appear to regard as 'classical' is Frede's reconstruction and interpretation of the definitions in Frede (1974) 107–117; the philological aspects of his work, in particular his reconstruction of the definitions provided by Diogenes Laërtius and of their interrelations, are largely taken over by Bobzien (1986) 45–50, but also by Mignucci (1978), although Mignucci's interpretation of the Stoic theory is in the end entirely different. Also Long/ Sedley (1987) II, 234 take over Frede's textual supplement and his interpretation of the definitions provided by Diogenes Laërtius. Cf. also Sorabji (1980) 272. I am afraid, however, that Frede's reconstruction has somewhat prematurely been taken over as 'standard'. In the first place it is based on a change in the text of Diogenes Laërtius' definition of the impossible into ὃ μή ἐστιν ἐπιδεκτικὸν τοῦ ἀληθὲς εἶναι ⟨ἢ ἐπιδεκτικὸν μέν ἐστι, τὰ δ' ἐκτὸς αὐτῷ ἐναντιοῦται πρὸς τὸ ἀληθὲς εἶναι⟩, thus introducing a reference to 'external circumstances'. This change is called forth by the claim that all definitions have to be coherently interconnected and is supported by a reference to the parallel definition in Boethius, which, as it stands, reads: 'impossibile .... quod nullam unquam suscipiat veritatem, aliis extra eventum ipsius prohibentibus' (cf. Frede (1974) 110). Now Frede acknowledges that, in this form, Boethius' definition does not warrant the above addition in Diogenes' text, because it says that what is impossible is so *exclusively by virtue of external factors*, whereas the logically required definition in Diogenes would, according to Frede, have to define the impossible as that which is *either* in itself impossible *or* excluded by external factors. Therefore he proposes to change also Boethius' definition into '...suscipiat veritatem <sive natura propria sive> aliis extra ...etc' ((1974) 113). All this is not very elegant from the point of view of textual criticism, and I prefer to assume, following a suggestion of Jaap Mansfeld, that in Boethius' text the definition of the non-necessary has been lost and that the words 'aliis extra eventum ipsius prohibentibus' in one way or another originally constituted the final part of that missing definition. Furthermore, as I shall argue below, it is not at all *a priori* clear that all preserved definitions have to be coherently interconnected. Rather, in view of the fact that, as I think can be shown, the Stoics used their modal terms in several senses and in view of the fact that most ancient sources were singularly unalert to this phenomenon, it is even *a priori* plausible that the preserved definitions *are not in all respects coherently interconnected.* I cannot go into these matters here at length, but the following remarks may at least serve to indicate some implausible consequences of Frede's reconstruction. Frede's starting point, the thesis that the genitives absolute in the definitions in Diogenes give *additional* requirements for something being possible, impossible etc., in combination with his failure to properly differentiate between what I shall here for the sake of convenience call 'logical' and 'physical'

definitions themselves have been described as circular, and not without reason.[76] In fact, the very variety of existing interpretations[77] should make us sceptical about whether there was only *one*

---

possibility, leads to very curious statements like '....möglich ist alles, was von nichts daran gehindert wird, der Fall zu sein, selbst wenn es nie eintreten wird. Nichts aber hindert das Gegenteil dessen, was vorherbestimmt ist, daran der Fall zu sein. Also ist es möglich, auch wenn es nicht eintreten wird'. That the opposite of what is fated is not prevented from coming about is *prima facie* implausible and seems to be contradicted by the two Chrysippean examples provided by Cic. *De Fato* 13 (= *SVF* II, 954 = *FDS* 989): this jewel not being broken and Cypselus' reign are—in a special sense—not *necessary*, but since they are nevertheless fated, i.e. caused by a complex of unchangeable causes, their opposites *are* prevented, viz. by fate, from coming about. Next, having interpreted Diogenes' definition of 'possible' as 'that which is in itself possible and is not prevented by external factors from being true', Frede adduces the description of the possible to be found in Alex. Aphr. (*Fat.* 176, 15–16 = *FDS* 989: τοῦτο ὃ ὑπ' οὐδενὸς κωλύεται γενέσθαι, κἂν μὴ γένηται) as a parallel to the definition in Diogenes as he interprets it. He obviously does not recognize that the opponents Alexander had in mind—in all probability not the orthodox Stoics—interpreted the κωλύεται in this phrase in an epistemic sense: 'that which is prevented by nothing from coming about' means, as Alexander himself (!) adds (176, 18–22), 'that of which we do not know the causes which prevent it from coming about', the implication being that the external factors in fact *do* play a role, though from the perspective of human action this is irrelevant since the factors are not known as such. So, contrary to what Frede suggests, it is not true that the possible is here defined as that which is *de facto* not prevented from coming about by external causes.

[76] Kneale/ Kneale (1962) 125.

[77] In particular the right interpretation of the genitives absolute in the definitions of Diogenes (and the corresponding ablatives absolute in those of Boethius) is disputed. Take, e.g., the definition of the possible as given by D. L. VII, 75 (= *SVF* II, 201 = *FDS* 914): δυνατὸν μὲν τὸ ἐπιδεκτικὸν τοῦ ἀληθὲς εἶναι τῶν ἐκτὸς μηδὲν ἐναντιουμένων. The question which here arises is whether the genitive absolute referring to external circumstances qualifies the words ἀληθὲς εἶναι or whether it qualifies the *definiendum* δυνατόν. In the former case the implication is that external circumstances only prevent the possible from becoming true, not from being possible. In that case that which is prevented by external factors from coming to be may still be possible. In the latter case, however, the genitive absolute adds a further necessary condition for something being possible. In that case, in other words, the possible is that which is *both* intrinsically possible *and* not prevented by external factors.

The former interpretation—which provides the Stoics with a use of the term 'possible' very much resembling that which was endorsed by the dialectician Philo, a contemporary of Chrysippus—has been opted for by Mates (1953) 40–41 and Reesor (1965). The latter interpretation, on the other hand, has been endorsed by Frede (1974) 107–117, and by a number of scholars (mentioned above, n. 75) who took over his position. A position between the two—though different from the position defended by Sharples, on whihc see the text above and the reference in the next footnote—has been taken by Kneale/ Kneale (1962) 123 ff., who distinguish between an absolute strictly defined sense in which the modal operators were used by the Stoics.

In this connection I want to start out from some observations made by R. Sharples, who has convincingly argued that the Stoic modalities were used differently in different contexts.[78] The Stoics were probably not interested in construing a system of modal operators for its own sake. Their interest in the proper meaning of modal terms arguably arose from physical and ethical considerations. In a physical context the required meanings of terms like possible and necessary were different from those required, for instance, in the context of human ethics in order to save moral responsibility. For these reasons I would suggest that, methodologically speaking, the texts which provide us with information about how the Stoics actually *used* their modal operators should be regarded as more informative than the surviving definitions.

These definitions themselves might be—and usually have been—read as suggesting that in the Stoic account of modalities two kinds of factors were relevant: what might be called 'internal factors' on the one hand and 'external factors' on the other.[79] From a physical point of view, there are, according to the Stoics, two factors which jointly determine whether it is possible for a particular thing (body) to change or to do something (for the sake of convenience I shall further speak of: to φ) or not, viz. internal disposition and external factors. It appears—though in this respect our sources appear to be embarrassingly confused—that the distinction here drawn between internal and external factors corresponds to the Stoic distinction between internal, 'perfect' causes (*autotelê aitia*), and the external 'auxiliary' causes' (*sunerga aitia*).[80] It

---

and a relative use of the modal operators, corresponding with the alleged distinction between simple and complex facts. For a critique of this view see now Bobzien (1986) 133.

[78] Sharples (1981). Sharples did not, as I do here, apply the distinction between physical and logical necessity—a distinction which I believe to be particularly rewarding in this context.

[79] It should be borne in mind, however, that depending on whether we read the Stoic modalities as *Axiomsmodi* or as *Sachverhaltsmodi* (on which see below, 295, esp n. 88), the term 'external factors' refers to different things. In the former case, external factors are the actual state of affairs in the outer world, as opposed to the inner (logical) coherence of the proposition. In the latter case, external factors are the causes external to the agent. I would argue, however, that in both views external factors can be disregarded for the purpose of establishing a person's moral responsibility for a particular act.

[80] I here commit myself to Frede's interpretation of the ancient Stoic

seems that in general the internal causes were given a higher status. In some cases, presumably, a perfect, internal cause, i.e. the way a thing is internally disposed, was thought to be sufficient to bring about the effect. In other cases a *sunergon* was thought to be needed in addition, the famous Chrysippean example in this respect of course being the cylinder, which rolls both in virtue of its own nature, i.e. because its form allows it to roll (*autoteles aition*), and because it is pushed (*sunergon aition*). From a physical point of view, of course, both internal and external causes are determined by fate (or perhaps we should simply say that they *are* fate).[81]

Accordingly, if we want to be sure that it is *physically* possible for something to φ, e.g. for a cylinder to roll, we will have to establish (1) that there are no internal factors preventing it from φ-ing, and (2) that there are no external factors preventing it from φ-ing. If both conditions are fulfilled, the physical state of affairs is such as to allow the thing at issue to φ—that, in fact, is what we mean when we say that it will be possible *in a physical sense* for our particular thing to φ. If, on the other hand, we want to be sure that it is physically speaking *impossible* for something to φ, we shall have to determine either (1) that there are internal causes which prevent it from φ-ing (if these internal causes are perfect causes we need not go on looking for an external preventing factor) or (2) that—internal preventing causes being absent or insufficient—there are external causes which prevent it from φ-ing.

Given the role of fate, what is prevented from happening is fated not to happen (i.e. it is in a physical sense impossible for it to happen), and what is *not* prevented from happening, i.e. what is physically possible, is fated and hence in the physical sense necessary. In other words, since both internal and external causes are physically speaking completely determined by (or co-exten-

aetiology as given in Frede (1980). For a somewhat different view see Sorabji (1980), esp. 260–261. For a fully different interpretation of the interrelations between the several kinds of causes, see now Görler (1987). A useful survey of the Stoic theory of causation, including an assessment of the principal testimonies, is provided by Duhot (1989). For our purpose, however, the details of the Stoic theory of causes (and hence also the scholarly controversies concerning the proper interpretation of those details) are immaterial except for two—to my knowledge uncontested—features, viz. (1) that there are internal and external causes, and (2) that in the context of determining moral responsibility only internal causes are relevant.

81 Cic. *De Fato* 43.

sive with) fate,[82] we may conclude that on the purely physical level the Stoics were committed to strong determinism, according to which only the necessary is possible (the non-necessary being, conversely, impossible). In fact, it seems they thought this strong determinism was even required in order to account for the ordered nature of the physical world and in order to allow a physical explanation of that world in terms of laws and causes.[83]

At the same time the extant evidence clearly indicates that the Stoics thought that, especially in the context of ethics, a different sense in which things are necessary or possible had to be introduced in order to save an agent's moral responsibility for his actions, in spite of the fact that they are completely determined from a physical point of view. It seems that in this context they stressed that only the internal causes were relevant.[84] Thus

---

82 For εἱμαρμένη being all-embracing see e.g. D. L. VII, 149 (= *SVF* II, 915 = *FDS* 998A), with references (*laudationes*) to specific works of Chrysippus, Posidonius and Boethus. For some lucid comments on this strong physical determinism see Long/ Sedley (1987) I, 343. See also Inwood (1985) 66–68 and Donini (1975) 220.

83 For the Stoic arguments for the existence of fate see Long/ Sedley (1987) I, 343, with references to relevant texts.

84 For a different view, however, see Long/ Sedley (1987) I, 235, who argue that responsibility implies opportunity and that 'if I want to claim credit for not smashing the jewel, I must in particular show that it was possible for me to smash it, in the sense that I had the opportunity to do so'. This, it is argued, implies that (1) the jewel was breakable, and (2) that circumstances did not prevent me from breaking it. There are several objections to be made to this interpretation. First of all I think in cases such as the one adduced by Long/ Sedley the Stoics would rather have focused on the possibility of the *agent* breaking the jewel than on the possibility of the jewel being broken. Secondly, we should make a difference (because the Stoics themselves apparently made a difference) between claiming credit for something and being responsible for something. Chrysippus argued—to judge from Plu. *SR* 1038f. (= *SVF* III, 211) that one cannot claim credit for not doing things which nobody would normally speaking ever (want to) do, such as having sex with an ugly old crone, or saying that '3 actually equals 4'—yet also in these cases the one who abstains from these actions *is* still to be called virtuous (and hence, I take it, responsible for his deeds); see on this Algra (1991). A more serious objection is that the common connotations of the word 'opportunity' together with the curious examples adduced by Long and Sedley tends to give a libertarian misrepresentation of the soft-determinist position the Stoics took in this respect.(Compare the interpretation provided by Nem. *De Nat. Hom.* p. 91–94 Matthaei (= *SVF* II, 991); See also Sharples (1981) 83). In the case of a jewel placed on the moon or, as Long/ Sedley suggest, locked away in a bank vault, the Stoics would probably have stated that it is *impossible* (in the morally significant, 'logical' sense) for the agent to break it (for, normally speaking, *nobody* could ever break such a jewel) so that the question of moral responsibility for not breaking it

Chrysippus argued that our assenting must be prompted by an impression as an external cause, but that in the end it will depend on our own mental state as an internal cause.[85] Thus the person who assents may himself be held responsible for doing so and cannot blame the circumstances—in this case, his impressions—for forcing him to do so. Neither, apparently, does the fact that our internal mental disposition is in the end itself fully physically determined play any role.

Whether or not such an attempt to save both physical determinism and moral responsibility may be regarded as successful has often been doubted. However, that should not concern us here.[86] What does concern us is that this curious ethical perspective forced the Stoics to maintain that

(1) there is a sense of 'possible' in which things which are *de facto* physically excluded may nevertheless be possible;
(2) in order to determine whether it is possible *in this sense* for something to φ, external factors are irrelevant;
(3) in order to determine whether it is possible in this sense for something to φ, its actual intrinsic *physical* disposition is also irrelevant.

---

need not even be put. The Stoic point rather seems to be that even if the jewel is close at hand, my not smashing it may be fated—or, to put it differently, my smashing it may be prevented by external circumstances. In view of such preventing external circumstances I may be said not to have the opportunity to smash the jewel (in the commonly accepted, libertarian sense of the word 'opportunity'). Yet the only thing which morally counts is my internal disposition, while the occurrence of fated and inevitable external circumstances seems to be morally immaterial. That the Stoics indeed thought that external preventing factors, even if playing a major role in determining the actual course of events, should be *discounted* in assessing moral responsibility, may also be inferred from the epistemic Stoic reading of 'not prevented', which was attacked by Alex. Aphr. *Fat.* 176, 14–27 (= *FDS* 1009; not in *SVF*; printed as LS fr. 38 H). According to this reading, the statement that something is 'not prevented' by external factors should be taken to mean that it may well be *de facto* prevented by them, but that we do not *recognize* these preventing factors. This implies that external factors *do* play a role, physically speaking, though they may be disregarded for moral purposes. Alhough this epistemic reading arguably reflects a later development in Stoic modal theory, one may safely assume that it relates to earlier Stoic attempts and that earlier Stoics were also of the opinion, *not* that the presence or absence of external preventing circumstances determines whether we are morally responsible for doing something, but that on the contrary these circumstances have to be disregarded in this connection.

85 Cf. Cic. *De Fato* 42–43 (= *SVF* II, 974).

86 For a good discussion of these problems see now Inwood (1985) 66 ff.

This raises the question what, then, it *is* that determines whether something is possible or impossible in this specific non-determinist sense. It is tempting to reply that it is *logical* possibility or impossibility which is at stake here. It may indeed be convenient to use this term, as I have in fact already been doing, in order to distinguish this kind of possibility from *physical* possibility. It should be pointed out, however, that the use of these labels is in itself not really informative unless it is shown how the distinction was actually accounted for by the Stoics themselves.[87] However, they seem to have made no terminological distinction between these two kinds of possibility *we* may recognize. Moreover, it is not immediately clear how the Stoics could give a unified and coherent semantical account of the notion of what I called 'logical' possibility.

A first complicating factor is the fact that for the Stoics what I called 'internal factors' apparently could pertain either to the physical fitness of a *body* (thing) to be, or to become, or to do something, or to the possibility of a certain *proposition* (*axiôma*) to be or become true. This is what induced Bobzien to speak of *Axiomsmodi* on the one hand and *Sachverhaltsmodi* on the other.[88] But even if we assume that all statements about something being possible for a *thing* can ultimately be translated into statements about the possible truth of a proposition, it is not easy to see how, technically speaking, logical possibility and impossibility were defined. To be sure, at least one example of an impossible proposition is fairly unproblematic, viz. the *axiôma* τέθνηκεν οὗτος ('this man is dead').[89] It is in virtue of the particular referring, *c.q.* signifying character of its elements, not in virtue of any external state of affairs, that it can never be true. For in those circumstances where we would expect it to be true (i.e. if οὗτος was formerly used to denote Dion and if Dion is now dead) the *axiôma* 'perishes' because the referent of οὗτος, a term which is strictly deictic, is

[87] The distinction between logical and physical possibility (or necessity etc.) is by no means universally accepted as self-evident. Cf. S. Haack, *Philosophy of Logics* (Cambridge 1978) 171, for a brief survey of these and connected issues and for further references.

[88] Bobzien (1986) 50–51. If we may trust our sources, a similar ambiguity between *Axiomsmodi* and *Sachverhaltsmodi* was already apparent in the examples used by Philo the dialectician. Cf. the *testimonia* adduced by Bobzien (1986) 40–41.

[89] Alex. Aphr. *In APr.* 177, 25 ff. (= *SVF* II, 202 a = *FDS* 994).

no longer there.[90] So in this case the *axiôma* is impossible, because it has no intrinsic capacity of becoming true.

The semantic background of the possibility (possible truth) of the *axiôma* 'this jewel will be smashed' seems to be more obscure. The underlying idea probably was that the general *concept* of jewel is such that the thought of a jewel being smashed is not absurd.[91] This may account for the example, given in Diogenes Laërtius' list of definitions, of a proposition which is 'impossible', viz. 'the earth flies'. Probably the concept of 'earth' was thought to be such that 'the earth flies' was considered to be an absurd statement.[92] In that case, however, there still seems to be a problem, since normally speaking general concepts seem to play no role in Stoic semantics.[93]

This being as it may, we can nevertheless now see, however dimly, in what sense Chrysippus thought that when we assent to a proposition, it is at the same time *possible* for us not to do so, even if our mental disposition and our impressions actually determine that, physically speaking, it is impossible that we do not assent. In this same sense it is possible for a jewel to be smashed, even if neither its own physical disposition nor external forces will ever cause it to be smashed.[94] In the sphere of human action this non-physical sense of the modal terms provided the Stoics with a means to argue for the moral responsibility of the agent in a fully

---

[90] For a good discussion of this curious 'perishing' of *axiômata* see Bobzien (1986) 18–21.

[91] It should be noted, however, that such a general concept was, according to the Stoics, by no means a hypostatized something, but, as an *ennoêma*, the result of a process of abstraction. Statements about 'universal jewels', e.g. whether they are liable to destruction, are in the end reducible to statements about token jewels. (This, by the way, seems to present problems of its own for the idea of logical necessity and possibility). For an excellent discussion of these matters see Long/ Sedley (1987) I, 179–183.

[92] That we are here dealing with the second, not strictly physical, non-determinist sense of 'impossible' is shown, I believe, by the fact that the definition as it stands (in D. L.) contains no reference to the possible role of external factors, which would be required, as we saw, if the problem of *physical* impossibility were at stake. It is true, the text has been emended by Frede so as to supply such a reference. However, see my remarks above, n. 75.

[93] See on these matters now the excellent discussion in Long/ Sedley (1987) I, 181–183 (on 'universals') and 199–202 (on *lekta*).

[94] Cic. *De Fato* 13 (= *SVF* II, 954 = *FDS* 989); see on this example Long/ Sedley (1987) I, 235, though I disagree with their view that according to Chrysippus the presence or absence of preventing external circumstances was morally significant.

determinist world, however futile we may judge this solution from a moral point of view.[95]

In this way, i.e. if we do not take the preserved definitions as our starting-point, but rather focus on the different contexts in which the Stoics used their modal terms, and on the actual examples of things possible etc. which are provided by the sources, a fairly coherent picture arises. From the point of view of physics the Stoics were committed to a strongly determinist theory, according to which the possible could be defined as that which is neither prevented by internal, *nor* by external factors, the impossible being that which is either prevented by internal, or by external factors. At least in contexts where human (rational) action and the possibility of saving human responsibility were at issue—but possibly also in other contexts (as may appear from the example of perishable jewels and the like) and possibly also for different reasons—they were committed to what might be called a variant of soft-determinism, purporting, for one thing, to save human responsibility. On this line of thought the possible was defined as a kind of intrinsic logical possibility, external factors here being irrelevant.

The coexistence of these two different perspectives—and, more particularly, the failure or unwillingness of many of our anti-Stoic sources to understand the subtleties of these coextensive perspectives—may explain the confused nature of many of our testimonies, including the two lists of definitions. From the (libertarian) point of view of the opponents of the Stoics this coexistence made no sense and they tended, accordingly, to represent the Stoic doctrine as if all kinds of necessity, possibility etc. could simply be identified.[96]

We are now in a position to consider Hahm's claim that the two fragments from *Peri dunatôn* do not contain Chrysippus' own

---

[95] Of course this soft-determinist sense of 'possible' and 'impossible', though applying to all kinds of things and events, only got a moral relevance in the case of rational human agents (on which see Inwood (1985) 66–91). Failure, on the side of the libertarian opponents of the Stoics, to understand this led to reproaches such as that voiced by Nem. *De Nat. Hom.* p. 293 Matthaei = p. 105 ff. Morani (= *SVF* II, 991) that if in this way our walking is 'up to us' (ἐφ' ἡμῖν), then also the burning of fire is up to the fire. Nemesius interestingly adds that this seems indeed to have been the opinion of the Stoic Philopator.

[96] Cf. Sharples (1981) 91.

positive views. According to Hahm they link up the eternity of the cosmos and its indestructibility (i.e. the *impossibility* of its destruction). This, he argues, cannot be what Chrysippus himself thought, since many things that are possible never happen, so that even if the cosmos is eternal (which Chrysippus admits by using the adverb αἰδίως) it might still be possible for it to be destroyed. Presumably Hahm assumed that the term 'possibility' was unequivocally used to denote what I have here called logical possibility. If, however, the view outlined in the present subsection is correct, it seems that Chrysippus could also very well be dealing with what I have called *physical* possibility. Whether this is so cannot *a priori* be assumed, but will have to be established by a closer investigation of the fragments at issue. This is what I shall try to do in the next subsection.

6.2.5 I want to argue that the Chrysippean quotations themselves primarily deal with the *physical impossibility* of the destruction of the cosmos. This is precisely, I suggest, why Chrysippus speaks of 'virtual indestructibility' (ὥσπερ ἀφθαρσίαν). At the same time I think there are indications that, in the context of these fragments Chrysippus also dealt with the *logical possibility* of the destruction of the cosmos. We may now consider his arguments.

At the beginning of the first quotation Chrysippus announces that he will deal with the question whether the cosmos should be called perishable, *tout court*. He then proceeds by adducing—as a first argument for the *virtual* indestructibility of the cosmos, i.e. its eternity—the fact that the cosmos occupies the centre of space. That this is only one out of two (or more) arguments is shown by the second quotation, in which we are informed that the imperishability of the cosmos is also otherwise secured (καθ' ἕτερον τρόπον). That Plutarch does not dwell on the other cause(s) for the indestructibility of the cosmos should, I presume, be ascribed to his polemical intentions. Nevertheless, the second quotation contains a clear indication as to the nature of what probably was the only other argument adduced by Chrysippus. It says that the cosmos takes its peculiar spatial position in virtue of the fact that it is 'thus qualified' (τοιάδε τις οὖσα). To my mind this unambiguously shows that in this same context Chrysippus discussed the fact that the indestructibility of the cosmos was due to the structure of the cosmos itself, in its 'classical' Chrysippean role.

It appears, then, that the original Chrysippean context contained two arguments intended to prove that the cosmos, physically speaking, cannot perish: (1) its internal structure, and (2) external factors, or rather, the absence of an independent spatial centre to which the parts of the cosmos could move.

It is the occurrence of the latter argument which has caused surprise among scholars[97] and which allowed Plutarch to regard the passage as one of the many Stoic self-contradictions. For the Stoic view that only bodies can act as a cause is widely attested.[98] Accordingly, space, whether as *topos*, *chôra*, or *kenon*, being incorporeal, cannot act in any way.[99] In other words, normally speaking the cosmos is *its own* 'perfect cause' (*autoteles aition*), so that in order to prove the physical incorruptibility of the cosmos it is in principle sufficient to point out that its structure is such as to preserve itself eternally. There are no (external) causes outside the cosmos itself. So at first blush we are inclined to grant Plutarch his point when he remarks that

> ... in thinking that the universe could not remain indestructible if by accident it had got become situated in another part of the void, he is evidently afraid lest the universe be dissolved and destroyed because the parts of substance move towards the middle. This he would not fear, however, did he not hold that bodies naturally move from all points towards the middle—the middle not of substance but of space that encompasses substance. Yet of this he has very frequently said that it is impossible and contrary to nature, because in the void there exists no difference by which bodies are drawn in one direction rather than another, but that the structure of the universe is responsible for the motion <of all the parts> moving from all points and tending towards its centre or middle.[100]

---

97 Cherniss (1976) 581 note c, reacts to Goldschmidt's interpretation—viz. that the position of the universe itself determines the centre of the void—with the objection that it is not clear how in that case Chrysippus could think that the position would make any difference to its indestructibility, presumably since the cosmos, on this line of thought, would everywhere create its own middle position.

98 To mention only a few instances: Ar. Did. fr. 18 Diels (= *SVF* I, 89) explicitly ascribes to Zeno among other things the view that τὸ μὲν αἴτιον σῶμα; Aët. IV, 20, 2 (= *SVF* II, 387; only in ps.-Plu. and in the *Aëtius Arabus*) ascribes to the Stoics in general the view that πᾶν γὰρ τὸ δρῶν ἢ καὶ ποιοῦν σῶμα and that πᾶν τὸ κινοῦν ... σῶμά ἐστι; S. E. *M.* 8, 263 (= *SVF* II, 363) has it that τὸ ἀσώματον κατ' αὐτοὺς (scil. the Stoics) οὔτε ποιεῖν τι πέφυκεν οὔτε πάσχειν.

99 Cf. Cleom. *Cael.* I, 1, 66 Todd: (κενοῦ) ... οὔτε τι πάσχοντος οὔτε ποιοῦντος; and I, 1, 99–100 Todd: τὸ μὲν κενὸν οὐδὲν ποιεῖ.

100 Plu. *SR* 1054d–e, (Cherniss' translation, with some small changes).

On closer scrutiny, however, there seems to be a way to make sense of Chrysippus' statement—and thus to forestall Plutarch's objection—without assuming, as Hahm does, that it does not contain his own views.

I would suggest that the situation, described in the first fragment, in which there is a centre of space independent of the (structure of the) cosmos which exerts a kind of attractive influence on the parts of the cosmos, thus causing its destruction, is not meant as a counterfactual situation of the *real*—i.e. the Stoic—cosmos, but as a situation of a *counterfactual cosmos*, or, more precisely, a situation occurring in a counterfactual physical world. In other words, Chrysippus argues that it is *conceivable* for there to be a cosmos which is not held together, as the actual (Stoic) one, by its centripetal force, but which has its parts moving naturally towards an independent centre of space, which may thus be said to act on the cosmos. This may very well be why he does not use the words 'if it *were* located elsewhere', but 'if it should be *imagined* to be elsewhere' (εἰ ἀλλαχῇ *νοηθείη ὤν*)—i.e. if the focusing point of the natural motion of the parts of the cosmos were conceived not to be the centre of the cosmos. In the counterfactual, but nevertheless conceivable case sketched here the destruction of the cosmos will be inevitable, although, as the argument purports to show, as things are (note the words εὐθὺς τοιάδε τις οὖσα) the cosmos is eternal. This means that in the second (logical) sense of 'possible' outlined in the previous subsection the cosmos *can* be destroyed (i.e. its destruction is conceivable), a thesis which was very probably explicitly put forward and defended in the same context. This hypothetical interpretation—which to me seems to be the only way to make sense of Chrysippus' actual words—may be seen to gain additional credibility from the following 'circumstantial evidence'.

In the first place, as I already noted, it explains the words *ὥσπερ* ἀφθαρσία in the first Chrysippean fragment: the cosmos is not *absolutely*—i.e. logically speaking—indestructible, but it is nevertheless *physically*, or *actually* indestructible. That the *logical* possibility of the cosmos' being destroyed was discussed in the context of our two fragments, may derive some initial plausibility from the fact that the next 'contradiction' to be discussed by Plutarch is precisely the fact that, according to the 'theory of the possible' (ὁ τῶν δυνατῶν λόγος) the same things are possible and impossible—

'impossible' in the sense of 'contrary to *heimarmenê*'—at the same time.[101] Moreover, if the information provided by Epictetus concerning Chrysippus' endorsing the thesis that something which neither is nor ever will be true might still be possible,[102] is derived from Chrysippus' *Peri dunatôn*, as the context seems to suggest,[103] we may conclude that the issue of the possibility of some things which will in fact never occur was dealt with in the work from which our two passages were quoted.

Next, that the Stoics thought that the cosmos, in the sense which is relevant here (not of course, a particular *diakosmêsis*) *can* be destroyed though it actually never *will* be, is also attested by two passages in Origen. In one of them, *Contra Celsum* I, 21 (= *SVF* II, 1053) we find the following statement about the cosmos (= God[104]):

> ... according to the Stoics also God is a body, and they are not ashamed to call him liable to be changed, changeable as a whole and alterable and simply liable to destruction (δυνάμενον φθαρῆναι) if there is someone to destroy him. By a happy coincidence, however, he is not actually destroyed, because there is nothing to destroy him.[105]

It is only one step from this argument to the famous argument put forward by Plato in the *Timaeus* concerning the indestructibility of the cosmos. There, at *Tim.* 32c, it is argued that though the cosmos, being material, is in principle destructible, it is in fact eternal by the will of the good Demiurge.[106] That the Stoics were committed to a comparable view of the theodicy is not in fact *a priori* implausible—though explicit evidence is lacking—since, as we know, they regarded God—being the active principle of the

---

101 Plu. *SR* 1055e: εἰ γὰρ οὐκ ἔστι δυνατὸν ὅπερ ἤ ἐστιν ἀληθὲς ἢ ἔσται κατὰ Διόδωρον, ἀλλὰ πᾶν τὸ ἐπιδεκτικὸν τοῦ γενέσθαι, κἂν μὴ μέλλῃ γενήσεσθαι, δυνατόν ἐστιν, ἔσται δυνατὰ πολλὰ τῶν μὴ καθ' εἱμαρμένην. On the probability of Plutarch having had recourse to the original text of Chrysippus' *Peri dunatôn*, see Babut (1969) 229.

102 Cf. Epict. *Diss.* II, 19, 1 ff. (= *SVF* II, 283, first part).

103 Cf. Epict. *Diss.* II, 19, 9 (= *SVF* II, 283, second part): γέγραφε δὲ καὶ Χρύσιππος θαυμαστῶς ἐν τῷ πρώτῳ περὶ Δυνατῶν.

104 On the equation of God and cosmos in early Stoic thought see D. L. VII, 137 (= *SVF* II, 526), Ar. Did. fr. 31 (= *SVF* II, 527) and Ar. Did. fr. 29 (= *SVF* II, 528).

105 ... καὶ ὁ θεὸς τοῖς Στωικοῖς ἐστι σῶμα, οὐκ αἰδομένοις λέγειν αὐτὸν τρεπτὸν καὶ δι' ὅλων ἀλλοιωτὸν καὶ μεταβλητὸν καὶ ἁπαξαπλῶς δυνάμενον φθαρῆναι, εἰ ἔχει τὸν φθείροντα· εὐτυχοῦντα δὲ μὴ φθαρῆναι παρὰ τὸ μηδὲν εἶναι τὸ φθεῖρον αὐτόν. Cf. also *In Evang. Ioann.* 13, 21 (= *SVF* II, 1054).

106 Cf. also the argument at *Tim.* 41b on the *aphtharsia* of the Gods.

cosmos—as a good and provident God. It might even be argued that, similarly to what is the case in the sphere of human ethics, also God's responsibility for the indestructibility of the cosmos could only be claimed if it were shown that the destruction of the cosmos was still in a sense possible.[107]

We may conclude that the two quotations from Chrysippus' *Peri dunatôn* preserved by Plutarch can, *pace* Hahm, be squared with what is otherwise known about the Stoic theory of modalities. Moreover, *pace* Plutarch (his second contradiction), they can also be squared with what is otherwise known about the Stoic doctrine of the inertness of space (*topos, chôra,* or *kenon*). Chrysippus does not imply that there could be within the Stoic system—i.e. *ceteris paribus*—an independent spatial centre with a dynamics of its own in attracting cosmic matter. The possibility of the cosmos being destroyed by the 'attractive' influence of an independent centre in space is merely a *logical* possibility. However, as things actually are in the Stoic physical world, the fact that the cosmic substance occupies the middle position is 'a concomitant (συντέτευχεν) of substance from the very fact that it is the kind of thing it is' (εὐθὺς τοιάδε τις οὖσα). In other words, in the actual physical world of Chrysippus space does get its determinations

---

[107] The statement that there is nothing to destroy the cosmos reminds us of Aristotle's argument *ex hypothesi* in his exoteric *Peri Philosophias* (fr. 19a), which argues that the cosmos is indestructible by virtue of the absence of both an internal and an external cause of destruction. We may add the testimony of Ar. Did. fr. 37 (= *SVF* II, 599, *in fine*) which informs us (after a discussion of the conflagration) that, according to the Stoics, the οὐσία of the cosmos as well as its διοικῶν are indestructible because the cosmos will not destroy itself, while there is no external factor to cause its destruction either (οὔτε αὐτῆς ἐξ αὐτῆς οὔτε ἔξωθέν τινος ἀναιρήσοντος αὐτὴν). In his pioneering 'Providence and the Destrucion of the Universe in Early Stoic Thought' (1979), Mansfeld has persuasively argued that the ideas of Zeno and other early Stoics concerning the theodicy were formed under the sway of the arguments put forward in this connection in Plato's *Timaeus* and Aristotle's *Peri Philosophias.* However, I think that Mansfeld's thesis that Zeno's theory of eternal recurrence should be regarded as a critical alternative to both Plato's and Aristotle's views should at least be qualified. One of Mansfeld's main claims, viz. that the theory of conflagrations and eternal recurrence allowed the early Stoics to defend the idea that the conflagration which ends every singular *diakosmêsis* is the best state for the world, has already been challenged by Long (1985). I would like to point out, in addition, that the texts here discussed seem to suggest that for the Stoics the problem of the theodicy primarily concerned the cosmos in the broader sense of the word (i.e. the cosmos *qua* οὐσία, or *qua* God) and that in this respect they subscribed to, rather than criticized, the arguments of Plato and Aristotle.

from the cosmos and in this way the cosmos so to speak always creates its own middle position.[108] It seems, then, that by thus taking the context of modal logic into account in interpretating the quotations from *Peri dunatôn*, we can show that the second contradiction detected by Plurtarch is harmless. From this point of view, then, there are as yet no reasons to regard the quotations as either confused or non-Chrysippean in character.

6.2.6 The first contradiction of which Plutarch accuses Chrysippus is yet to be discussed: how could Chrysippus both maintain that the void has no up, down, or centre, and that the cosmos rests at the centre of space? If we may for the moment disregard the question how Chrysippus actually called the space of which the cosmos took the central part, the following observations may be made.

As I noted, Chrysippus held that the extracosmic infinite void can have no spatial determinations *of its own*. This, according, to Plutarch, is exactly why he objected against the Epicurean theory of the downward motion of the atoms through the void. The same emphasis on what I would like to call the *isotropy* of the void is to be found in Cleomedes *Cael.* I, 1, 150–153 Todd, where it is argued that 'being infinite and incorporeal [the void] has no up nor down, nor front nor rear, nor right nor left, nor centre. For those directions are found in the case of bodies'.[109]

This however, does not necessarily preclude the possibility that Chrysippus thought that the spatial determinations of the cosmos itself could be extended to the surrounding void. In that case also an infinite void might be said to acquire certain spatial determinations, by virtue of its containing a unique spherical cosmos

---

108 Note that the meaning of the words συντέτευχεν and συντυχία in the second quotation can be inferred from the phrase συντέτευχεν ἀιδίως τὸν μέσον κατειληφυῖα τόπον εὐθὺς τοιάδε τις οὖσα. I have opted for using the word 'concomitant' in my translation (rather than Cherniss' 'accident') because it clearly illustrates that the position of the cosmos is *inextricably* bound up with its internal structure. It is obvious that Plutarch's glossing of συντέτευχεν as 'happened accidentally' (συμπέπτωκεν αὐτομάτως, *SR* 1055d) and as 'by coincidence' (ἐκ τύχης 1055a) is a polemical distortion.

109 *Cael.* I, 1, 150–153 Todd: ἄπειρον τοίνυν ἅμα καὶ ἀσώματον ὄν οὔτε ἄνω ἄν τι ἔχοι, οὔτε κάτω, οὔτε ἔμπροσθεν οὔτε ὄπισθεν, οὔτε ἐκ δεξιῶν οὔτε ἐξ εὐωνύμων, οὔτε μέσον· αὗται γὰρ αἱ σχέσεις ἐπτὰ οὖσαι περὶ σώματα θεωροῦνται. For the connection between the isotropy of space and the immobility and coherence of the cosmos see Algra (1988a) 158–159.

which, so to speak, projects its own spatial determinations (up, down, centre) into the void. In fact, since there is no other way of making sense of Chrysippus' words, we shall have to draw the general conclusion that on his line of thought there was a difference between the void *qua* empty space *as such*, and the void *qua* space surrounding the cosmos. Yet this leaves open several ways in which this conclusion may be further fleshed out.

Perhaps the most obvious one is to argue that the void is only isotropic when regarded in and by itself, not when regarded as surrounding the unique Stoic cosmos. Chrysippus' argument against Epicurus—in whose physical world there is no unique spherical cosmos imparting its own determinations to the void—could in that case be explained as a mere *ad hominem* argument.[110] However, a serious objection against such an interpretation would be that it is not so much because it is *void* that the void is said to exhibit no spatial differentiation, but rather because it is *infinite*.[111] In fact Plutarch *SR* 1055c (= *SVF* II, 550) provides us with another *verbatim* quotation from Chrysippus in which he argues that an *intracosmic* void, if real (*quod non*), would exhibit the same spatial differentiations as the cosmic body.[112] Now it might still be thought that the idea behind this was that the void is only *in itself* infinite, but that it 'loses' its infinity when partly occupied by a cosmos. However, also this, I believe, won't do. For Chrysippus not only maintained that void in itself is infinite, but also that the void which *de facto* extends outside our cosmos is infinite.[113] Moreover, also the sum total of cosmos and void, *to pan*, was thought to be infinite.[114] So our second conclusion may be that, unless Chrysippus was contradicting his own position, he must have meant that the space of which the cosmos occupies the central part is a *finite*

---

[110] This view seems to have been endorsed by Goldschmidt (1953) 29–30, and 43.

[111] Cf. Plu. *SR* 1054b ὅτι τοῦ κόσμου κενὸν ἐκτὸς ἄπειρόν ἐστι τὸ δ' ἄπειρον οὔτ' ἀρχὴν οὔτε μέσον οὔτε τελευτὴν ἔχει πολλάκις ὑπ' αὐτοῦ λέγεται. Cleom. *Cael.* I, 1, 150–153 Todd, adduces both the void being ἀσώματον and its being ἄπειρον as reasons for its not exhibiting any spatial differentiation.

[112] This text is quoted, and briefly discussed, in the appendix to the present chapter, below, 338–339.

[113] Cf. Ar. Did. fr. 25 (= *SVF* II, 503 = text (1) in section 6.1. of this chapter): τὸ μὲν οὖν κενὸν ἄπειρον εἶναι λέγεσθαι· τὸ γὰρ ἐκτὸς τοῦ κόσμου τοιοῦτ' εἶναι ...etc.

[114] Cf. S. E. *M.* 9, 332 (= *SVF* II, 524). On the possibly Chrysippean origin of the notion of τὸ πᾶν see above n. 34.

space. In this connection we may note that the words ἐν ἀπείρῳ only occur in Plutarch's polemical paraphrase and that Chrysippus himself does not in any way imply that that of which the cosmos occupies the centre is infinite.

As a third step, we may now tentatively connect this second conclusion with our previous observations on the Chrysippean notion of *chôra*. For apart from not speaking of the cosmos being at the centre of an *infinite* space, the Chrysippean quotations in Plutarch do not even use the word *kenon* to denote this space. In the first quotation Chrysippus speaks, rather enigmatically, of the indestructibility of the cosmos being secured by ἡ τῆς χώρας κατάληψις, οἷον διὰ τὸ ἐν μέσῳ εἶναι. To some extent I have of course read my own interpretation into the text by translating 'the part of *chôra* that it has occupied', and it should be admitted that the Greek phrase is ambiguous. It may mean either that what the cosmos *de facto* occupies is *chôra*, and that it is in the middle of something else, or that it occupies a *part* of *chôra* and that it is in the centre of this same *chôra*. Plutarch's paraphrase at *SR* 1054 is also ambiguous in this respect. For at 1054d he speaks of an ἐν ἀπείρῳ *μέσον* τινὰ *τόπον* καὶ *μέσην χώραν* which might seem to suggest the former interpretation.[115] However, at *SR* 1054e he paraphrases Chrysippus as stating that the parts of the cosmos would move towards the centre of the *chôra* which surrounds (or encompasses) it (ἐπὶ τὸ μέσον ... *τῆς περιεχούσης* τὴν οὐσίαν *χώρας*).

Now, as we saw in the previous section, Chrysippus' definitions of *chôra* seem to suggest that the *chôra* of the cosmos should be regarded as a larger but finite space encompassing the cosmos but being able to contain 'more of it'. If we assume that it is this *chôra* which Chrysippus had in mind in the text quoted by Plutarch, we can disprove the alleged self-contradictory nature of Chrysippus' thought. In that case it is Plutarch who distorts Chrysippus' original intention in order to be able to construct the required contradiction. He does so by glossing Chrysippus' ἡ τῆς χώρας κατάληψις not only (correctly) by the μέσον ... τῆς χώρας but also (wrongly) by the words ἐν ἀπείρῳ (1054 d) and ἐν τῷ κενῷ (1054

[115] Note, however that in ordinary Greek idiom the expression μέση χώρα could also mean 'the centre *of* χώρα' (cf. *LSJ*, *s.v.*. μέσος, I, a), in which case this phrase could also be translated 'a middle place in the infinite and a centre *of space*'.

e), thus suggesting that the cosmos is said to be at the centre of the (infinite) void. If this admittedly conjectural interpretation is accepted, we may conclude that the finite space of which the cosmos occupies the centre, and which is *de facto* bounded by the working sphere of the *tonos* of the cosmos, was something for which Chrysippus could apply the separate spatial concept of *chôra*.

Such an application of the concept of *chôra* might seem odd and in some respects it certainly is odd and problematic. First of all, do we have to assume that when the cosmos itself expands in the act of a conflagration its *chôra* expands as well? This is what one would naturally infer from Chrysippus' statement in the second quotation that the cosmos is *eternally* at the centre. But perhaps this is only a manner of speaking intended to cover both the actual or usual situation in which the cosmos is indeed at the centre of a larger spherical *chôra* and the eventual situation in which the cosmos will temporarily fully occupy this space (which after all will still be spherical). This brings us to a more serious drawback of this application of the concept of *chôra*. For where does this *chôra* end and where does the rest of the infinite void begin? It is perhaps because of such perplexities that other Stoics preferred to take *chôra* as just the infinite sum total of space. Nevertheless, given the fact that according to Chrysippus only something finite could exhibit determinations, and given the additional fact that Chrysippus thought, as we saw, that also a counterfactual *internal* void space would have to be included in the working sphere of the cosmic *tonos*, it is on second thoughts perhaps not that odd that he considered the possibility that the spatial determinations of the cosmic sphere also extend to part of the extracosmic void.

Yet even if such a 'cosmic' application of the Chrysippean concept of *chôra* would thus in itself seem to make sense, it will have to be conceded to Plutarch that the way Chrysippus apparently used it to prove the stability of the cosmos is odd. After all, he did not really need it. The Stoic theses that all changes of the cosmos will have to be caused and determined by the cosmos itself, that the cosmos moves towards its own centre, and that incorporeals (including space) cannot act as a cause, were together sufficient to establish the coherence of the cosmos in the infinite void.[116] This

---

116 On which see Algra (1988a) *passim*.

also may have induced later Stoics to drop the Chrysippean concept of *chôra.*

6.2.7 At this point it should be stressed that the interpretation of the two Chrysippean quotations which has been developed in the previous subsections is not entirely novel. Indeed, that some such interpretation as the one here provided is needed, has been felt by scholars such as Sambursky and Furley.[117] To quote Sambursky: '(the quotations) .... actually refer to the spherical shape of the cosmos, i.e. the symmetrical grouping of the material world around a central point'. However neither Furley nor Sambursky further worked out this suggestion. Accordingly, their apodictic statement was an easy target for Hahm who argued that 'it is hard to see how anyone can think that with these words Chrysippus refers to motion towards the middle of the cosmos or to the even distribution of matter around its centre'.[118] I hope that the present section has shown that the two Chrysippean fragments make sense and are coherent in a way much resembling that proposed by Sambursky and Furley.

Inevitably, of course, much of what has been said, and in particular my interpretation of Chrysippus' concept of *chôra,* and of its application at a 'cosmic scale', remains speculative to a certain extent. One thing, however, seems to have been fairly securely established. Chrysippus thought that the spatial determinations of the cosmos were at least to a certain extent also projected on the extracosmic space. Regardless whether this resulted in the conception of an infinite spherical space or in the conception of a finite spherical space, and whether or not in that case it was called *chôra,* the fact remains that here we witness a curious feature of the early Stoic theory of space. On the one hand space or void considered as such—being incorporeal—is isotropic and inert. On the other hand, in so far as it is considered as dependent on the cosmos it takes over its spatial determinations. This raises the question whether there actually *is* such a thing as 'space as such' or 'void as such'. This question can only be answered by relating the Stoic theory of space to Stoic ontology (and epistemology) in general. This is what I shall try to do in the next section.

---

117 Sambursky (1959) 112; Furley (1966) 21.
118 Hahm (1977) 260.

6.3 *The ontology of place and void in early Stoic thought*

6.3.1 In the present section I shall investigate, in so far as the fragmentary nature of the evidence permits, how the Stoic theory of space (*topos, chôra, kenon*) was embedded in Stoic ontology and epistemology. I shall first try to make clear what exactly the Stoics thought to be the ontological status of place and void and how they conceived of the way in which these incorporeals were related to physical reality. This will inevitably involve some excursions into the difficult and much discussed problem of the relationship between language, thought and reality in Stoic philosophy. After a discussion of the epistemological and semantic aspects of the Stoic concept of *topos* in 6.3.2, I shall turn to show in 6.3.3 that the required physical role of the void (*kenon*) and, in fact, of *space as such*, forced the Stoics to endow it with an exceptional status among the incorporeals.

6.3.2 Stoic ontology sharply distinguishes corporeal being from incorporeals. Both body and incorporeals belong the general class of 'somethings' (*tina*).[119] To be a something probably means nothing more than 'to be part of the world as we conceive it and to be a proper subject for discourse'.[120] Both place and void are listed as incorporeals.[121] For the rest, the class of incorporeals comprises a rather heterogeneous number of things which, though not being 'real', i.e. corporeal, are nevertheless an integrate part of the world or at least of the way in which we conceive of the world. There is some disagreement on what exactly is to be included in the class of incorporeals, but at least the inclusion of four items—sometimes referred to as the 'canonical' incorporeals—is undisputed: *lekta,*

---

[119] The most comprensive recent discussion of the Stoic theory of the 'something' as the supreme genus is Brunschwig (1988), of which an English version is now printed in Brunschwig (1992) 92–158 (but references here will be to the original French version). For a critique of some aspects of Brunschwig's reconstruction, see Mansfeld (1992) 100–102. On the heterogeneous character of the class of Stoic incorporeals see Brunschwig (1988) 90–109, esp. 91.

[120] It is now usually agreed that general concepts (the equivalents of Platonic Ideas) as well as non-referential objects of thought do not pass this test, and that the Stoics excluded them from the class of ἀσώματα to subsume them under the heading of οὔτινα ('nothings' or 'not-somethings'), on which see Hadot (1968) 161 ff; Wurm (1973) 178 ff; Pasquino (1978) 375 ff.; Long/ Sedley (1987) I, 162–166; Brunschwig (1988) 22 and 40.

[121] Cf. S. E. *M.* 10, 218 (= *SVF* II, 331).

place (*topos*), void (*kenon*) and time.[122] The statement 'there are *asômata*' does not involve any ontological implications in a strict sense—i.e. it does not imply that the *asômata* are beings in a strict sense (i.e. corporeal). They do not exist, but only subsist (ὑφίσταναι). This, of course, immediately raises the question what they are.

In so far as they might seem to be mere objects of thought (*noêmata*) without having an independent existence in the outer world, it might be tempting to treat them, in a subjective-idealist fashion, as *our* contribution, as the ordering schemes governing our thinking about actual reality. In this way scholars have more than once endowed the ancient Stoic theory of place and space with subjective-idealist features. Thus we find statements like: 'What they mean by calling space, place and void and time incorporeals seems clear enough when we accept their postulates ... talk about space, place and void is our contribution' or: '... void is a construct of the mind'.[123] However, it can be shown that at least for the four 'canonical' incorporeals such an 'idealist' interpretation is untenable. Two observations are useful in this connection. First, it is worthwhile to take account of the way in which the early Stoics used the term ὑπάρχειν in connection with the incorporeals, because this informs us of the ontological status which they were willing to accord to the latter. Secondly, also their own accounts of the formation of concepts and of the particular way in which we arrive at the concepts of incorporeals clearly shows that according to the Stoics the *asômata* are more than just 'impositions of the mind'.

Our sources indicate that the Stoics used the term ὑπάρχειν at least in connection with both *lekta* and time. *Lekta* which are truly predicated (like 'περιπατεῖ' in the *axiôma* 'Δίων περιπατεῖ', when at the moment of utterance Dion is actually walking) were thought to correspond with a part or aspect of material reality. As has been plausibly shown by Long and Graeser, the term ὑπάρχειν is used to indicate precisely this correspondence between an incorporeal —e.g. a particular *lekton*—and something in corporeal reality. This means that the word ὑπάρχειν which is usually equivalent to

[122] On other possible candidates (notably ἐπιφάνεια) see Brunschwig (1988) 28–30.
[123] Watson (1966) 38 and 86. See also Bréhier (1907) 50.

εἶναι, has, when applied to *lekta*, a truthfunctional rather than an ontological meaning.[124] It is of course in virtue of this very correspondence between *lekta* and corporeal reality that, whatever they are, they cannot be said to be mere constructs of thought. A similar use of ὑπάρχειν also plays a role in the early Stoic philosophy of time, as is evident from Arius Didymus, fr. 26 Diels (Stobaeus, *Ecl.* 1, 106, 5 = *SVF* II, 509 = LS 51B), where the following view on time is ascribed to Chrysippus:

> He also says that only the present belongs (ὑπάρχειν); the past and the future subsist (ὑφεστάναι), but belong in no way, just as only predicates which are [actual] attributes are said to belong, for instance, walking around belongs to me when I am walking around, but it does not belong when I am lying down or sitting.[125]

Here also we may conclude that the subjective-idealist interpretation—apparent, for instance, in Christensen's statement that time and space, as well as the other incorporeals, 'can be conceived as meanings or meaning-constituents, to which man has assigned a name, signifying but not denoting'—is inadequate.[126] The notion of correspondence which is here invoked, and which is expressed by the term ὑπάρχειν, involves obvious ontological implications. Apparently the non-corporeal character of time is in itself not sufficient to make us conclude that it is just an 'impositions of the mind'. As is also shown by its definition ('the *diastêma* of the motion of the cosmos'), time does not depend on our experience or our consciousness. Rather, it seems to be an abstraction of an aspect of corporeal reality. We shall presently see that the remaining two canonical incorporeals, place and void, are probably to be regarded in a similar way.

This brings us to the subject of concept-formation. The (*ennoêma*) of an incorporeal must be the intentional object of a 'conception' (*ennoia*), which, as a mental state, is corporeal.[127] Such an *ennoia* has to be a *phantasia*.[128] To the possible objection that *asômata*

---

124 Cf. Long (1971) 89–90; Graeser (1971) 299 f.

125 Μόνον δὲ ὑπάρχειν φησὶ τὸν ἐνεστῶτα, τὸν δὲ παρῳχημένον καὶ τὸν μέλλοντα ὑφεστάναι μέν, ὑπάρχειν δὲ οὐδαμῶς φησιν. ὡς καὶ κατηγορήματα ὑπάρχειν λέγεται μόνα τὰ συμβεβηκότα οἷον τὸ περιπατεῖν ὑπάρχει μοι ὅτε περιπατῶ, ὅτε δὲ κατακέκλιμαι ἢ κάθημαι οὐχ ὑπάρχει ... (transl. Long/ Sedley).

126 Christensen (1962) 25.

127 On the use of 'concept' and 'conception' as translation of ἔννοια and ἐννόημα see Long/ Sedley (1987) I, 182

128 Cf. Plu. *Comm. Not.* 1084f.: φαντασία γάρ τις ἡ ἔννοιά ἐστιν and esp. D.

cannot act on a body and that it is for that reason difficult to see how an *asômaton* like place could cause a *phantasia*, the Stoics had an interesting reply which is attested by Sextus *M.* 8, 409–410 (= *SVF* II, 85 = LS 27E):

> For they [the Stoics] say, just as the trainer or drill-sergeant sometimes takes hold of the boy's hands to drill him and to teach him to make certain motions, but sometimes stands at a distance and moves to a certain drill, to provide himself as a model for the boy—so too some impressions touch, as it were, and make contact with the commanding-faculty to make their printing in it, as do white and black, and body in general; whereas others have a nature like that of the incorporeal sayables (*lekta*), and the commanding-faculty is impressed *in relation to them*, not *by* them.[129]

Though—as Sextus (*ibidem*) is only too quick to retort—the analogy does not go on all fours, the upshot of the argument is more or less clear. The *phantasiai* of incorporeals come about in an *indirect* manner, not *by* the incorporeals, but only 'in accordance with' or 'in relation to' them. The same question is treated, this time not from the perspective of the origin of mental states (*phantasiai*) but from the perspective of concept-formation (*nooumena*), by Diogenes Laërtius VII, 53 (= *SVF* II, 87). Diogenes—or rather Diocles of Magnesia, his probable source—there offers a list of mental operations by means of which concepts are established (e.g. analogy, abstraction). About both *topos* and *lekta* it is said that they νοεῖται κατὰ μετάβασιν which I would translate: 'they are conceived of by way of abstraction (or: transition)'.[130] In the case

---

L. VII, 51, where of all φαντασίαι some are said to be αἰσθητικαί, directly issuing from sense perception, whereas others are διὰ τῆς διανοίας καθάπερ τῶν ἀσωμάτων καὶ τῶν ἄλλων τῶν λογῷ λαμβανομένων.

[129] S. E. *M.* 8, 409–410: ὥσπερ γάρ, φασίν, ὁ παιδοτρίβης καὶ ὁπλομάχος ἔσθ' ὅτε μὲν λαβόμενος τῶν χειρῶν τοῦ παιδὸς ῥυθμίζει καὶ διδάσκει τινὰς κινεῖσθαι κινήσεις, ἔσθ' ὅτε δὲ ἄπωθεν ἑστὼς καί πως κινούμενος ἐν ῥυθμῷ παρέχει ἑαυτὸν ἐκείνῳ πρὸς μίμησιν, οὕτω καὶ τῶν φανταστῶν ἔνια μὲν οἱονεὶ ψαύοντα καὶ θιγγάνοντα τοῦ ἡγεμονικοῦ ποιεῖται τὴν ἐν τούτῳ τύπωσιν, ὁποῖόν ἐστι τὸ λευκὸν καὶ μέλαν καὶ κοινῶς τὸ σῶμα, ἔνια δὲ τοιαύτην ἔχει φύσιν, τοῦ ἡγεμονικοῦ ἐπ' αὐτοῖς φαντασιουμένου καὶ οὐχ ὑπ' αὐτῶν, ὁποῖά ἐστι τὰ ἀσώματα λεκτά. Translation: Long/ Sedley.

[130] Bréhier (1907) 38 wrongly translates: 'par la transition' referring to the fact that 'l'intervalle peut être constitué tantôt par un corps tantôt par un autre ... le lieu (ainsi que les exprimables) est conçu par la transition (μετάβασις). Le lieu est le point de passage de plusieurs corps qui s'y succèdent comme un συνήμμενον est le passage d'une proposition à une autre'. Much better: Watson (1966) 25: 'transference from things clear', or Christensen (1962) 58: 'transferring or transforming presentations to other fields of

of *lekta* it is fairly obvious that this means that a, physically speaking, inseparable aspect of corporeal reality is separated and isolated from its 'material substratum', e.g. Dion's walking from the material Dion-in-a-state-of-walking. Analogously, the concept of *topos* (*topos* being in itself mere dimension) may have been thought to be arrived at by abstracting the dimension from a three-dimensional body.[131]

If concepts like *topos* and *kenon* are thus arrived at by a *phantasia*, the Stoic criterion of truth can be applied: in the case of each particular *phantasia* we can ask whether it is a καταληπτικὴ φαντασία.[132] Such a καταληπτικὴ φαντασία is best described as a *phantasia* which by its own nature is such as to command assent (*synkatathesis*). As such it *ipso facto* has to correspond with something in physical reality. Now the way this assent is given differs. In some cases it is directly induced by sensation, in others a more complex mental act is required: in the latter case, the *phantasia* has to be measured against other *phantasiai*, by *logos*.[133]

It is not immediately clear in which of the two ways assent to the *phantasia* of *topos* is given. On the basis of the analogy, drawn in our sources between the formation of the concept of *topos* on the one hand and of *lekta* on the other, it could perhaps be plausibly inferred that, since it is part and parcel of a *phantasia* of a body

---

inquiry'.

[131] In that case place would clearly be non-absolute. It is tempting to add the testimony of Iamblichus preserved by Simp. *In Cat.* 361, 10–12 Kalbfleisch (= *SVF* II, 507 = Iamb. fr. 112 Dalsgaard Larssen) where we read that according to the Stoics place παρυφίσταται τοῖς σώμασιν. On closer view, however, we appear to be dealing with one lemma of a *diaeresis* made against the background of Iamblichus' own, typically neoplatonic, 'intellectual theory of place', according to which place is taken to be 'formative place': it has a form-like priority to sensible substances. From this point of view the statement that the alternative conception of place (rejected by Iamblichus) takes it that place παρυφίσταται τοῖς σώμασιν need imply no more than that on this conception place is *not* formative and prior to substances. It could just mean that place, like body, exists at the non-intelligible level, in which case the statement could also cover e.g. the Epicurean conception of (absolute) place. On the passage in Diogenes Laërtius see now Long/ Sedley (1987) I, 199–202, esp. 200.

[132] See e.g. S. E. *M.* 7, 247 ff. (= *SVF* II, 65 = LS 40 E) for definitions of the καταληπτικὴ φαντασία.

[133] Cf. S. E. *M.* 7, 37 (= *SVF* II, 107): ἔοικε .... τῷ δὲ ζυγῷ καὶ κανόνι ἡ αἴσθησις καὶ ἡ διάνοια δι' ἧς γίνεται τὰ τῆς κρίσεως and D. L. VII, 52 (= *SVF* II, 84): ἡ δὲ καταλήψις γίνεται ... δι' αἰσθήσει μὲν ... λογῷ δὲ τῶν δι' ἀποδείξεως συναγομένων, ὧσπερ τὸ θεοὺς εἶναι.

(*sôma*) that the body is three-dimensional,[134] the kataleptic *phantasia* of *topos* presents itself together with the kataleptic *phantasia* of a particular body. But it may also be that the Stoics thought that in this case assent depends on an elementary *logos*. Such a *logos* could be provided in the form of some arguments adduced by Cleomedes. In what purportedly is a series of proofs for the existence of the void, although it actually deals with place or space,[135] he argues that we recognize that in a vessel filled with water and a stone, the water will fill up the *topos* of the stone as soon as the stone is taken out. This same phenomenon is witnessed, he adds, in the case of air.[136] In this way, in other words, we may be induced to assent to the *phantasia* of an *independent* place. It should be noted, however, that in that case the analogy with the *lekta* does not seem to hold any longer, for a *lekton* is not (or: does not subsist as) an independent something in physical reality. It should also be noted that in that case place appears to be quite unlike time in that the latter is given a clearly derivative subsistence: it is the 'length' (*diastêma*) of the motion of the cosmos. There is no such thing as time in itself, independently of the corporeal cosmos.

The problem which we are here confronted with is the very one we encountered previously, when dealing with Aristotle's presentation of *topos* as an item in the category of quantity. In chapter 4, I argued in this connection that it was virtually impossible to make sense of the conception of *topos* as a property. Yet it is precisely some such conception which seems to be required by Stoic ontology. When regarded from this perspective, Stoic place, being three-dimensionally extended, has to be an aspect of—and as such has to be ontologically dependent on—corporeal reality. In other words it has to belong to what in chapter 1 was called the (a) kind of place. Nevertheless, if the concept of place is to play any role in the explanation of locomotion, a concept of place as an *independent* three-dimensional extension (our (c) concept of place) seems to be what is required. That the Stoics gave in to the latter requirement appears not only from the passages in Cleomedes I just referred to, but also from the way in which *topos* was actually

[134] Note that the common Stoic definition of corporeal being was τὸ τριχῇ διαστατὸν μετὰ ἀντιτυπίας. (Cf. Galen *De Qual. Inc.* 1, vol. XIX, 464 Kühn = *SVF* II, 502).

[135] *Cael.* I, 1, 25–33 Todd.

[136] *Cael.* I, 1, 33–38 Todd.

defined. It was defined as a subspecies of space (τὸ οἷόν τε κατέχεσθαι ὑπὸ σώματος), viz. as space *de facto* occupied by body. The formula opted for ('that which can be occupied by being and which is *de facto* occupied')[137] suggests that place was conceived of as *independent* of the body which occupies it.

It could be suggested that the Stoics thought these two perspectives were in the end compatible in so far as the cosmos is regarded as a whole. The parts of the cosmos may change their individual places, which are nevertheless—because always *de facto* occupied by body—inseparable from the bodily mass of the cosmos. In the case of the void, however, such an escape cannot be provided. There we witness a definite shift towards a kind of 'absolute' space.

6.3.3 When discussing the ontological status of the void, one would like to begin by studying the way the Stoics thought the concept of void was acquired. Unfortunately, however, we do not have any explicit description of this process of concept-formation. Yet it is perhaps legitimate to infer from the (purely negative) characteristics the void is usually invested with, that among the mental operations listed in Diogenes Laërtius (see previous subsection) 'privation' (στέρησις) is the most likely candidate. Being is, so to speak, stripped of all attributes save extension.[138] However this may be, whether our *phantasia* of the void also reveals an aspect of reality is a question which cannot possibly be decided by a simple appeal to sense experience. The statement 'there is void' obviously belongs to the class of *phantasiai* which has to be measured by *logos*. As such, it is not dissimilar from the example adduced by Sextus in connection with the group of *phantasiai* which are to be judged by *logos*, viz. 'that there are gods' (τὸ θεοὺς εἶναι).[139] Whether or not there are gods, and whether or not there 'is' a void—though the meaning of 'there is' might in each case be different—are questions which have to be decided by adducing external evidence, i.e. other indubitable facts. An instance of such a 'proof' of the existence of gods is provided by Cicero *ND* II, where

---

137 For the different versions of this definition see above, section 6.1.

138 Cf. Cleom. *Cael.* I, 1, 65 ff. Todd: ἔστι δὲ ἁπλουστάτη αὐτοῦ ἡ ἐπίνοια ἀσωμάτου τε καὶ ἀναφοῦς ὄντος, καὶ οὔτε σχῆμα ἔχοντος οὔτε σχηματιζομένου etc.

139 See above, n. 133.

it is first shown that there must be gods (esse deos, § 4 ff.) and subsequently what they are like (quales sint, § 45 ff.). In a rather similar way an instance of a Stoic proof of the 'existence' of an extracosmic void is provided by Cleomedes *Cael.* I, 1, 20 ff. Todd, where it is first argued that there is a void (εἰ ἔστι τὸ κενόν, *Cael.* I, 1, 20 ff. Todd) and subsequently that this void is infinite (ὅτι ἄπειρον τὸ κενόν, *Cael.* I, 1, 112 ff. Todd).

Cleomedes gives three arguments for the existence of a void. The first (*Cael.* I, 1, 20–38 Todd) actually amounts to a series of arguments for the existence of an independent place. This need not surprise us, since Cleomedes, as we saw in section 1, used the term *kenon* also to denote space as such. The second argument (*Cael.* I, 1, 39–43 Todd), is based on a thought experiment: we can imagine the cosmos moving from its place. The space into which it would thus move would have to be empty. The third argument provided (*Cael.* I, 1, 42–54 Todd) is the 'classical' one that, when there will be a conflagration, the cosmic mass will occupy more than ten thousand times the volume of the cosmos in its present state.[140] Since this space is at the present moment largely unoccupied, it is void. Cleomedes goes even further. He argues that even if one is not prepared to accept that a conflagration might occur, one still has to assume that there really is an infinite void extending outside the cosmos, for one could at least *imagine* a dilatation of the cosmic mass, and that into which the cosmos would expand would have to be void.[141]

Now from the perspective of ontology one might have expected a treatment of void as the *potential* extension of the cosmic body.[142] In this connection we might draw an analogy with the way past

---

140 See also Aët. II, 9, 2 (= *SVF* II, 609).

141 *Cael.* I, 1, 52–53 Todd: Αὐτὸ ἂν τοῦτο, εἰς ὃ τῇ ἐπινοίᾳ χωροίη κατὰ τὴν ἔκτασιν, κενὸν ἂν εἴη. This argument appears to be directed primarily against the Peripatetics, as is shown by the words that follow: ὅθεν οἱ λέγοντες ἔξω τοῦ κόσμου μηδὲν εἶναι φλυαροῦσιν κτλ.

142 An interpretation along these lines of Cleomedes' conception of the void has been provided by by Todd (1982). Todd argues that the void *has no extension* (!) and, if I understand him correctly, that as such it is a kind of infinite extensional potentiality. This interpretation, however, is based on a mistaken inference of the void having *no extension* from its having *no shape* (*op. cit.* 130). Moreover it is positively refuted by Cleomedes' statement that the void 'surrounds' the cosmos (*Cael.* I, 1, 99 Todd: τὸ μὲν *περιέχον* αὐτὴν [scil.: τὴν οὐσίαν]) as well as by the nature of the arguments against an intracosmic void (*Cael.* I, 1, 68 ff. Todd).

and future time were treated by Chrysippus. In fact the analogy seems to have been drawn by Chrysippus himself, to judge from Arius Didymus fr. 26 and from Plutarch *Comm. Not.* 1081 ff., where it is said that in his 'On the Void' (*Peri Kenou*) Chrysippus discussed the difference between the present, past, and future.[143] Just as past and future correspond to non-real temporal situations of the material cosmos, whereas only the present corresponds ('belongs') to its real temporal situation (even if only roughly speaking, κατὰ πλάτος),[144] so also the void might be taken to be the one-time spatial extension of the cosmos, whereas place (*topos*) is its actual spatial extension. However tempting this analogy might seem to be, I believe it does not hold for at least the following two reasons.

First, far from being just a potential attribute of the cosmos, something we can only think of but which is not yet real, the void is in fact always represented as if it were self-subsistent. It is there, extending infinitely outside the cosmos, awaiting, so to speak, its dilatation. Just like place it is a subspecies of space (τὸ οἷόν τε κατέχεσθαι ὑπὸ ὄντος), more precisely it is that part of space which is unoccupied. It thus definitely belongs to the furniture of the outer world and as such it seems to be accorded a quite privileged position among the incorporeals. The Stoics were apparently unable to imagine a conflagration with the concomitant dilatation of the cosmic sphere otherwise than as taking place into a space that was already there. They could in principle have taken refuge to the idea of a cosmos *creating its own larger extension* in the act of conflagration,[145] but such a conception would probably have been too unheard-of to be plausibly maintained.

As we saw, Cleomedes goes on to argue for the *infinity* of the extra-cosmic void. And here we run into a second problem. For there is yet another respect in which the analogy between space and time does not hold, viz. the way in which the infinity of either is motivated. The subsistence of an infinite past and of an infinite future can be traced back to the cosmos being—*qua* dynamical continuum—eternal. In this particular sense the cosmos has always existed and will go on existing forever so that

[143] The analogy has been worked out by Goldschmidt (1953) 39.
[144] Cf. Ar. Did., fr. 26 (= *SVF* II, 509).
[145] On the interpretation worked out along these lines by Todd (1982), see above, n. 142.

time, being defined as the extension of the cosmic motion (διάστημα τῆς τοῦ κόσμου κινήσεως),[146] is infinite. However, though the cosmos will at times increase its volume (conflagration) it will *never* actually be *spatially* infinite.[147]

It might be objected that the Chrysippean void, even if conceived as an actually subsistent extension, could nevertheless be regarded as an attribute of the cosmos, albeit as a *possible* attribute, in that it represents not the extension which *will* eventually be occupied, but only the extension which *can* be occupied ('can' here being used in the peculiarly Stoic sense of what we might call logical possibility).[148] This may indeed have been how Chrysippus *cum suis* saw things, witness the fact that the void was defined as 'that which can be occupied by being (τὸ οἷόν τε ὑπὸ ὄντος κατέχεσθαι), but which is not occupied'. Cleomedes also appears to have reasoned along these lines when he argued, in a statement referred to earlier, that even if one is not prepared to accept that a conflagration might occur, one still has to assume that there really is an infinite void extending outside the cosmos, for one could at least *imagine* a dilatation of the cosmic mass, and that into which the cosmos would expand will have to be void. Yet all this appears to be merely another way of putting the problem, for it leaves the infinite void with its exceptional status among the incorporeals.[149]

It is interesting to see by what arguments the early Stoics thought the infinity of the void could be defended. On the one hand this infinity was regarded more or less as a natural consequence of the reified status of the void. In other words, once an extra-cosmic void had been proved 'to be there', the most commonsensical thing to assume was apparently that this void was infinite. Indeed the evidence indicates that the (early) Stoics

---

146 Cf. *SVF* II, 510.

147 Also Brunschwig (1988) 100–101 points to this weakness of the analogy between void and time: 'il n'existe aucun temps "vide" qui serait au-delà des limites de celui qu'occupe le mouvement du monde'.

148 Cf. Bréhier (1907) 49: 'au fond le vide n'est qu' un attribut des corps, non pas un attribut réel, mais un attribut possible'.

149 We might ask why other potential attributes of the cosmos are not given a similar exceptional thing-like status. We may add, moreover, that it is difficult to conceive how, even in logical or conceptual sense of possibility, the Stoics could have maintained that it is possible for such a cosmos to be infinite in the first place—given the fact that their cosmos exhibits form and order.

subscribed to a kind of common sense argument which was also endorsed by other philosophers. According to this argument everything limited must have something next to it. Since there is nothing next to the void, the void cannot be limited.[150] In addition, however, we are fortunate to have what appears to have been a more technical argument, preserved by Arius Didymus (fr. 25 = *SVF* II, 503, quoted above, section 6.1, text 1, (4)), which attempts to defend the infinity of the void *not* in terms of common sense beliefs, but through specifically Stoic arguments. It claims that

> Just as the corporeal is finite, so the incorporeal is infinite; for time and the void are infinite. For just as the nothing constitutes no limit, so also is there no limit (or: is it no limit) to the nothing, e.g. to the void. For by its own nature (κατὰ γὰρ τὴν αὐτοῦ ὑπόστασιν) it is infinite; but it is being limited when it is filled up (περατοῦται δ' αὖ τοῦτο ἐκπληρούμενον); but when that which fills it is taken away, you cannot conceive of its boundary.[151]

What exactly is the gist of this (no doubt condensed) argument? The claim that only the corporeal is limited, whereas the incorporeal, *qua talis*, has no limit may be baffling at first sight. Place, for one thing, is most definitely limited. And if speaking of limited or 'unlimited' makes any sense at all in the case of λεκτὰ, it is unclear why λεκτὰ should be said to be unlimited (not even when this term is taken in the sense of 'indefinite'). It is probably for this reason that the thesis is immediately narrowed down to time and the void. Yet the scope of the argument arguably also includes place, for the idea is probably that the incorporeals place, void and time are *of themselves* without limitation and that any limit they may have is derived from the corporeal. This indeed seems to be what is supposed by the sentence dealing with place,

---

[150] Various versions of this argument as (allegedly) used by the Stoics are recorded by Simp. *In Cael.* 284, 28 ff. Heiberg (partly in *SVF* II, 535) and Cleom. *Cael.* I, 1, 112–149 Todd (partly in *SVF* II, 538). It should be noted, incidentally, that this evidence has been dismissed, to my mind on insufficient grounds, by Inwood (1991) 261–263. Alex. Aphr. *Quaest.* III, 101, 10 ff. Bruns attacks the argument as an (Epicurean?) argument for the infinity of τὸ ὄν, while explicitly adding that ὁ λόγος [...] τὴν πιθανότητα ἔχει παρὰ τῆς φαντασίας καὶ τῆς αἰσθήσεως. Versions of the argument are also to be found in Archyt. 47 A 24 DK; Eudem. fr. 65 Wehrli; Epicur. *Ep. Hdt.* 41.

[151] Ar. Did. fr. 25, last part (= *SVF* II, 503): [...] καθάπερ δὲ τὸ σωματικὸν πεπερασμένον εἶναι, οὕτως τὸ ἀσώματον ἄπειρον· ὅ τε γὰρ χρόνος ἄπειρος καὶ τὸ κενόν. ὥσπερ γὰρ τὸ μηδὲν οὐδέν ἐστι πέρας, οὕτως οὐδὲ τοῦ μηδενός, οἷόν ἐστι τὸ κενόν. κατὰ γὰρ τὴν αὐτοῦ ὑπόστασιν ἄπειρόν ἐστι, περατοῦται δ' αὖ τοῦτο ἐκπληρούμενον· τοῦ δὲ πληροῦντος ἀρθέντος οὐκ ἔστιν αὐτοῦ νοῆσαι πέρας.

which immediately precedes this quotation and which claims that τὸν δὲ τόπον πεπερασμένον διὰ τὸ μηδὲν σῶμα ἄπειρον εἶναι. In that case the argument may be reconstructed as follows: the void, if limited, is either (a) limited by itself, or (b) limited by an actually occupying body (in which case it would of course, strictly speaking, cease to be void). Since neither (a) nor (b), the void must be infinite.

In spite of the *prima facie* appeal of this argument—the main point in its favour of course being that it is hard to think of a bounding surface of the void—it is less powerful than Chrysippus would have us believe. Thus it might be objected that the disjunction 'either (a) or (b)' is in the case at issue here not as exclusive as seems to be supposed. Could it not be argued, on the principles of Stoic ontology and physics, that the void is limited by what in the event of a conflagration will be the limit (*peras*) of the cosmic body? After all, the void is the void *of* the cosmos, the receptacle needed for the eventual increase of the cosmic volume. Hence the cosmic body is in the end the *raison d' être* of the void. Why then shouldn't it also be the void's *raison d'être fini*?

It might in fact be asked whether Chrysippus' speaking instead of a void which is infinite *in its own right* (κατὰ τὴν αὑτοῦ ὑπόστασιν)—independently, that is, of any possible space-filler (τοῦ δὲ πληροῦντος ἀρθέντος)—did not bring him too close to the Epicurean conception of the void as a self-subsistent space. At any rate he appears to have credited the void with a rather exceptional status among the incorporeals. In spite of the fact that the Stoics would agree that void is *like place* in that both are subspecies of space (τὸ οἷόν τε κατέχεσθαι ὑπὸ σώματος), and in spite of the fact that Chrysippus apparently argued that void is in some respects *like time*, little stress was laid on the void's derivative status as an incorporeal. Instead—to judge from Ar. Did. fr. 25—Chrysippus presented the void *as if* it were something self-subsistent. By not regarding the void as simply and solely corresponding to the one-time extension of the cosmos (an extension which will be very large, but finite), but by taking it instead to be infinite *in itself* (κατὰ γὰρ τὴν αὑτοῦ ὑπόστασιν) and *qua incorporeal* (τὸ ἀσώματον ἄπειρον) he appears definitely to have severed the link between the void and the corporeal cosmos. The question is whether or not this rather exceptional, more or less self-subsistent, status of the void should be seen as a serious problem.

The answer to this question will depend on our view of what it means, ontologically speaking, to 'be' (or rather: to subsist as) an incorporeal. If we assume that, in view of the exclusivity of corporeal being, the mode of existence or subsistence of incorporeals should be a *derivative* one—as is indeed suggested by the way time and *lekta* (and perhaps also place) are being presented—a self-subsistent void will appear to be a strange element within the class of incorporeals. In that case we are once more faced (as we were also in the case of *topos*, discussed in the previous subsection) with conflicting claims of Stoic ontology on the one hand and Stoic physics on the other. Stoic ontology, on this interpretation, did not really allow for an *infinite* void as a reified feature of the outer world. Stoic physics, on the contrary, needed an extracosmic void as a necessary condition for the expansion of the cosmos at a conflagration. Quite in line with the common sense conception of space of the (c) kind (on which see chapter 1), the Stoics were apparently only able to think of such a void as 'being already there'. Once such a conception of void was admitted, it was, for the reasons outlined above, only natural to assume that it was infinite.

If, on the other hand, we assume that the criterion for being an incorporeal was just to be a part of the outside world (i.e. to enjoy some kind of extra-mental reality) while at the same time not being corporeal,[152] there is nothing wrong with the class of incorporeals comprising both (abstracted) aspects or attributes of corporeal reality (like time, or *lekta*) and independent non-corporeal things (like the void). In that case, however, one is struck by the utter diversity of the items contained in the class of incorporeals, and one might still ask whether a void subsisting *independently* of the corporeal cosmos—note, again, that Arius' account envisages the situation where the occupying body is thought away (τοῦ δὲ πληροῦντος ἀρθέντος)—would not be better at home in an ontology that could treat it as an existent in its own right, on a par with corporeal being.[153] I think, therefore, that we are justified in concluding that on both interpretations—i.e. on both conceptions of the class of incorporeals—the early Stoic

[152] This seems to be the opinion expressed in Brunschwig (1988) 76–77.

[153] In other words, one may well wonder whether it is reasonable to deny the void existence for the sole reason that it is not corporeal, and as such unable to act. Compare the equally incorporeal and inert Epicurean void which is nonetheless an existent *sui generis* next to the atoms.

conception of a self-subsistent infinite extra-cosmic void, as well as the arguments used to support it, were at least susceptible to doubt or criticism. In the next section I shall try to show that such doubts or criticisms may well have induced Posidonius to take a different stance on this subject.

6.4 *Posidonius and the Stoic theory of the extra-cosmic void*

6.4.1 We have seen that according to orthodox Stoic physics the unique spherical cosmos was surrounded by an infinite void, a void which had to be there for the cosmos to expand into at the periodical conflagration. We have also seen that, in addition, its infinite extent was proved by separate arguments, some of which are also familiar from other ancient philosophers. Our sources suggest, however, that at this point the Stoics were not altogether unanimous. For one thing, Aëtius II, 9 claims that Posidonius differed from the Stoics in general in holding that the void was just large enough to be occupied by the cosmos in the eventual conflagration. For another, chapter 8 of a work commonly referred to as Achilles' *Isagoge* (an introduction to Aratus) ascribes the tenet of a finite extra-cosmic void—quite cavalierly, it seems—to the Stoics in general.[154] The same ascription occurs in ch. 3 of yet another, anonymous introduction to Aratus, appended to the text of Achilles in the Codex Vaticanus 191, and printed by Maass as 'Anonymus I'.[155] These shreds of evidence present us with various problems. What value, for a start, should we attach to the testimony of Aëtius in the absence of any other evidence concerning Posidonius' alleged heterodox conception of the extra-cosmic void? This question has been given various answers. The majority of scholars took Aëtius' testimony at face value (using Diels' reconstruction of the text as a basis), and accepted Posidonius' heterodoxy without giving it much further thought.[156] Some added the rider that the view ascribed to Posidonius is odd and

---

[154] Cf. E. Maass, *Commentariorum in Aratum Reliquiae* (Berlin 1958[2]) (hereafter: *CAR*), 38 (partly printed as *SVF* II, 610).

[155] Maass, *CAR* 92–93 (not in *SVF*).

[156] Cf. E. Zeller, *Die Philosophie der Griechen in ihrer geschichtlichen Entwicklung*, III, 1 (Leipzig 1923[5]) 596 n. 3; A. Schmekel, *Die Philosophie der mittleren Stoa in ihrem geschichtlichen Zusammenhange dargestellt* (Berlin 1892, repr. Hildesheim 1974) 240; Sorabji (1988a), 129; Algra (1988b).

that we shall probably never know its background.[157] Other scholars, however, went further still and rejected the evidence of Aëtius because the tenet of a finite extra-cosmic void was thought to be 'intrinsically implausible' and because it is not attested anywhere else.[158] For these reasons Ian Kidd even proposed to emend the text in order to bring Aëtius' testimony on Posidonius in line with the demands of Stoic orthodoxy.[159] But also the testimonies of Achilles and Anonymus give rise to a number of questions. Most importantly, how do they relate to each other and to the testimony of Aëtius, and is their testimony of any relevance to the question of Posidonius' conception of the void? I have discussed these issues in detail elsewhere.[160] In the present section I shall confine myself to a brief presentation of the evidence (6.4.2) and an attempt to assess this evidence against the background of the early Stoic theory of space and of the void (6.4.3).

6.4.2 As a starting point in discussing the testimony of Aëtius on Posidonius I shall take the version of ps.-Plutarch (*Plac.* II, 9), which is more complete and provides the better text, although, as we shall see, some textcritical problems still remain. I here render the text as I have reconstructed it, largely on the basis of the Greek text of Mau's Teubner edition. For the sake of convenience I have numbered the lemmata and smoothened the translation into something like a neatly running text. The chapter is headed 'On what is outside the cosmos, whether there is void' (Περὶ τοῦ ἐκτὸς τοῦ κόσμου εἰ ἔστι κενόν):

---

[157] Cf. K. Reinhardt, *Poseidonios* (München 1921) 43: 'Aber was in diesen Dogmen wirkte, ob nur die Stellungnahme, ob nur die Vermittlung zwischen dem begrenzten Weltraum des Aristoteles und dem unendlichen des Chrysipp, oder noch etwas anderes: wie wollte man das erfahren?'. Also Theiler (1982) 179, in his commentary on fr. 302, speaks of a 'seltsame Auffassung ... mit der er der trivialstoischen Lehre vom ἄπειρον κενόν entgegentritt'. Note, incidentally, that the texts which Theiler here adduces as parallels are not really parallels. They speak of the very large, though not infinite, place, or part of the void, which will be occupied by the cosmos in conflagration; they do not speak of the extent of the void as such.

[158] Thus E. Müller, *De Posidonio Manilii auctore* (Leipzig 1901) 35 (*non vidi*), followed by Rehm (1922) and Pohlenz (1948–49) 108; also sceptical are M. Laffranque, *Poseidonios d'Apamée* (Paris 1964) 310–311; Kidd (1988) 391–394.

[159] Cf. Edelstein/ Kidd (1988^2), 102 (fr. 97 a *in app.*); and Kidd (1988) 393–394.

[160] See Algra (1993).

1. The followers of Pythagoras <say that> outside the cosmos there is a void in which the cosmos exhales and out of which <it inhales>.
2. The Stoics <say that outside the cosmos there is a void> into which <the cosmos> is dissolved at the conflagration; it is infinite.
3. Posidonius <maintains that> it is not infinite, but just as large as is sufficient for the dissolution <of the cosmos>. In the first book of the 'On the Void'.
4. Aristotle said that there is no void.
5. Plato <said that> there is neither a void outside nor inside the cosmos.[161]

There are almost identical versions of this chapter in Eusebius,[162] and Qosta ibn Luqa (i.e. the so-called Aëtius Arabus),[163] and a very garbled one in ps.-Galen,[164] all of them dependent on ps.-Plutarch. There is also an independent version of this Aëtian chapter in Stobaeus.[165] The Posidonius lemma has of course been included

---

161 1. Οἱ μὲν ἀπὸ Πυθαγόρου ἐκτὸς εἶναι τοῦ κόσμου κενὸν εἰς ὃ ἀναπνεῖ ὁ κόσμος καὶ ἐξ οὗ. 2. Οἱ δὲ Στωικοὶ εἰς ὃ κατὰ τὴν ἐκπύρωσιν ἀναλύεται [τὸ] ἄπειρον. 3. Ποσειδώνιος οὐκ ἄπειρον ἀλλ' ὅσον αὔταρκες εἰς τὴν διάλυσιν ἐν τῷ πρώτῳ περὶ κενοῦ. 4. Αριστοτέλης ἔλεγεν μηδὲν εἶναι κενόν. 5. Πλάτων μήτ' ἐκτὸς τοῦ κόσμου κενὸν εἶναι μήτ' ἐντός Some textcritical remarks: 2] τὸ ἄπειρον MSS et Aet. Arab. ('was unendlich ist'): ἄπειρον Eusebius; 3] ἀλλ' ὅσον MSS: καθ' ὅσον Kidd *in app.*; [ἐν τῷ πρώτῳ περὶ κενοῦ.] del. Diels, Mau; 4] [ἔλεγεν] del. Mau. Note, moreover, that Diels mistakenly read ἔλεγεν εἶναι κενόν (i.e. he missed the word μὴ) and commented (*Prol.* to *DG*, 9) that Plutarch 'Aristoteli ... indoctissime simul et audacissime contrariam ipsi sententiam attribuit ... atque ultima quoque misere turbavit'.

162 For information about the various witnesses of Aëtius and on the way in which they interrelate see now the convenient survey in the first pages of of Runia (1989). Eus. *PE* XV, 40, is headed Περὶ τοῦ ἐκτός and has the following version (I give the text as printed by Mras): Οἱ μὲν ἀπὸ Πυθαγόρου ἐκτὸς εἶναι τοῦ κόσμου κενόν, εἰς ὃ ἀναπνεῖ ὁ κόσμος καὶ ἐξ οὗ. Οἱ δὲ Στωϊκοί εἰς ὃ καί τῇ ἐκπυρώσει ἀναλύεται, ἄπειρον. Ποσειδώνιος οὐκ ἄπειρον ἀλλ' ὅσον αὔταρκες εἰς τὴν διάβασιν. Πλάτων, 'Αριστοτέλης μήτ' ἐκτὸς τοῦ κόσμου διάκενον εἶναι μήτ' ἐντός.

163 Qosta Ibn Luqa in Daiber (1980) 147: "Ueber das, was außerhalb der Welt ist, ob es ein Vakuum ist: (1) Die Anhängerschaft des Pythagoras glaubte, daß außerhalb der Welt ein Vakuum ist; in es und aus ihm atmet die Welt. (2) Die Stoiker glaubten daß außerhalb der Welt ein Vakuum ist, in welches sich auflöst, was unendlich ist. (3) Poseidonius glaubte, daß es kein Unendliches gibt, sonder soviel wie zur Auflösung benötigt wird. (4) Platon und Aristoteles glaubten daß es überhaupt kein Vakuum gibt, weder außerhalb noch innerhalb der Welt".

164 The garbled version of Ps.-Gal. *Phil. Hist.* 52 (Diels *DG*, 623) is headed Περὶ τοῦ ἐκτός and, after a lacuna which was apparently caused by *saut du même au même*, continues as follows: *** ἐκτὸς εἶναι τοῦ κόσμου, εἰς ὃ ἀναπνεῖν τὸν κόσμον καὶ ἐξ οὗ. οἱ δὲ Στωικοὶ εἰς ὃ ἀναπνεῖν μόνον. Πλάτων δὲ μήτε ἐκτὸς εἶναι τοῦ κόσμου μήτε ἐντὸς κενόν, ἐκτός δὲ μικρόν τι εἰς ὃ ἀναπνεῖν

165 Stob. *Ecl.* I, 18, p. 160, 8 Wachsmuth (= fr. 97 b Edelstein/ Kidd

(either in the ps.-Plutarchean version or in both versions) in collections of Posidonian fragments, viz. as frs. 84 and 97a Edelstein/Kidd, and fr. 302 Theiler.[166]

Of course we are only dealing with a doxographical testimony which in principle may give simplified or distorted information, and which may exaggerate differences of opinion, in this case between the orthodox Stoics and Posidonius. In fact not even the reference to the title ἐν τῷ πρώτῳ περὶ κενοῦ can be taken to lend additional credibility to the testimony of Aëtius.[167] And indeed, as I already indicated, for various reasons scholars have thrown doubt on the reliability of this testimony. One of them is Ian Kidd who in the commentary accompanying his edition of the Posidonian fragments even expresses the view that, for reasons which I shall presently discuss, Posidonius *cannot* have defended the heterodox position which the third lemma in this Aëtius chapter ascribes to him. He accordingly suggests that 'if ἀλλ' ὅσον [...] is changed to καθ' ὅσον the sentence makes sense: Posidonius said that the void outside the cosmos was not infinite *in so far as* being sufficient for the dissolution' (my italics). According to Kidd, in other words, Posidonius was committed to nothing more than the common Stoic dogma that the void was in itself infinite, but that the part of it eventually to be occupied by the cosmos at a conflagration was finite.[168] Kidd adds that 'the error occurred from the form οὐκ ἄπειρον which invited ἀλλ' ὅσον.'

This suggestion, however, is in fact an attempt to restore the wording *of Posidonius*. It is therefore rather unfortunate that he presents it in the form of an emendation of the text *of Aëtius*, or even merely of ps.-Plutarch's version of Aëtius (the suggestion

---

(1988²), not in Theiler). For a full discussion of the textcritical aspects and of the various witnesses for Aëtius' testimony see Algra (1993) 476–485.

[166] In fact, the text of this lemma is printed two times in the collection of Edelstein/ Kidd (1988²), viz. as fr. 84 (as a *dubium* among the fragments of named books in virtue of the reference ἐν τῷ πρώτῳ περὶ κενοῦ) and as fr. 97 a (i.e. among the fragments of unnamed books, because the parallel version of Stobaeus lacks the reference). Theiler has printed it (fr. 302) including the reference to the title.

[167] On the title, which is probably an interpolation and which may originally well have been ment to go with the lemma 'Aristotle', see Algra (1993) 478–480.

[168] I find myself unable to go along with Inwood (1991) 254–266, who argues that the Chrysippean void was not infinite but rather 'unlimited', in the sense that it was of indefinite extent. His argument is partly based on a wrong assessment of the evidence of Achilles, on which see below, n. 181.

'fortasse καθ' ὅσον' is printed in the *apparatus* of fr. 97 a, i.e. ps.-Plutarch's version, but not in that of fr. 97 b, i.e. Stobaeus' version). Whatever Posidonius himself may have written, both ps.-Plutarch (and derived versions) and Stobaeus have ἀλλ' ὅσον. So there is little reason to doubt that this is what stood in the Aëtian original.[169] Of course Aëtius may have misrepresented Posidonius' original intention. But also in that case an attempt to emend the reconstructed text of Aëtius, or the text of ps.-Plutarch, is not what is called for.

What is more, the structure of the chapter as a whole strongly suggests that on the topic of the extent of the extra-cosmic void Aëtius detected a real discrepancy between Stoic orthodoxy on the one hand and Posidonius on the other. The second lemma winds up with the word ἄπειρον in an 'isolated' (and, hence, prominent) position—countered by an equally abrupt Ποσειδώνιος *οὐκ ἄπειρον* at the beginning of the third lemma.[170] This makes clear that we are here dealing with a *diaeresis* on the basis of the opposition ἄπειρον-οὐκ ἄπειρον. The chapter as a whole exhibits the structure typical of the *placita* literature in which tenets are what matters primarily, rather than the persons who held them (or, for that matter, their chronological order).[171] In the present case the basic division is of course one between (1) those who posited the existence of an extra-cosmic void and (2) those who denied it (Plato + Aristotle). Among the former category a primary subdivision is made between the views that (1a) there is a void for the

---

[169] Note, in addition, that on Kidd's line of thought we shall have to suppose, rather implausibly, that Aëtius provided Posidonius with a separate lemma, next to the previous one dealing with the Stoics in general, in virtue of a *doxa* which is not only quite compatible with the *doxa* contained in the previous lemma, but which also happens to have been shared by all Stoics. It is possible, however, that Aëtius or his source *mistook* a statement by Posidonius resembling the reading Kidd suggests for the statement that the void is not infinite *tout court*. In that case Kidd's suggestion might be accepted, *not* as an attempt to restore the original text of Aëtius, but as a pointing out one of the various possible ways in which an original remark by Posidonius may have been mistakenly condensed into the doxographic testimony that according to Posidonius the void was finite. On the other hand, it is not very likely that the *doxographicum* in Aëtius presents us with a *verbatim* quotation (*pace* Kidd (1988) 336, who speaks of the 'Posidonius sentence')

[170] See on this, and on Stobaeus' more smoothly running version, Algra (1993) 481–482.

[171] On this and other characteristics of the placita as a genre see now Mansfeld (1990a) esp. 3058; and Runia (1989) 269.

cosmos *to inhale* (Pythagoras) and that (1b) there is a void for the cosmos *to expand into* at the eventual conflagration (Stoics).[172] As I noted above, group (1b) is further subdivided along the lines of the opposition *finite* void/*infinite* void. Schematically, then, the overall structure of the chapter is as follows:

1a – extra-cosmic void for the cosmos *to inhale*
1b – extra-cosmic void for the cosmos *to expand into*, be it
  1b1 – infinite, or
  1b2 – finite;
2 – no void, neither outside nor inside the cosmos

Since the overall structure of the chapter thus strongly suggests that Aëtius *opposed* the Posidonian *doxa* to the common Stoic one, we have all the more reason to stick to the reading ἀλλ' ὅσον, which is found in all MSS and in all versions, as the correct (i.e. the original) one.

That the *doxa* of a finite void was present as an option in what we might broadly refer to as the placita tradition is confirmed by the evidence of Achilles and Anonymus *In Aratum*, both of which are in one way or another dependent on this placita tradition. As I indicated, these two treatises ascribe the view at issue to the Stoics in general. The work usually referred to as Achilles' *Isagoge* (but which in fact is an *epitomê* compiled *from* a work by Achilles) contains a chapter dedicated to the question 'whether there is a void outside <the cosmos>' (εἰ ἔστι τι ἐκτὸς κενόν).[173] It ascribed the following view to 'the Stoics':

> The Stoics, who say that there is a conflagration at certain fixed intervals of time, say that there is a void, but that is not infinite, but just large enough to be able to contain the all when it is dissolved.[174]

---

[172] Note that here also the parallelism in both ps.-Plutarch and Stobaeus between the two lemmata ((1a) and (1b)) of this *diaeresis* is underlined by the presentation. In Stobaeus we find two similarly constructed sentences (κενὸν εἰς ὃ ἀναπνεῖ ὁ κόσμος καὶ ἐξ οὗ in the lemma on Pythagoras, and κενὸν εἰς ὃ ... ἀναλύεται ὁ κόσμος in the lemma on the Stoics), whereas ps.-Plutarch stresses the symmetry by leaving out of the second lemma what can be substituted in thought from the first (viz. the words ἐκτὸς εἶναι τοῦ κόσμου κενὸν and, as we noticed above, ὁ κόσμος).

[173] This is chapter 8, Maass *CAR* 38.

[174] *CAR* 38: Οἱ δὲ Στωικοὶ ἐκπύρωσιν λέγοντες κόσμου κατά τινας ὡρισμένους χρόνους εἶναι, κενὸν μέν, οὐ μὴν ἄπειρόν φασιν ἀλλὰ τοσοῦτον ὅσον χωρῆσαι λυθὲν τὸ πᾶν.

I have discussed this chapter as a whole as well as this particular ascription in more detail elsewhere, and I here confine myself to summarizing the results of that discussion.[175] First of all, the chapter is based on the following *diairesis*:

(1) infinite extracosmic void (Epicurus)

(2) limited extracosmic void (Stoa)

(3) no extracosmic void (no name label; instead, a number of *arguments* are added)

Secondly, the ascription of the theory of a finite extra-cosmic void to the Stoics not only flatly contradicts what is otherwise known about the Stoic theory of the void, but it also runs counter to the information provided by Achilles himself at *Isag.* 9 (*CAR* 38), where, representing a scholastic-doxographical outline of a Stoic proof of the immobility of the cosmos, he speaks of the immobility of the cosmos in an *infinite* void. This suggests that Achilles' treatise was patched together, apparently in a rather mechanical way, from materials deriving from various sources. In view of the structural resemblance between the *diaireseis* in Achilles and in Aëtius II, 9, and in view of the fact that also elsewhere Achilles shows signs of being dependent on the placita tradition, we had best assume that Achilles here represents a specimen of the placita vulgate that was different from, though not unrelated to the Aëtian placita (we might speak of 'cousin writings').[176] In that case the odd contents of the lemma on 'the Stoics' can best be explained as a garbled and abridged version of what originally were two lemmata on the Stoics and Posidonius respectively.[177]

What Maass printed as 'Anonymus I' in his *Commentariorum in Aratum Reliquiae* is another, this time anonymous, introduction to Aratus' *Phainomena*. Its date cannot be established with any

---

[175] Cf. Algra (1993) 485–490.

[176] The details of the way Achilles relates to the placita tradition are not easy to establish. In his *Doxographi Graeci* Diels listed Achilles among the *testimonia Plutarchi*, thus making the very specific claim that, in spite of the differences, a number of passages in Achilles are to be traced back to the ps.-Plutarchean placita (a view also endorsed by Maass, witness the *marginalia* of his edition of Achilles). Later, however, he retracted this view, on which see G. Pasquali, 'Doxographica aus Basiliusscholien', *Nachr. der kön. Ges. der Wiss. zu Göttingen, philol.-hist. Kl.* 1910, 194–228, esp. 221, n. 1. David Runia has suggested to me that Achilles may have had direct access to the so-called *vetusta placita*.

[177] For a reconstruction of how this garbled version may have originated see Algra (1993) 487–488.

certainty, or so it seems. It contains a chapter entitled 'whether the earth is stationary or whether it moves' (εἰ ἕστηκεν ἡ γῆ ἢ κινεῖται, *CAR* 92–93), to which the following brief survey concerning the issue of the extra-cosmic void is appended:

> Whether there is a void outside the heavens, is not a futile (or: superfluous) question for us to ask. The Stoics say that there is, for in that respect the whole differs from the all, but that it is not infinite, as Epicurus and the others maintain.[178]

Now this is only a rudimentary account, and for that reason we should be cautious in trying to establish its provenance. Yet there are resemblances between this passage and the account in Achilles: in both cases, for example, we find the opposition Stoics *versus* Epicureans, and in both cases the Stoics are mistakenly credited with a theory of a finite void. Indeed, further study of the two treatises reveals that there are quite a number of passages where Anonymus I clearly parallels Achilles, although Achilles in those cases invariably offers a more complete version.[179] The conclusion seems warranted that Anonymus used Achilles (whether the extant *epitomê* or the original work), or that the passages here discussed are to be traced back to a common source used by both Achilles and Anonymus.

We may conclude that our passages from Achilles and Anonymus are best regarded as two versions of a doxographical account which was very similar (and related) to the account of Aëtius, an account originally comprising both 'the Stoics' and 'Posidonius'. This means that there is little reason to take them seriously as a 'dissident voice' concerning the Stoic theory of the void,[180] let alone as containing useful and reliable evidence on the *orthodox* early Stoic view.[181] Neither, however, should we assume that either of them contains information which, without inter-

---

178 *CAR* 92–93: εἰ δὲ ἔστι τι κενὸν ἔξωθεν τοῦ οὐρανοῦ, οὐ περίεργος ἡμῖν ἡ ζήτησις. πλὴν οἱ Στωικοὶ λέγουσιν εἶναι. ταύτῃ γὰρ διαφέρειν τὸ ὅλον τοῦ παντός, ἀλλ' οὐκ ἄπειρον, ὡς Ἐπίκουρος καὶ οἱ λοιποί.

179 For further details see Algra (1993) 488–489.

180 This is how Kidd (1988) 391, describes the account of Achilles.

181 This very odd claim is made, with respect to Achilles, by Inwood (1992) 259–260 (n. 44); apart from being based on what I take to be a mistaken analysis of the Chrysippean conception of infinity (but I am afraid I cannot go into that here), Inwood's interpretation fails to take account of the indirect doxographical nature and the garbled state of the evidence presented by Achilles.

mediary, derives from Posidonius.[182] In view of their debt to the placita tradition, they cannot count as independent witnesses in the required sense.[183] Yet, as was already noted, they do testify to the fact that the *doxa* of a finite void was present as an option in the placita literature, which gives us all the more reason not to go along with Kidd in attempting to eliminate that *doxa* from the text of Aëtius II, 9 by emendation.

All in all we may conclude that the explicit evidence for Posidonius' alleged heterodox conception of the extent of the extra-cosmic void is limited to the testimony of Aëtius—the accounts in Achilles and Anonymus I being additional witnesses only in a very weak sense. This provides us with little firm ground to stand on. Of course Aëtius' presentation may amount to a distortion of Posidonius' own intentions. Yet the evidence as such provides no reason to assume *a priori* that this is the case. The question whether Aëtius' testimony is reliable will therefore have to be decided —if it can be decided at all—by means of other, external, criteria. It is to such external criteria that we shall now direct our attention.

6.4.3 In his commentary to the testimony of Aëtius II, 9 Ian Kidd comes up with two arguments in favour of his thesis that we are dealing with a garbled and unreliable report. First of all, he claims that it is significant that none of the reporters who provide information about the Stoic theory of the void (Arius Didymus, Philo, Aëtius, Plutarch, Alexander of Aphrodisias, Diogenes Laërtius, Cleomedes and Sextus Empiricus) mentions the fact that any Stoic denied the infinity of the void. On closer scrutiny, however, this *argumentum ex silentio* hardly carries any weight.[184] Some of the sources at issue show no acquaintance at all with Posidonius or Posidonian physics (this goes for Philo, for Plutarch's

---

182 This suggestion can be found in Kidd (1988) 392: ' [...] the question must be asked whether the statement could derive from Posidonius (who was known to Achilles)'. As evidence that Achilles knew Posidonius Kidd adds Frs. 128 (this, incidentally, should be: 129), 148, and 209 Edelstein/ Kidd (1988^2); but see the next footnote.

183 We may add that Kidd's claim that Posidonius 'was known to Achilles' should be qualified anyway. The 'Posidonian fragments' provided by Achilles are all arguably derived from intermediate sources (such as Eudorus and Diodorus). For this see Kidd's own comments (1988) 549–550.

184 For a more detailed discussion of the value of the *argumentum ex silentio* in this case, see Algra (1993) 490–495.

source in *SR* and *Comm. Not.*, and to some extent for Alexander).[185] But even if our sources elsewhere do show knowledge of Posidonius' philosophy, the structure and purpose of the passages in which they speak of the Stoic void may be such as not to make us expect any reference to a Posidonian heterodoxy (this goes for Sextus, Cleomedes, Alexander and Diogenes Laërtius).[186] In one case the fragmentary nature of the preserved evidence precludes that we treat the silence on Posidonius as significant (Arius Didymus).[187]

So the 'problem' of the silence of the other sources appears to be a very small problem indeed, if a problem at all. It should be pointed out, in addition, that even if the silence of each and all of these reporters did turn out to be significant, the most we could infer would be that Posidonius probably never put forward the conception of a finite extra-cosmic void as a full-fledged alternative to the Chrysippean orthodoxy. It does not necessarily mean that the *doxa* in Aëtius had no basis whatsoever in Posidonius' work. It is conceivable that what Aëtius thought to be a *doxa* worthwhile to include may in its original setting have been hardly more than a suggestion on Posidonius' part. After all, the less strongly Posidonius was committed to this view, the less radically opposed to the Chrysippean orthodoxy it would have appeared to others, and the less surprising it is that it did not reach other ancient scholastic writings or doxographies. Stronger evidence appears to be needed, therefore, if the testimony of Aëtius is to be rejected.

This brings us to Kidd's second reason to distrust the evidence

---

185 The specifically anti-Stoic writings in which Plutarch discusses the Stoic void, *De Stoicorum Repugnantiis* (*SR*) and *De Communibus Notitiis* (*Comm. Not.*) *exclusively* focus on the early Stoics: no Stoic later than Antipater is referred to (for this reason von Arnim argued that the source of *SR* and *Comm. Not.* was Clitomachus, who compiled the arguments of Carneades. Cf. his introductory remarks at *SVF* I, xii–xiv). Among the attested fragments and testimonies of Posidonius only one is to be found in a surviving work of Alexander (fr. 133 Edelstein/ Kidd (1988²)), and it characteristically deals with such a special subject as theories of halos, so we are not to allowed to infer that also the more general aspects of Posdidonian physics were known to Alexander (moreover even this information was probably only derived from Geminus' *Epitomê* of Posidonius' *Meteorologica*, on which see Algra (1993) 493, n. 48).

186 For the specific reasons in each of these cases see Algra (1993) 491–494.

187 See Algra (1993) 494–495.

of Aëtius: once an extra-cosmic void had been accepted, "it would be a very peculiar concept to hold that it was finite".[188] But is this true? Yes and no. Yes, in the sense that it would seem to run counter to common sense, in particular to the common sense argument which was endorsed by the early Stoics as well as by other philosophers: everything limited must have something next to it, and since there is nothing next to the void, the void cannot be limited. We shall have to admit that, by contrast, the alternative position ascribed to Posidonius is rather counter-intuitive. When seen from a different point of view, however, the position ascribed to Posidonius is certainly not as peculiar as Kidd would have us believe. For, as I have also tried to establish in the previous section, the early Stoic conception of an infinite void was itself not entirely unproblematic from a systematic point of view. Accordingly, we cannot exclude the possibility that Posidonius advocated a revision of this part of Stoic physics *for theoretical reasons*. In that case, the fact that his alternative conception of a finite void was rather counter-intuitive may have been thought to be of minor importance. In this connection it should be noted that there are many concepts in physics, ancient and modern, which are at first sight odd or which run counter to common conceptions, but which may nevertheless be defended on good theoretical grounds. Now what theoretical reasons could Posidonius have had for taking the dissident position ascribed to him by Aëtius?

Two possible reasons have already been mentioned in the previous section. First of all, Chrysippus' 'technical' argument as preserved in Arius Didymus fr. 25 appeared to be unsatisfactory. According to Chrysippus the void was not limited by itself. Since it was obviously not limited by an actually occupying body either (for then it would no longer be void), it was concluded that it was infinite. Yet, as I noted, why couldn't we opt for a third possibility and argue that the void is limited by what in the event of a conflagration will be the limit of the cosmic body? If the cosmic body is the *raison d'être* of the void, why shouldn't it also be its *raison d'être fini*? Secondly, I have pointed out that the Stoic conception of an extra-cosmic void subsisting in its own right might be said to come too close to the Epicurean conception of a self-subsistent absolute void existing on a par with corporeal being. At any rate

188 Kidd (1988) 393.

such a void would seem to enjoy a much more independent status than any of the other canonical incorporeals.

With an eye to our discussion of the Chrysippean quotations in Plutarch in section 6.2, we may now add a third argument, an argument which concerns the problem of the *isotropy* (i.e. the lack of any spatial determinations) of the void. As we saw, Plutarch (at *SR* 1054 b–1055 a) charged Chrysippus with inconsistency at precisely this point.[189] According to Plutarch, Chrysippus defended the thesis that the extra-cosmic infinite void can have no spatial determinations *of its own.* That is why Chrysippus objected against the Epicurean theory of the downward motion of the atoms through the void. Plutarch also clearly implies, however, that we are not merely dealing with an *ad hominem* argument against Epicurus, but with an argument which would also hold for the Stoic void. The inconsistency, according to Plutarch, is to be located in the fact that Chrysippus also appears to have argued that the spatial determinations of the cosmos could be extended at least to some extent to that extra-cosmic void. That is why he could state, in his Περὶ δυνατῶν, that the cosmos is at the centre of the surrounding space—a space which, as I argued, was probably called the χώρα of the cosmos. Now whether or not I have been correct in making the latter claim concerning the precise denomination (χώρα) is rather immaterial for our present purpose. For, as I pointed out in subsection 6.2, in either case the fact remains that Chrysippus held *both* that the void, when taken independently, exhibits no spatial determinations *and* that (part of) this same void when seen as the surrounding finite receptacle of the cosmos (possibly called χώρα) *derives* spatial determinations from the cosmos. As I indicated earlier in this chapter, such a position was bound to call forth two questions, or rather objections. First, *to what extent* may the spatial determinations of the cosmos be extended into the void. Apparently the surrounding space of which Chrysippus speaks is finite, since the cosmos is said to occupy its centre. Even so, however, the idea that outside this *finite* and *anisotropic* space an *infinite* and *isotropic* void begins is, to say

189 Note that if it is true that Plutarch merely reproduced an existing argument of Academic provenance—on which see von Arnim's opinion as referred to above, n. 185—Posidonius may very well have been familiar with it.

the least, odd. Secondly, we are facing a problem, concerning the ontological status of the void, which we ran into earlier: does it make sense at all to speak of the void being isotropic 'in itself' or 'taken independently' any more than of e.g. time 'in itself'? Thus, although the second contradiction detected by Plutarch in the two fragments from *Peri dunatôn* quoted in section 6.2 (viz. the fact that Chrysippus sometimes argued that the void is isotropic, while at other times he defended the thesis that it has a centre) may—if the interpretation of Chrysippus' concept of *chôra* proposed in section 6.1 and 6.2 is correct—not be as glaring as Plutarch pretends it to be, it does reveal a fundamental ambiguity in the Stoic theory of space.

These observations may serve to show, once more, that the early Stoic conception of an infinite extra-cosmic void, as well as the arguments used to support it, were at least susceptible to doubt or criticism. Against this background it is conceivable that Posidonius took the Chrysippean notion of a finite spherical space encompassing and surrounding the cosmos (a space probably called χώρα, and just as large as is needed for the conflagration) as a basis for a cautious suggestion that it made more sense to drop the idea of the infinity of the void altogether. But he may well have gone beyond a mere suggestion. Dissatisfaction with the Chrysippean argument for the infinity of the void may have stimulated him to reject the orthodox position outright and to postulate that just as place (τόπος) gets its limit from the occupant body, so also the void is limited by the maximum extent of its one-time occupant (i.e. the cosmos at the event of a conflagration). Thus there would be no need to speak of the void having no limit *of its own*, or exhibiting no spatial differentiations *of its own*.

From the point of view of Stoic physics, in other words, the move which the placita tradition ascribed to Posidonius would not be utterly odd. On the contrary, it would arguably have the merit of rendering the Stoic theory of space more coherent in bringing the conception of the void in line with that of place (place deriving both its limit and its determinations from the cosmos). In addition it would rid Stoic physics of an element (an infinite isotropic void) which was perhaps better at home in the rival theory of Epicurus. At the same time it would seem to make sense from the point of view of Posidonius' own peculiar brand of Stoicism, since it might be regarded as an attempt to conciliate the requirement that there should be an extra-cosmic void for the cosmos to expand into with the fundamentally finitist nature of Stoic, but also of Platonic and Aristotelian, cosmology.[190] Posidonius' rather vigorously anti-Epicurean attitude as well as his recorded attempts to incorporate elements from the systems of Plato and Aristotle into Stoicism may thus lend additional credibility to our hypothetical reconstruction.[191] On balance, therefore, I believe the testimony of Aëtius should be accepted.

In that case we are left with one final point to explain. It is obvious that the conception of a finite extra-cosmic void did not carry the day among the Stoics, and we may well ask why. Also on this point we can of course only speculate, but it is worth pointing out some possible explanatory factors. First, as I already noted, in the area of the *principia physica* Stoicism appears to have been remarkably conservative. The system of Stoic physics as taught and discussed even in later antiquity was on the whole the system as constructed by Chrysippus. As I noted, even Posidonius' alleged heterodox conception of the void may have been put forward not so much as a radical alternative to the Chrysippean orthodoxy, but rather as a qualifying suggestion in order to bridge the gap between Chrysippus' concepts of χώρα and κενόν.[192] Secondly, for all its possible merits in terms of the systematic coherence of the Stoic theory of space, the idea of a finite extra-cosmic void was probably too intuitively abhorrent to be plausibly defended. After all common sense tends to view space as a self-subsistent three-dimensional extension and in that case the

---

[190] We may note that on the line of thought thus ascribed to Posidonius the situation of the cosmos at the time of a conflagration is analogous to that of a Platonic or Aristotelian cosmos: it is finite and yet not surrounded by anything else. The difference only becomes apparent *after* the conflagration when the cosmic mass shrinks, leaving a still *finite* void space behind.

[191] For Posidonius' anti-Epicurean polemics see fragments 22 (theology), 46 (mathematics, against Zeno of Sidon), 47 (idem, against Zeno of Sidon and Epicurus), 149 (psychology) in Edelstein/ Kidd (1988[2]). Although Galen, in *PHP* probably overstresses the affinity between Posidonius on the one hand and Plato and Aristotle on the other, it must be acknowledged that in the area of psychology Posidonius' debt to Plato in particular was considerable (cf. T 97 Edelstein/ Kidd (1988[2]), where Posidonius is said to have called Plato 'divine' and to have praised his theories of the passions and of the faculties of the soul).

[192] Note, incidentally, that Posidonius' correction would still enable him to adhere to the Chrysippean *definitions* of the spatial terms τόπος, χώρα and κενόν.

arguments adduced by Epicurus and others to prove that an extra-cosmic void space must be infinite are rather convincing.

Conclusions

In this chapter I have tried to provide a more or less coherent interpretation of the scanty evidence concerning the early Stoic philosophy of space. This interpretation was based on the analysis, in section 6.1 and 6.2 respectively, of the preserved Stoic definitions of spatial concepts and of two *verbatim* quotations from Chrysippus rendered by Plutarch *SR* 1054c–d. Having established, in as far as this is possible, the provenance of the several definitions, I concentrated on the Chrysippean material provided by Arius Didymus. It was shown that, although Chrysippus' concepts of *topos* and *kenon* became canonical among orthodox Stoics, his concept of *chôra* apparently lost its appeal. I argued that both surviving Chrysippean definitions suggested that *chôra* was thought to be a space which is larger than the body for which it provides room: it is that which can contain more than, or more of, that which it contains. I suggested that this concept of *chôra* may also have been applied in what I called a 'cosmic' setting, the *chôra* of the cosmos being the finite spherical space which encompasses and surrounds the cosmos and which provides the room it will need at a periodical conflagration.

In section 6.2. I have tried to show that this interpretation could be helpful in making sense of two quotations from Chrysippus' *Peri dunatôn* discussed by Plutarch. One of the two alleged contradictions detected by Plutarch can be eliminated, provided that we pay proper attention to the context from which the two fragments have been lifted, viz. a context concerned with modal logic. In addition, I argued that Chrysippus can be defended against the other charge too, if our interpretation of *chôra* as a finite spherical 'receptacle' of the cosmos is accepted. On any account, however, the two fragments were seen to lay bare an interesting ambiguity in Chrysippus' philosophy of space: apparently Chrysippus made a distinction between the void *qua* empty space *as such* (being infinite and isotropic) and the void *qua* space surrounding the cosmos (being at any rate anisotropic).

In section 6.3 the ontological status of place and void in Stoic thought was at the focus of attention. It was shown that the occurrence of an *ekpurôsis* forced the Stoics to have recourse to a conception of the void as something self-subsistent, a pre-existing space of what in chapter 1 was called the (a) kind, into which a body can move or expand. The void was thus given a quite privileged status among the incorporeals and it could well be asked whether this did not bring Chrysippus too close to the Epicurean position, according to which empty space existed next to (and ontologically on a par with) corporeal being.

In section 6.4, finally, I discussed the surviving evidence on Posidonius' conception of the (extent of the) extra-cosmic void. I concluded that, on balance, this evidence should be accepted, so that we may credit Posidonius with having claimed that the extra-cosmic void is not infinite. Making use of the findings of sections 6.2 and 6.3, I suggested that he may have had some good arguments for taking this position. His curious move may have been motivated by dissatisfaction with the Chrysippean arguments for the infinity of the void as well as by the desire to bridge the gap between *chôra* and *kenon*—and thus to adapt Stoic physics, in spite the inevitable extra-cosmic void, to the finitism of Aristotelian and Platonic cosmology.

Appendix: the Stoic theory of space and Aristotelian Physics

By way of an *excursus*, I want to add some observations on how the early Stoics conceived of the physical *role* of space, and on the question whether this view was influenced by Aristotle. The observation made earlier in this chapter, viz. that space *qua* void is a necessary condition for the expansion of the cosmos, is not novel. However, the conclusion arrived at in subsection 6.3.2, mainly on the basis of Cleomedes *Cael.* I, 1, 20–38 Todd, viz. that place, and in general space as such, could according to the Stoics be regarded as a necessary condition for *motion* of individual parts of the cosmos, has, as far as I know, not usually been recognized. I admit that explicit evidence concerning the role of *topos* in this respect—at least in so far as the *early* Stoics are concerned—is lacking.[193] Nevertheless, the nature of the arguments for an

193 See, however, Sen. *Ep.* 65, 11, where Plato and Aristotle are criticized for having introduced a *turba causarum*, in particular for treating as a cause what strictly speaking is just a necessary condition. *Locus* is given as one of the examples. See on this passage Frede (1980) 228.

independent place in Cleomedes, as well as the fact that the common denominator of both the place *into which* a body could *move*, and the void *into which* the cosmos can *expand* is called 'that which *can receive* body', sufficiently indicate that the common denominator, space, was regarded as a necessary condition for motion as well as for expansion. It is there, awaiting, so to speak, the arrival of body (A) and the departure of body (B) (as *topos*) or the expansion of the cosmos (as *kenon*).

This is an important conclusion from an historical point of view, since it allows us to refute Hahm's claim that Chrysippus took over Aristotle's argument (presented at *Phys.* Δ 215 a 21 ff.) that owing to the isotropy and inertness of the void, a body cannot move in such a void. According to Hahm, Chrysippus—at least at a certain point in his career—used this argument in order to prove the immobility of the cosmos in the surrounding void.[194] It may now easily be shown that Aristotle's arguments were such that they would have been unacceptable to a Stoic.

In the fourth book of the *Physics* Aristotle is concerned—among other things—with refuting the basic presupposition of those who posited a self-subsistent void (primarily the early atomists as it appears), viz. that such a self-subsistent empty space is a necessary condition for motion.[195] To this purpose he frames a number of *reductiones ad absurdum*. One of these turns the concept of the isotropy and inertness of the void into an argument against the very existence of such a void in that it shows that it involves the absurd consequence of motion being made impossible because in such a void there is no reason for a body to move one way rather than the other.

Now first of all, as I have already argued elsewhere in this book, Aristotle here presents us with a bad argument which does not appear to do justice to the early atomists' position.[196] For he argues here as if the atomist void was meant to be a *sufficient cause*, instead of merely a *necessary condition* for motion to occur.[197] But

[194] Hahm (1977) 122: 'The basis for Chrysippus' solution seems to have been Aristotle'. This, by the way, was already the opinion of Bréhier (1907) 47.

[195] Cf. *Phys.* Δ 215 a 21 ff. On Aristotle's arguments see Furley (1976) 92 ff. On the Stoic arguments for the immobility of the cosmos in the void see Algra (1988a) *passim*.

[196] See above, ch. 5, 206–207 and 250–252.

[197] It should be noted that the early atomists in all probability accepted

that need not concern us here. For the moment it is sufficient to note that this bad argument left an obvious way out for those who were ready to argue that a corporeal entity, though not moved by the void, can still be moved by itself or by another corporeal entity—the void, or rather space, merely serving as a necessary condition. If it can be shown—as I think has just been done—that this was the view the Stoics were committed to,[198] there is no reason to assume that they subscribed to Aristotle's unsatisfactory argument against the atomists and that, consequently, they believed that the void actually *precludes* motion. In fact, it is only a natural consequence of the Stoic ontology that the void, being incorporeal and thus unable to act as a cause, cannot *preclude* the possibility of motion (and so *cause* rest) either.

This interpretation seems to be corroborated by the Chrysippean *Gedankenexperiment* referred to by Plutarch *SR* 1055c (= *SVF* II, 550), which states that a part of cosmic matter, even if it found itself surrounded by an intracosmic void, would still continue its natural motion:

> It is reasonable that the way in which each of the parts moves when cohering with the rest is also the way in which it moves by itself, even if, for the sake of argument, we should in imagination suppose it to be in a void within this universe, for as it would be moving to the middle when surrounded from all sides by matter (συνεχόμενον πάντοθεν) so will it continue in this motion, even if for the sake of argument all about it suddenly comes to be void.[199]

In this connection we may also adduce Cleomedes' statement that 'we can imagine the cosmos itself moving from the place which

---

motion as an empirically given fact, not requiring an explanation in terms of causes, but rather forcing us towards a cosmology in which motion is at least *made possible* by a void. See on this problem e.g. the convenient survey in Steckel (1970) 203 ff.; Guthrie (1967–81) II, 396 ff. goes a long way in the same direction but, to my mind, spoils his case by still treating the atomist void as a sufficient cause for motion (*op. cit.* 398–399) on the basis of the testimony of Aristotle's *Physics*. Note, however that Aristotle himself provides a clue for the sense in which the void is an αἴτιον: αἴτιον κινήσεως οἴονται εἶναι τὸ κενὸν οὕτως ὡς ἐν ᾧ κινεῖται (214 a 24) and οὐ γὰρ ἂν δοκεῖν εἶναι κίνησιν εἰ μὴ εἴη κενόν. τὸ γὰρ πλῆρες ἀδύνατον εἶναι δέξασθαί τι (213 b 5–7).

198 See also my paper Algra (1988a) 155–180.

199 *SR* 1055c: ὃν τρόπον δὲ κινεῖται ἕκαστον τῶν μορίων συμφυὲς ὂν τῷ λοιπῷ, εὔλογον οὕτως καὶ καθ' αὑτὸ κινεῖσθαι, καὶ εἰ λόγου χάριν νοήσαιμεν αὐτὸ καὶ ὑποθοίμεθα εἶναι ἐν κενῷ τινι τοῦ κόσμου τούτου· ὡς γὰρ ἂν συνεχόμενον πάντοθεν ἐκινεῖτο ἐπὶ τὸ μέσον, μενεῖ ἐν τῇ κινήσει ταύτῃ, κἂν λόγου χάριν ἐξαίφνης περὶ αὐτὸ γένηται κενόν. Cherniss' text and translation.

it is now occupying'.[200] Whatever the role of the void in these two examples, it certainly does not preclude motion. It seems pointless, therefore, to assume that the Stoics were directly influenced by Aristotle's *Physics* in this respect.

In general, there seems to be little or no reason to connect the early Stoic theory of space with Aristotle. The only possible indication of such a connection to be found, seems to be the fact that, as Arius, Aëtius and Cleomedes suggest, the early Stoics used the analogy of a vessel in connection with space, an analogy which prominently figured in Aristotle's *Phys.* Δ. However, we may as well assume that this image had become stock in the Aristotelian oral tradition. The early Stoics' using it need not imply that they actually read the *Physics.* Though I would not dare to argue, like Sandbach, that the early Stoics did not know Aristotle's physical writings at all, I think most, if not all, resemblances pointed out by Hahm[201] could also very well be explained by a common reliance on everyday concepts, the common subject matter, and the more general familiarity of the Stoics with the Aristotelian school tradition. Direct influence of the Aristotelian school writings should only be assumed in those cases where the resemblance is too specific to be otherwise accounted for.

---

200 *Cael.* I, 1, 39–40 Todd: Καὶ μὴν καὶ τὸν κόσμον αὐτὸν δυνάμεθα ἐπινοῆσαι κινούμενον ἐκ τοῦ τόπου, ὃν νῦν τύγχανει κατειληφώς.

201 Cf. Hahm (1977) 103–105. As might be expected Hahm plays down the differences between the Stoic and the Aristotelian theories of space and place. Thus, concerning the fact that the Stoic cosmos, very much unlike its Aristotelian counterpart, is surrounded by a void, he argues that 'the difference is, on the face of it, merely one of definition' (!) (*op. cit.* 103). The only significant difference between the Chrysippean and the Aristotelian concept of place is, according to Hahm, 'that Chrysippus ignores Aristotle's technical definition of place as 'the immediate continent of a body' or the inner surface of the container, and instead emphasizes the other side of Aristotle's notion, namely, that place is 'something occupied by body''(*op. cit.* 105). Accordingly, the 'similarities' are made rather heavy weather of: 'Chrysippus shows himself to be in Aristotle's debt by considering void in the context of place' (*op. cit.* 105). But see my remarks above, n. 28, on the different ways in which void and place are related in Aristotelian and Stoic physics.

BIBLIOGRAPHY

This list contains all titles referred to more than once. Of all other books and articles full references are given in the footnotes

AA. VV. (1961), *Aristote et les problèmes de méthode*: Communications présentées au Symposium Aristotelicum, tenu à Louvain du 24 août au 1er septembre 1960; avant-propos de Suzanne Mansion; Aristote, Traductions et Études (Louvain/ Paris).

Ackrill, J. L. (1963), *Aristotle's* Categories *and* De interpretatione, Clarendon Aristotle Series (Oxford, repr. w. corr. 1966).

Adam, C. – Tannery, P. (edd.) (1905), *Oeuvres de Descartes*, vol. 8.1: *Principia philosophiae* (Paris, repr. 1974).

Adorno, F. (1980), 'Epicuro: *Epistola a Erodoto* 39, 7–40, 3', *Elenchos* (1) 245–275.

Algra, K. A. (1988a), 'The Early Stoics on the Immobility and Coherence of the Cosmos', *Phronesis* (33) 155–180.

——, (1988b), *Concepts of Space in Classical and Hellenistic Greek Philosophy* (diss. Utrecht)

——, (1990), 'Chrysippus on Virtuous Abstention from Ugly Old Women (Plutarch, *SR* 1038 E – 1039A)', *CQ* (40.2) 450–458.

——, (1992), ''Place' in Context: On Theophrastus Fr. 21 and 22 Wimmer', in: Fortenbaugh – Gutas (1992) 141–165.

——, (1993), 'Posidonius' Conception of the Extra-Cosmic Void: The Evidence and the Arguments', *Mnemosyne* (46.4) 473–505.

Allen, R. E. (ed.) (1965), *Studies in Plato's Metaphysics*, International library of philosophy and scientific method (London/ New York).

——, (1969), 'Individual Properties in Aristotle's Categories', *Phronesis* (14) 31–39.

Anderson, W. E. (1976), 'Cartesian Motion', in: Machamer – Turnbull (1976) 200–223.

Archer-Hind, R. D. (1888), *The Timaeus of Plato* (London, repr. New York 1973).

Ashbaugh, A. F. (1988), *Plato's Theory of Explanation*: A Study of the Cosmological Account in the *Timaeus* (Albany, NY).

Ast, F. (1835–38), *Lexicon Platonicum sive vocum Platonicarum index*, 3 vols. (Leipzig, various reprints).

Babut, D. (1969), *Plutarque et le Stoicisme*, Publications de l'Université de Lyon (Paris).

Baeumker, C. (1890), *Das Problem der Materie in der griechischen Philosophie: eine historisch-kritische Untersuchung* (Münster, repr. Frankfurt am Main 1963).

Bailey, C. (1928), *The Greek Atomists and Epicurus* (Oxford, repr. New York 1964).

Barnes, J. – Schofield, M. – Sorabji, R. (edd.) (1975–79), *Articles on Aristotle*, 4 vols.: I. Science. II. Ethics and Politics. III. Metaphysics. IV. Psychology and Aesthetics (London).

Bobzien, S. (1986), *Die stoische Modallogik*, Epistemata, Reihe Philosophie 32 (Würzburg).

Boehner, P. (ed.) (1944), *The Tractatus de successivis Attributed to William Ockham*, edited with a Study on the Life and Works of Ockham, Franciscan Institute Publications, Philosophy Series 1 (New York).

Bolton, R. (1991), 'Aristotle's Method in Natural Science: Physics I', in: L. Judson (ed.), *Aristotle's Physics: a Collection of Essays* (Oxford) 1–29.

Bossier F. – Steel, C. (1972), 'Priscianus Lydus en de *In de Anima* van Pseudo(?)-Simplicius', *Tijdschrift voor Filosofie* (34) 761–822.

Bourgey, L. (1955), *Observation et expérience chez Aristote*, Bibliothèque d'histoire de la philosophie (Paris).

Bréhier, E. (1907), *La théorie des incorporels dans l'ancien stoïcisme* (Paris, [3]1962 in series Bibliothèque d'histoire de la philosophie).

Brisson, L. (1974), *Le même et l'autre dans la structure ontologique du Timée de Platon: un commentaire systématique du Timée de Platon*, Publications de l'Université de Paris X Nanterre. Lettres et Sciences humaines. Sér. A, Thèses et travaux 23 (Paris).

Brunschwig, J. (ed.) (1978), *Les Stoïciens et leur logique*. Actes du colloque de Chantilly, 18–22 septembre 1976, Bibliothèque d'histoire de la philosophie (Paris).

——, (1988), 'La théorie stoïcienne du genre suprême et l'ontologie platonicienne', in: J. Barnes – M. Mignucci (edd.), *Matter and Metaphysics*, Elenchos 14 (Napoli), 19–127 (english version in Brunschwig (1994) 92–157).

——, (1994), *Papers in Hellenistic Philosophy* (Cambridge).

Bunge, M. (1978), 'Physical Space', in: M. Svilar – A. Mercier (edd.), *L' Espace*, Institut International de Philosophie, Entretiens de Berne 12 – 16 sept. 1976 (Bern etc.) 133–149.

Burnyeat, M. F. (1984), 'The Sceptic in his Place and Time', in: R. Rorty – J. B. Schneewind – Q. Skinner (edd.), *Philosophy in History*, Essays on the Historiography of Philosophy (Cambridge etc.) 225–254.

Carroll, W. E. (ed.) (1985), J. A. Weisheipl: *Nature and Motion in the Middle Ages*, Studies in Philosophy and the History of Philosophy 1 (Washington DC).

Charlton, W. (1970), *Aristotle's* Physics*: Books I and II*, Clarendon Aristotle Series (Oxford, repr. 1984).

Cherniss, H. F. (1935), *Aristotle's Criticism of Presocratic Philosophy* (Baltimore, repr. New York 1964).

——, (1944), *Aristotle's Criticism of Plato and the Academy* (Baltimore, New York [2]1962).

——, (1954), 'A Much Misread Passage of the *Timaeus* (*Timaeus* 49 c 7–50 b 5)', *AJPh* (75, 2) 113–130.

——, (1957), 'The Relation of the *Timaeus* to Plato's Later Dialogues', *AJPh* (78) 225–266, repr. in: L. Tarán (ed.), H. Cherniss: *Collected Papers* (Leiden 1977) 298–339.

——, (1976), *Plutarch's Moralia* vol. 13, Pt. 2, Loeb Classical Library (Cambridge Mass./ London).

Christensen, J. (1962), *An Essay on the Unity of Stoic Philosophy*, Scandinavian University Books (København).

Claghorn, G. S. (1954), *Aristotle's Criticism of Plato's 'Timaeus'* (Den Haag).

Cornford, F. M. (1936), 'The Invention of Space', in: J. A. K. Thomson (ed.), *Essays in Honour of Gilbert Murray* (London) 215–235, repr. in: A. C. Bowen (ed.) *Selected Papers of F. M. Cornford*, Greek and Roman Philosophy 10 (New York 1987) 199–219.

——, (1937), *Plato's Cosmology: the* Timaeus *of Plato* translated with a running commentary; International Library of Psychology, Philosophy and Scientific Method (London, various repr.).

Crombie, I. M. (1962–63), *An Examination of Plato's Doctrines*, 2 vols.: I. Plato on

Man and Society. II. Plato on Knowledge and Reality; International Library of Philosophy and Scientific Method (London/ New York).

Daiber, H. (ed.) (1980), *Aëtius Arabus: Die Vorsokratiker in arabischer Überlieferung*, Veröffentlichungen der Orientalischen Kommission, Akademie der Wissenschaften und der Literatur in Mainz XXXIII (Wiesbaden).

——, (1992), 'The *Meteorology* of Theophrastus in Syriac and Arabic Translation', in: Fortenbaugh - Gutas (1992) 166–293.

Davidson, H. A. (1969), *Averroes Middle Commentary on Porphyry's* Isagoge, *and on Aristotle's* Categoriae; [respectively] transl. from the Hebrew and Latin versions, [and] transl. from the original Arabic and the Hebrew and Latin versions; Mediaeval Academy of America 79; Corpus philosophorum medii aevi; Corpus commentariorum Averrois in Aristotelem, versio Anglica 1, a, 1–2 (Cambridge Mass. etc.).

Diels, H. (1879), *Doxographi Graeci* (Berlin, [4]1976 and later repr.).

——, (1893), 'Ueber das physikalische System des Straton', *Sitzungsber. kgl. Pr. Akad. Wiss. Berlin* 101–127; repr. in: W. Burkert (ed.) (1969), H. Diels: *Kleine Schriften* (Darmstadt) 239–265.

Dijksterhuis, E. J. ([4]1980), *De Mechanisering van het Wereldbeeld* (Amsterdam).

Driscoll, J. (1979), 'The Platonic Ancestry of Primary Substance', *Phronesis* 24 (1979) 253–269.

Donini, P. L. (1975), 'Fato e voluntà umana in Crisippo', *AAT* (109) 187–230.

Duhem, P. (1913), *Le système du monde. Histoire des doctrines cosmologiques de Platon à Copernic*, vol. 1 (Paris, repr. 1957).

Duhot, J. J. (1989), *La conception stoïcienne de la causalité* (Paris).

Düring, I. (1957), *Aristotle in the Ancient Biographical Tradition*, Studia Graeca et Latina Gothoburgensia 5 (Göteborg) 241–246.

——, (1966), *Aristoteles: Darstellung und Interpretation seines Denkens*, Bibliothek der klassischen Altertumswissenschaften, N.F. Reihe 1, H. 2 (Heidelberg).

Edelstein, L. - Kidd, I. G. (edd.) (1972), *Posidonius*, vol. 1: *The Fragments*, Cambridge Classical Texts and Commentaries 13 (Cambridge, rev. ed. 1988).

Einstein, A. (1949), 'Autobiographisches/ Autobiographical Notes', in: P. A. Schilpp (ed.), *Albert Einstein: Philosopher-Scientist*, vol. 1, The Library of Living Philosophers (Evanston, IL, various reprints) 1–95.

Ellis, J. (1988), 'The Aporematic Character of Theophrastus' *Metaphysics*', in: Fortenbough - Sharples (1988), 216–223.

Evans, J. D. G. (1977), *Aristotle's Concept of Dialectic* (Cambridge etc.).

Feyerabend, P. K. (1962), 'Explanation, Reduction and Empiricism', in: H. Feigl - G. Maxwell (edd.), *Scientific Explanation, Space, and Time*, Minnesota Studies in the Philosophy of Science 3 (Minneapolis, various repr.) 28–97.

Finley, M. I. (1975), 'Generalizations in Ancient History', in: M. I. Finley, *The Use and Abuse of History* (London) 60–74.

Flashar, H. (1983), 'Aristoteles', in: H. Flashar (ed.), *Grundriss der Geschichte der Philosophie*, begr. von F. Ueberweg, vol. 3 (Basel/ Stuttgart) 177–185.

Fortenbaugh, W. W. - Sharples, R. W. (edd.) (1988), *Theophrastean Studies; On Natural Science, Physics and Metaphysics, Ethics, Religion and Rhetoric*, Rutgers University Studies in Classical Humanities 3 (New Brunswick/ Oxford).

Fortenbaugh, W. W. - Gutas, D. (edd.) (1992), *Theophrastus: His Psychological, Doxographical and Scientific Writings*, Rutgers University Studies in the Classical Humanities 5 (New Brunswick/ London).

Frede, M. (1974), *Die stoische Logik*, Abhandlungen der Akademie der Wissenschaften in Gottingen. Philol.-hist. Kl. Folge 3, Nr. 88 (Göttingen).

——, (1980), 'The Original Notion of Cause', in: Schofield – Burnyeat – Barnes (1980) 217–249; repr. in: Frede (1987b) 225–243.

——, (1987a), 'Individuals in Aristotle', in: Frede (1987b) 49–71.

——, (1987b), *Essays in Ancient Philosophy* (Oxford).

——, (1987c), 'The Title, Unity, and Authenticity of the Aristotelian *Categories*', in: Frede (1987b) 11–28 (originally published as 'Titel, Einheit und Echtheit der Kategorien', in: P. Moraux – J. Wiesner (edd.), *Zweifelhaftes im Corpus Aristotelicum* (Berlin 1983) 1–30).

Furley, D. J. (1966), 'Lucretius and the Stoics', *BICS* (13) 13–33.

——, (1967), *Two Studies in the Greek Atomists*, 1. Indivisible Magnitudes; 2. Aristotle and Epicurus on Voluntary Action (Princeton, NJ).

——, (1976), 'Aristotle and the Atomists on Motion in a Void', in: Machamer –Turnbull (1976) 83–100.

——, (1985), 'Strato's Theory of the Void', in: J. Wiesner (ed.), *Aristoteles, Werk und Wirkung: Paul Moraux gewidmet*, vol. 1: Aristoteles und seine Schule (Berlin/ New York) 594–609.

——, (1982), 'The Greek Commentators' Treatment of Aristotle's Theory of the Continuous', in: N. Kretzmann (ed.), *Infinity and Continuity in Ancient and Medieval Thought* (Ithaca/ London) 17–36.

——, (1987), *The Greek Cosmologists*, I. The Formation of the Atomic Theory and its Earliest Critics (Cambridge etc.).

——, (1989), *Cosmic Problems: Essays on Greek and Roman Philosophy of Nature* (Cambridge) .

Gatzemeier, M. (1970), *Die Naturphilosophie des Straton von Lampsakos: zur Geschichte des Problems der Bewegung im Bereich des frühen Peripatos*, Monographien zur Naturphilosophie 10 (Meisenheim am Glan).

Görler, W. (1987), 'Hauptursachen bei Chrysipp und Cicero', *RhM* (130) 254–274

Goldschmidt,V. (1953), *Le système stoïcien et l'idée de temps* (Paris, [4]1979).

Gotthelf, A. (1987), 'Aristotle's Conception of Final Causality', in: A. Gotthelf – J. G. Lennox, *Philosophical Issues in Aristotle's Biology* (Cambridge) 204–242

Gottschalk, H. B. (1965), 'Strato of Lampsakos: Some Texts', *Proceedings of the Leeds Philosophical and Literary Society*, Lit. and Hist. Sect. (11.6) 95–182.

——, (1987), *Aristotelian Philosophy in the Roman World from the Time of Cicero to the End of the Second Century AD*, in: W. Haase – H. Temporini (edd.), *Aufstieg und Niedergang der römischen Welt* II, Band 36. 2 (Berlin/ New York) 1079–1174.

Goulet, R. (ed.) (1980), Cléomède: *Théorie élémentaire* («De motu circulari corporum caelestium»), Histoire des doctrines de l'antiquité classique 3 (Paris).

Graeser, A. (1971), 'A Propos ΥΠΑΡΧΕΙΝ bei den Stoikern', *ABG* (15) 299–305.

Grant, E. (1981a), *Much Ado about Nothing: Theories of Space and Vacuum from the Middle Ages to the Scientific Revolution* (Cambridge).

——, (1981b), 'The Medieval Doctrine of Place: Some Fundamental Problems and Solutions' in: Maierù – Bagliani (1981) 57–79.

Guthrie, W. K. C. (1957), 'Aristotle as Historian', *JHS* (77) 35–41, repr. in: D. J. Furley – R. E. Allen, *Studies in Presocratic Philosophy*, vol. 1: *The Beginnings of Philosophy*, International Library of Philosophy and Scientific Method (London 1970) 239–254.

——, (1967–1981), *A History of Greek Philosophy*, 6 vols. (Cambridge etc., various repr.).

Hacking, I. (1983), *Representing and Intervening: Introductory Topics in the Philosophy of Natural Science* (Cambridge).
Hadot, I. (1978), *Le problème du néoplatonisme Alexandrin; Hiérocles et Simplicius* (Paris).
Hadot, P. (1968), *Porphyre et Victorinus,* 2 vols. (Paris).
Hahm, D. E. (1977), *The Origins of Stoic Cosmology* (Columbus, Ohio).
Happ, H. (1971), *Hyle: Studien zum aristotelischen Materie-Begriff* (Berlin/ New York).
Heath, Th. (1949), *Mathematics in Aristotle* (Oxford, repr. 1970).
Hoffmann. P. (1979), 'Simplicius: Corollarium de Loco' in: *L' astronomie dans l' antiquité classique*; Actes du Colloque tenu à l' Université de Toulouse-Le Mirail, 21–23 octobre 1977, Collection d Études anciennes (Paris) 143–161.
Huby, P. (1978), 'Epicurus' Attitude to Democritus', *Phronesis* (23) 80–86.
Hussey, E. (1983), *Aristotle's* Physics*: books iii and iv,* Clarendon Aristotle series (Oxford/ New York).
Inwood, B. (1981), 'The Origin of Epicurus' Concept of Void', *CPh* (76) 273–285.
——, (1985), *Ethics and Human Action in Early Stoicism* (Oxford).
——, (1991), 'Chrysippus on Extension and the Void', *RIPh* 178 (1991.3) 245–266.
Irwin T. (1988), *Aristotle's First Principles* (Oxford).
Jaeger, W. (1912), *Studien zur Entstehungsgeschichte der Metaphysik des Aristoteles* (Berlin).
——, (1923), *Aristoteles. Grundlegung einer Geschichte seiner Entwicklung* (Berlin, numerous [revised] reprints and translations).
Jammer, M. (1954), *Concepts of Space: the History of Theories of Space in Physics,* Foreword by Albert Einstein (Cambridge Mass., [2]1969).
Kahn, Ch. (1973), *The Verb 'Be' in Ancient Greek,* vol. 16 of: J. W. M. Verhaar (ed.), *The Verb 'Be' and its Synonyms, Philosophical and Grammatical Studies,* Foundation of Language, Suppl. Series 16 (Dordrecht/ Boston).
Keyt, D. (1961), 'Aristotle on Plato's Receptacle', *AJPh* (82) 291–300.
Kidd, I. G. (1988), *Posidonius,* vol. 2: *The Commentary,* Cambridge Classical Texts and Commentaries 14 A&B (Cambridge etc.).
King, H. R. (1950), 'Aristotle's Theory of ΤΟΠΟΣ', *CQ* (44) 76–96.
Kneale, W. – Kneale, M. (1962), *The Development of Logic* (Oxford, repr. with corr. 1975).
Koyré, A. (1957), *From the Closed World to the Infinite Universe* (Baltimore, repr. 1987).
Koyré, A. – Cohen, I. B. – Whitman, A. (edd.) (1972), *Isaac Newton's Philosophiae naturalis principia mathematica.* The third edition (1726) with variant readings, 2 vols. (Cambridge, MA).
Kuhn, Th. S. (1962), *The Structure of Scientific Revolutions,* Phoenix Books (Chicago, 2nd enl. ed. 1970, various repr.).
Kullmann, W. (1974), *Wissenschaft und Methode.* Interpretationen zur Aristotelischen Theorie der Naturwissenschaft (Berlin/ New York).
Kung, J. (1988), 'Why the Receptacle is not a Mirror', *AGPh* (70) 167–178.
Lang, H. S. (1992), 'Why Fire goes Up', in: H. S. Lang, *Aristotle's Physics and its Medieval Varieties,* SUNY series in ancient Greek philosophy (Albany) 63–84.
Le Blond, J.-M. (1939), *Logique et méthode chez Aristote: étude sur la recherche des principes dans la physique aristotélicienne,* Bibliothèque d'histoire de la philosophie (Paris (diss. 1938), [3]1973).

Lee, E. N. (1966), 'On the Metaphysics of the Image in Plato's *Timaeus*', *The Monist* (50) 341–368.

Löbl, R. (1976), *Demokrits Atome*: eine Untersuchung zur Überlieferung und zu einigen wichtigen Lehrstücken in Demokrits Physik; Habelts Dissertationsdrucke. Reihe klassische Philologie 24 (Bonn).

Long, A. A. (1971), 'Language and Thought in Stoicism', in: A. A. Long (ed.), *Problems in Stoicism* (London) 75–113.

——, (1985), 'The Stoics on World-Conflagration and Everlasting Recurrence', in: R. H. Epp (ed.), *Spindel Conference 1984: Recovering the Stoics*. Southern Journal of Philosophy, suppl. (23) 13–37.

Long, A. A. – Sedley, D. N. (1987), *The Hellenistic philosophers*, 2 vols.: I. Translations of the principal sources with philosophical commentary; II. Greek and Latin texts with notes and bibliography (Cambridge).

Maass, E. (1958²), *Commentariorum in Aratum Reliquiae* (Berlin).

Machamer, P. K. – Turnbull, R. G. (1976), *Motion and Time, Space and Matter: Interrelations in the History of Philosophy and Science* (Columbus, Ohio).

Machamer, P. K. (1978), 'Aristotle on Natural Place and Natural Motion', *Isis* (69) 377–387.

Maier, A. (²1952), *An der Grenze von Scholastik und Naturwissenschaft*: Die Struktur der Materiellen Substanz; Das Problem der Gravitation; Die Mathematik der Formlatituden (Roma).

Maierù, A. – Bagliani, A. P. (edd.) (1981), *Studi sul XIV secolo in memoria di Anneliese Maier* (Roma).

Mansfeld, J. (1979), 'Providence and the Destruction of the Universe in Early Stoic Thought', in: M. J. Vermaseren (ed.), *Studies in Hellenistic Religions*, EPRO 78 (Leiden) 129–188, repr. in: Mansfeld (1989) Study I.

——, (1985), 'Historical and Philosophical Aspects of Gorgias' 'On What Is Not'', in: L. Montoneri – F. Romano (edd.), *Gorgia e la Sofistica*, Siculorum Gymnasium 38 (Catania) 243–271, repr. as Study IV in: Mansfeld (1990b) 97–125.

——, (1986), 'Aristotle, Plato, and the Preplatonic Doxography and Chronography', in: G. Cambiano (ed.), *Storiografia e dossografia nella filosofia antica*, [atti del convegno internazionale su 'La storiografia filosofica antica', Stresa, 27 e 28 settembre 1984], Biblioteca storico-filosofica 2 (Torino) 1–59, repr. as Study II in: Mansfeld (1990b) 22–83.

——, (1989), *Studies in Later Greek Philosophy and Gnosticism* (London).

——, (1990a), 'Doxography and Dialectic: The *Sitz im Leben* of the 'Placita'', in: W. Haase – H. Temporini, *Aufstieg und Niedergang der römischen Welt*, Part II, vol. 36.4 (Berlin/ New York 1990) 3056–3229.

——, (1990b), *Studies in the Historiography of Greek Philosophy* (Assen).

——, (1992), *Heresiography in Context: Hyppolytus'* Elenchos *as a Source for Greek Philosophy*, Philosophia Antiqua 51 (Leiden etc.).

Mates, B. (1953), *Stoic Logic*, University of California Publications in Philosophy 26 (Berkeley/ Los Angelos, 1961, microfilm repr. Ann Harbor 1992).

McGuire, J. E. (1982), 'Space, Infinity and Indivisibility: Newton on the Creation of Matter', in: Z. Bechler (ed.), *Contemporary Newtonian Research*, Studies in the History of Modern Science 9 (Dordrecht) 145–190.

Mendell, H. (1987), 'Topoi on Topos: The Development of Aristotle's Concept of Place', *Phronesis* (32) 206–231.

Mignucci, M. (1978), 'Sur la logique modale des Stoïciens', in: Brunschwig (1978) 317–346.

Mohr, R. D. (1985), *The Platonic Cosmology*, Philosophia Antiqua 42 (Leiden).

Moraux, P. (1951), *Les listes anciennes des ouvrages d' Aristote*; préface par A. Mansion; Aristote: traductions et études (Louvain).

——, (1973–84), *Der Aristotelismus bei den Griechen*: von Andronikos bis Alexander von Aphrodisias, 2 vols.: I. Die Renaissance des Aristotelismus im I. Jh. v. Chr. II. Der Aristotelismus im I. und II. Jh. n. Chr.; Peripatoi 5–6 (Berlin/ New York).

Moreau, J. (1965), *L'espace et le temps selon Aristote*, Saggi e testi 4 (Padova).

Morsink, J. (1982), 'The Mandate of *Topics* I, 2', *Apeiron* (16) 102–128.

Mueller, I. (1970), 'Aristotle on Geometrical Objects', *AGPh* (52) 156–171, repr. in: Barnes – Schofield – Sorabji (1975–79) III, 96–107.

Nussbaum, M. C. (1986a), *The Fragility of Goodness: Luck and Ethics in Greek Tragedy and Philosophy* (Cambridge).

——, (ed.) (1986b), *Owen, G. E. L.: Logic, Science and Dialectic*: Collected Papers in Greek Philosophy (London).

Owen, G. E. L. (1953), 'The Place of the *Timaeus* in Plato's Dialogues', *CQ* N.S. (3) 79–95, repr. in: Nussbaum (1986b) 65–84.

——, (1961), 'Τιθέναι τὰ φαινόμενα /Tithenai ta Phainomena', in: AA. VV. (1961) 83–103; repr. in: Barnes – Schofield – Sorabji (1975–79) I, 113–126; repr. in: Nussbaum (1986b) 239–251.

——, (1965), 'Inherence', *Phronesis* (10) 97–105; repr. in: Nussbaum (1986b) 252–258.

——, (1966), 'The Platonism of Aristotle', *PBA* (51) 125–150; repr. in: Strawson, P. F. (ed.) (1968), Studies in the Philosophy of Thought and Action (Oxford) 147–174; in: Barnes – Schofield – Sorabji (1975–79) I, 14–34 and in: Nussbaum (1986b) 200–220.

——, (1976), 'Aristotle on Time', in: Machamer/ Turnbull (1976) 3–27; repr. in: Barnes – Schofield – Sorabji (1975–79) I, 140–158 and in: Nussbaum (1986b) 295–314.

——, (1986), 'Aristotelian Mechanics', in: A. Gotthelf (ed.), *Aristotle on Nature and Living Things*: Philosophical and Historical Studies Presented to David M. Balme on his Seventieth Birthday (Pittsburgh/ Bristol 1986) 227–245, repr. in: Nussbaum (1986b) 315–333.

Owens, J. (1951), *The Doctrine of Being in the Aristotelian 'Metaphysics'*: a Study in the Greek Background of Mediaeval Thought (Toronto, 3rd rev. ed. 1978).

Pasquino, P. (1978), 'Le statut ontologique des incorporels dans l'ancien stoïcisme', in: Brunschwig (1978) 375–386.

Patricios, N. N. (1971), 'The Spatial Concepts of the Ancient Greeks', *AClass* (14) 17–36.

Pohlenz, M. (1948–49), *Die Stoa: Geschichte einer geistigen Bewegung*, 2 vols. (Göttingen, ^{4}1970–1972).

Quine, W. V. O. (1953), 'Two Dogmas of Empiricism', in: W. V. O. Quine, *From a Logical Point of View*: 9 Logico-Philosophical Essays (Cambridge, ^{2}1961, 1971) 20–46.

Raalte, M. van (1988), 'The Idea of the Cosmos as an Organic Whole in Theophrastus' *Metaphysics*', in: Fortenbough – Sharples (1988) 189–215.

Reesor, M. E. (1965), 'Fate and Possibility in Early Stoic Philosophy', *Phoenix* (19) 285–297.

Regenbogen, O. (1940), 'Theophrastos', in: Pauly-Wissowa, *Real-Enzyclopädie der classischen Altertumswissenschaft*, Supp. Bd. VII (Stuttgart) 1354–1562.

Rehm, A. (1922), 'Kleomedes', in: Pauly-Wissowa, *Real-Enzyclopädie der classischen Altertumswissenschaft*, Bd. XI (Stuttgart) 679–694.

Rijk, L. M. de (1980), 'On Ancient and Mediaeval Semantics and Metaphysics (3)', *Vivarium* (18) 1–62.

——, (1986), *Plato's Sophist*: a Philosophical Commentary, Verhandelingen der Koninklijke Nederlandse Akademie van Wetenschappen. Afd. Letterk. N.R., dl. 133 (Amsterdam etc.).

Rist, J. M. (1972), *Epicurus, an Introduction* (Cambridge etc.).

Robinet, A. (ed.) (1957), *Correspondance Leibniz-Clarke*; presentée d'après les manuscrits originaux des bibliothèques de Hanovre et de Londres; Bibliothèque de philosophie contemporaine. Histoire de la philosophie et philosophie generale (Paris).

Rosen, S. (1983), *Plato's Sophist*, The Drama of Original and Image (New Haven/ London).

Ross, W. D. (1923), *Aristotle* (London, [5]1949, various repr.).

——, (1924), *Aristotle's Metaphysics*, 2 vols. (Oxford, 1953 (with corr.), 1958).

——, (1936), *Aristotle's Physics*, 2 vols. (Oxford, 1955 (with corr.), 1979).

——, (1951), *Plato's theory of ideas* (Oxford and later repr.).

Runia, D. T. (1989), 'Xenophanes on the Moon: A *Doxographicum* in Aëtius', *Phronesis* (34) 245–269.

Sachs, E. (1917), *Die fünf platonischen Körper*, History of Ideas in Ancient Greece (Berlin, repr. New York 1976).

Sambursky, S. (1956), *The Physical World of the Greeks*; transl. from the Hebrew by Merton Dagut (London, [2]1960).

——, (1959), *Physics of the Stoics* (London).

——, (1962), *The Physical World of Late Antiquity* (London).

——, (1982), *The Concept of Place in Late Neoplatonism*, The Israel Academy of Sciences and Humanities (Jerusalem).

Schmekel, A. (1892), *Die Philosophie der mittleren Stoa in ihrem geschichtlichen Zusammenhange dargestellt* (Berlin, repr. Hildesheim 1974).

Schofield, M. – Burnyeat, M. F. – Barnes, J. (edd.) (1980), *Doubt and Dogmatism*: Studies in Hellenistic Epistemology: [papers of a conference on Hellenistic philosophy held at Oriel College, Oxford, March 1978] (Oxford).

Schulz, D. J. (1966), *Das Problem der Materie in Platons 'Timaios'*, Abhandlungen zur Philosophie, Psychologie und Pädagogik 31 (Bonn).

Sedley, D. N. (1982), 'Two Conceptions of Vacuum', *Phronesis* (27) 175–193.

——, (1987), 'Philoponus' Conception of Space', in: Sorabji (1987b) 140–153.

Shapere, D. (1966), 'Meaning and Scientific Change', in: R. G. Colodny (ed.), *Mind and Cosmos*: Essays in Contemporary Science and Philosophy, University of Pittsburg Series in the Philosophy of Science 3 (Pittsburg) 41–85.

——, (1984), *Reason and the Search for Knowledge*: Investigations in the Philosophy of Science; Boston Studies in the Philosophy of Science (Dordrecht etc.).

Sharples, R. W. (1981), 'Necessity in the Stoic Doctrine of Fate', *SO* (56) 81–97.

Skemp, J. B. (1960), 'ΥΛΗ and ΥΠΟΔΟΧΗ', in: I. Düring – G. E. L. Owen (edd.), *Aristotle and Plato in the Mid-Fourth Century*: Papers of the Symposium Aristotelicum held at Oxford in August, 1957, Studia Graeca et Latina Gothoburgensia 11 (Göteborg) 201–212.

Solmsen, F. (1960), *Aristotle's System of the Physical World: a Comparison with his Predecessors*, Cornell Studies in Classical Philology 33 (Ithaca, NY).

——, (1977), 'Epicurus on Void, Matter and Genesis', *Phronesis* (22) 263–281, repr. in: F. Solmsen, *Kleine Schriften*, vol. 3 (Hildesheim etc. 1982) 333–351.

Sorabji, R. (1980), 'Causation, Laws and Necessity', in: Schofield – Burnyeat – Barnes (1980) 250–282.

——, (1983), *Time, Creation and the Continuum: Theories in Antiquity and the Early Middle Ages* (London 1983).
——, (1987a), 'John Philoponus', in: Sorabji (1987b) 1–40.
——, (1987b) (ed.), *Philoponus and the Rejection of Aristotelian Science* (London/ Ithaca).
——, (1988a), *Matter, Space and Motion: Theories on Antiquity and their Sequel* (Ithaca NY).
——, (1988b), 'Theophrastus on Place', in: Fortenbaugh – Sharples (1988) 139–166.
Steckel, H. (1970), 'Demokritos', in: Pauly-Wissowa, *Real-Enzyclopädie der classischen Altertumswissenschaft*, Suppl. Bd. 12 (Stuttgart) 191–223.
Steinmetz, P. (1964), *Die Physik des Theophrastos von Eresos*, Palingenesia 1 (Bad Homburg etc.).
Szlezák, T. A. (ed.) (1972), Pseudo-Archytas: *Über die Kategorien*, Texte zur griechischen Aristoteles-Exegese, Peripatoi 4 (Berlin/ New York).
Tarán, L. (1965), *Parmenides*: a text with translation, commentary and critical essays (Princeton, NJ).
Taylor, A. E. (1928), 'Aristotle's Doctrine of Space', = Appendix 2 in: A. E. Taylor, *A Commentary on Plato's Timaeus* (Oxford, repr. 1962) 664–677.
Theiler, W. (1982), *Poseidonios: Die Fragmente*, 2 vols., Texte und Kommentare 10 (Berlin/ New York).
Todd, R. B. (1982), 'Cleomedes and the Stoic Concept of the Void', *Apeiron* (16) 129–136.
Vlastos, G. (1975), *Plato's Universe* (Oxford).
Voelke, A. J. (1990), 'Vide et non-être chez Leucippe et Démocrite', *RThPh* (122) 341–352.
Wagner, H. (1967), *Aristoteles Physikvorlesung*, Aristoteles Werke in deutscher Übersetzung 11 (Darmstadt, repr. 1979).
Wardy, R. (1991), 'Transcendental Dialectic', *Phronesis* (36, 1) 88–106.
Watson, G. (1966), *The Stoic Theory of Knowledge* (Belfast).
Wehrli, F. (^{2}1983), 'Der Peripatos bis zum Beginn der römischen Kaiserzeit', in: F. Ueberweg, *Die Philosophie der Antike*, vol. 3 (Ältere Akademie, Aristoteles, Peripatos), herausgegeben von H. Flashar, Grundriss der Geschichte der Philosophie (Basel/ Stuttgart) 459–599.
Weil, E. (1975), 'The Place of Logic in Aristotle's Thought', in: Barnes – Schofield – Sorabji (1975–79) I, 88–112 (originally published as 'La place de la logique dans la pensée aristotélicienne', *RMM* 56 (1951) 283–315).
Weisheipl, J. A. (1985a), 'The Principle *Omne quod movetur ab alio movetur* in Medieval Physics', in: Carroll (1985) 75–97 (repr. from: *Isis* 56 (1965) 26–45).
——, (1985b) 'The Spectre of *motor coniunctus*' in: Carroll (1985) 99–120 (repr. from Maierù – Bagliani (1981) 81–104).
Wieland, W. (1962), *Die aristotelische Physik*: Untersuchungen über die Grundlegung der Naturwissenschaft und die sprachlichen Bedingungen der Prinzipienforschung bei Aristoteles (Göttingen, ^{2}1970).
Wolff, M. (1987) 'Philoponus and the Rise of Preclassical Dynamics', in: R. Sorabji (ed.), *Philoponus and the Rejection of Aristotelian Science* (London) 84–121.
Wolfson, H. A. (1929), *Crescas' Critique of Aristotle: Problems of Aristotle's Physics in Jewish and Arabic Philosophy*, Harvard Semitic Studies VI (Cambridge MA).
Wurm, K. (1973), *Substanz und Qualität*: ein Beitrag zur Interpretation der plotinischen Traktate VI 1, 2 und 3; Quellen und Studien zur Philosophie 5 (Berlin/ New York).

Zeller, E. – Mondolfo R. (1967), *La filosofia dei greci nello suo sviluppo storico*, 1, 3 'Eleati', a cura di G. Reale (Firenze).

——, (1974), *La filosofia dei greci nello suo sviluppo storico*, 2, 3, 1 'Platone e l'Accademia antica', a cura di M. Isnardi Parente (Firenze).

——, (1966), *La filosofia dei greci nel suo sviluppo storico*, 2, 5 a cura di A. Plebe, (Firenze).

INDICES

(I) Index Locorum
(II) General Index

(I) Index locorum

(II) General Index

PHILOSOPHIA ANTIQUA

A SERIES OF STUDIES ON ANCIENT PHILOSOPHY

EDITED BY

J. MANSFELD, D.T. RUNIA AND J.C.M. VAN WINDEN

Available in this series:

50. SIMPLICIUS. *Commentaire sur les Catégories.* Traduction commentée sous la direction de I. Hadot. I: Introduction, première partie (p. 1-9, 3 Kalbfleisch). Traduction de Ph. Hoffmann (avec la collaboration d'I. et P. Hadot). Commentaire et notes à la traduction par I. Hadot avec des appendices de P. Hadot et J.-P. Mahé. 1990. ISBN 90 04 09015 0
51. SIMPLICIUS. *Commentaire sur les Catégories.* Traduction commentée sous la direction de I. Hadot. III: Préambule aux Catégories. Commentaire au premier chapitre des Catégories (p. 21-40, 13 Kalbfleisch). Traduction de Ph. Hoffmann (avec la collaboration d'I. Hadot, P. Hadot et C. Luna). Commentaire et notes à la traduction par C. Luna. 1990. ISBN 90 04 09016 9
52. MAGEE, J. *Boethius on Signification and Mind.* 1989. ISBN 90 04 09096 7
53. BOS, E.P. and MEIJER, P.A. (eds.) *On Proclus and His Influence in Medieval Philosophy.* 1992. ISBN 90 04 09429 6
54. FORTENBAUGH, W.W., et al. (eds.) *Theophrastes of Eresos.* Sources for His Life, Writings, Thought and Influence. 1992. ISBN 90 04 09440 7 *set*
55. SHANKMAN, A. *Aristotle's* De insomniis. A Commentary. ISBN 90 04 09476 8
56. MANSFELD, J. *Heresiography in Context.* Hippolytos' *Elenchos* as a Source for Greek Philosophy. 1992. ISBN 90 04 09616 7
57. O'BRIEN, D. *Théodicée plotinienne, théodicée gnostique.* 1993. ISBN 90 04 09618 3
58. BAXTER, T.M.S. *The Cratylus.* Plato's Critique of Naming. 1992. ISBN 90 04 09597 7
59. DORANDI, T. (Hrsg.) *Theodor Gomperz. Eine Auswahl herkulanischer kleiner Schriften (1864-1909).* 1993. ISBN 90 04 09819 4
60. FILODEMO. *Storia dei filosofi. La stoà da Zenone a Panezio* (PHerc. 1018). Edizione, traduzione e commento a cura di T. Dorandi. 1994. ISBN 90 04 09963 8
61. MANSFELD, J. *Prolegomena.* Questions to be Settled Before the Study of an Author, or a Text. 1994. ISBN 90 04 10084 9
62. FLANNERY, S.J., L.K. *Ways into the Logic of Alexander of Aphrodisias.* 1995. ISBN 90 04 09998 0
63. LAKMANN, M.-L. *Der Platoniker Tauros in der Darstellung des Aulus Gellius.* 1995. ISBN 90 04 10096 2
64. SHARPLES, R.W. *Theophrastus of Eresus.* Sources for his Life, Writings, Thought and Influence. Commentary Volume 5. Sources on Biology (Human Physiology, Living Creatures, Botany: Texts 328-435). 1995. ISBN 90 04 10174 8
65. ALGRA, K. *Concepts of Space in Greek Thought.* 1995. ISBN 90 04 10172 1